D0969093

PERSPECTIVES IN WORLD FOOD AND AGRICULTURE 2004

PERSPECTIVES IN WORLD FOOD AND AGRICULTURE 2004

EDITED BY

COLIN G. SCANES

JOHN A. MIRANOWSKI

THE WORLD FOOD PRIZE

Iowa State Press
A Blackwell Publishing Company

COLIN G. SCANES is a Professor of Animal Science at Iowa State University. He was formerly Executive Associate Dean of the College of Agriculture and founding Interim Director of the Plant Sciences Institute at Iowa State University. He was educated in the United Kingdom with a B.Sc. from Hull University and Ph.D. from the University of Wales. He was formerly on the faculty at the University of Leeds and Rutgers—The State University of New Jersey, where he was department chair. He has published extensively with more than 10 books and 500 papers. He has received numerous awards, including Honorary Professor at the Agricultural University of Ukraine.

JOHN A. MIRANOWSKI holds both an A.M. and a Ph.D. in Economics from Harvard University and a B.S. from Iowa State University. He is presently a Professor of Agricultural Economics at Iowa State University. He was formerly Director, Resources and Technology Division at the USDA's Economic Research Service and Executive Coordinator of the Secretary of Agriculture's Policy Coordination Council, Special Assistant to the Deputy Secretary of Agriculture, and Head of the Department of Economics at Iowa State University.

"The State of World Food Security" © Food and Agriculture Organization of the United Nations, used with permission.

© 2004 Iowa State Press
A Blackwell Publishing Company
All rights reserved

Iowa State Press
2121 State Avenue, Ames, Iowa 50014

Orders: 1–800–862–6657
Office: 1–515–292–0140
Fax: 1–515–292–3348
Web site: www.iowastatepress.com

Authorization to photocopy items for internal or personal use, or the internal or personal use of specific clients, is granted by Iowa State Press, provided that the base fee of $.10 per copy is paid directly to the Copyright Clearance Center, 222 Rosewood Drive, Danvers, MA 01923. For those organizations that have been granted a photocopy license by CCC, a separate system of payments has been arranged. The fee code for users of the Transactional Reporting Service is 0–8138–2021–9/2004 $.10.

Library of Congress Cataloging-in-Publication Data

Perspectives in world food and agriculture, 2004 / edited by Colin G. Scanes and John A. Miranowski.
    p. cm.
  ISBN 0-8138-2021-9 (alk. paper)
  1. Agriculture. 2. Food supply. 3. Agriculture and state. I. Scanes, C. G. II. Miranowski, J. A.
    S439.P44 2003
    338.1'8—dc21
                                                    2003009569

The last digit is the print number: 9 8 7 6 5 4 3 2 1

This book is dedicated to

Dr. Norman E. Borlaug

Scientist,
Agriculturalist,
Humanitarian,
Father of the green revolution,
Nobel Peace Prize Laureate,
Founder of the World Food Prize

Dr. Norman E. Borlaug (center) meeting the editors, John Miranowski (left) and Colin Scanes (right) at the Iowa capitol building at the announcement of Norman Borlaug Day in 2002.

CONTENTS

FOREWORD

The partnership between the World Food Prize and the series editors was born of a conversation at the World Food Prize Foundation headquarters in Des Moines in 2001.

At that time, Colin Scanes and John Miranowski told me of their vision for a volume that would fill the void they saw at the nexus of research and public policy on food and agriculture. They explained that their plan was to design an annual publication with contemporary relevance, but one which would also serve as an historical archive to trace the statistical changes in key measurements of both the food supply and human needs and deprivation.

The editors' vision for their publication fit well with the World Food Prize's search, at the time, for an effective vehicle that could disseminate the rich and diverse presentations made each October at its International Symposium held in Des Moines near or on World Food Day. At the turn of the 21st century, food and agricultural policy had been thrust onto the main agenda of the international political system. Yet, few international actors were well versed in the essential details of food production or the extent of human suffering possible as the world's population expands in the coming decades. Moreover, the linkages of poverty eradication to countering terrorism were only beginning to become apparent. I sensed an incipient partnership.

Founded by Nobel Peace Prize Laureate Dr. Norman E. Borlaug, the World Food Prize prides itself on bringing together global experts, policymakers, academic specialists, and business leaders to address cutting edge issues at the intersection of agricultural development, human deprivation, and international security. Topics recently explored at the World Food Prize International Symposium often were harbingers of issues that later became center stage in global diplomatic forums, and at the heart of regional disputes. Those topics included: the role of genetically modified crops in feeding developing countries (2000); the impact of HIV/AIDS on food production in Africa (2001); the threat of agroterrorism (2001); global water insecurity and its implications for peace in the Middle East (2002); and the essential role of the United Nations in implementing the Millennium Development Goals (2003). The Scanes and Miranowski volume seemed the perfect vehicle to distribute and preserve the research and insights of the experts we had assembled to address these issues.

In addition to containing the papers given at our symposium, this annual publication could also incorporate the address given by our World Food Prize Laureate following the presentation of our $250,000 prize, providing a continuum of rich overviews of the major challenges facing our planet in the struggle to ensure adequate food for all in the 21st century.

This rich trove of information, commentary, and prescriptions for change has been available to only those in attendance at our symposia—until now. With the publication of Dr. Scanes's and Dr. Miranowski's first volume, in what assuredly will be a long and distinguished series, the insights and analyses of the leading figures on the frontiers of feeding the hungry will now be available around the world. Of particular significance is the inclusion of Norman Borlaug's *tour d'horizon* on feeding a world of 10 billion people, given on October 21, 2003 at Iowa State University, where both editors hold distinguished faculty positions.

Fifty years from now, a student somewhere in the world may pick up this volume and read the analysis by and dictums of Dr. Borlaug—the individual credited with saving more lives than any other person who has ever lived. And, motivated by this Nobel Laureate's words, maybe, just maybe, this student will be inspired to try to achieve a similar scientific breakthrough that will result in feeding the world's burgeoning population.

If that does occur, it will be because Dr. Borlaug's *summa agricultura* lecture has been preserved thanks to the initiative of Colin Scanes and John Miranowski, who had the vision to create this series, and the foresight of Iowa State Press and Blackwell Publishing, who made it a reality. The World Food Prize is proud to be a partner with them in this endeavor.

Kenneth M. Quinn
Ambassador (ret.)
President, The World Food Prize

PREFACE

Perspectives in World Food and Agriculture: 2004 has as its goal bringing together essays and reviews on frontiers in the food system, agricultural research, and agricultural policy in North America, Europe, and the developing world, together with a summary of statistical data on world agriculture. The volume is a partnership with the World Food Prize, and we are delighted to include the World Food Prize logo on the cover.

The United Nations Food and Agriculture Organization (FAO) estimates that there are over 800 million people in the world who are undernourished (covered in detail in chapter 1). Food security throughout the world is a moral imperative, such that all people receive sufficient calories, protein, vitamins, and minerals (macro- and trace) for not only the absence of hunger but also productive and healthy lives. Lack of food violates human rights and is both a cause and effect of poverty. Moreover, there is considerable potential for regional conflicts over food and water and links between endemic poverty and terrorism. In this increasingly interdependent global environment, overcoming hunger and poverty is critical on ethical grounds; enlightened self-interest from the developing world is also imperative.

Agriculture represents a crucial mechanism for reducing poverty and stimulating economic activity while protecting the environment. In the United States, food production/processing represents ~16 percent of the GNP. In many developing countries, food and agriculture represent more than 50 percent of the GNP, and more than half the population lives in rural areas. The movement to cities (driven by economic considerations) represents twin challenges for food and water security to the urban dwellers and to enhancing rural economic development. Policymakers in multinational and national governmental and nongovernmental development agencies are recognizing, rediscovering, or reinvigorating the importance of food, agriculture, and water.

This volume is divided into the following sections:

Section I   Frontiers in World Food
Section II   Frontiers in Food

Section III   Frontiers in Animal Agriculture
Section IV   Frontiers in Water
Section V   Frontiers in Policy and Ethics
Section VI   Statistics and Trends in World Agriculture

Each section consists of a series of invited essays or reviews and is introduced by a contextual statement.

"Frontiers in World Food" considers the world food situation with examples of successful science-based approaches. The impact of the World Food Prize and its laureates is stressed—individuals *can* and *have made* the difference. Norman E. Borlaug received the Nobel Peace Prize as the father of the "green revolution," which has done so much to reduce hunger, particularly in Asia. He has contributed an essay on the challenge of feeding a world population of 10 billion people. The section also contains a biography of Dr. Borlaug by Ambassador Kenneth Quinn. Also included is a review by Pedro Sanchez (winner of the 2002 World Food Prize) on a success story in agricultural research and development in the developing world. This is based on the recognition of the declining fertility of African soils and includes a successful approach to overcome this problem.

"Frontiers in Food" considers the issues of food safety and marketing in the U.S.

"Frontiers in Animal Agriculture" considers livestock as a means to reduce poverty in developing countries, the impact of horses for rural economic development in North America, and the thorny issue of animal waste, which affects agricultural producers, neighbors, and the physical environment (potentially contaminating air, water, and soil).

"Frontiers in Water" is based on the 2002 World Food Prize symposium. Lack of adequate supplies of high-quality water, particularly in the Middle East, represents an obstacle to development and the alleviation of hunger and poverty and a source of friction between countries.

"Frontiers in Policy and Ethics" includes an essay on the ethics of food and agriculture by Paul Thompson together with a perspective on the effects of China's accession to the World Trade Organization and an extensive discussion of agroterrorism. The issue of agricultural biotechnology in developing countries is addressed by a cutting edge review by David Zilberman and his colleagues.

"Statistics and Trends in World Agriculture" provides statistical information on trends in agriculture and hence factors that affect world agriculture.

It is hoped that this volume will be useful to a diverse readership, from policymakers (national and international), to officials and administrators (government, nongovernment organizations, universities, and research institutions), to farmer/commodity organizations, to agricultural leaders, to agricultural-related businesses, to agricultural scientists (universities, research institutes, and industry), to agricultural educators, to undergraduate and graduate students of agriculture (particularly in perspectives/capstone courses), and to the general public interested in food and agricultural issues. It is anticipated that this volume will be the first of a series.

Acknowledgments

The editors of this volume are indebted to the Advisory Council members for their invaluable assistance. The Advisory Council consists of a group of eminent agriculturists and agricultural researchers from across the World, and is comprised of the following individuals:

Dr. Ronald Cantrell, Director-General, International Rice Research Institute. Dr. Cantrell is a prominent plant scientist. He is presently Director-General, International Rice Research Institute. He was formerly Head of Agronomy at Iowa State University and with the wheat and maize breeding center, CIMMYT.

Dr. Jiaan Cheng, Vice President, Zhejiang University, China. Dr. Cheng is the President Elect for the International Consortium of Agricultural Universities.

Dr. Csaba Csaki, World Bank. Dr. Csaki is a prominent agricultural economist. He is presently with World Bank.

Dr. Eddy Decuypere, Katholieke Universiteit, Leuven, Belgium. Dr. Decuypere is a prominent animal scientist.

Dr. Ralph Hardy, National Agricultural Biotechnology Council, U.S.A. Dr. Hardy is a prominent plant scientist. He is presently President of the National Agricultural Biotechnology Council and was formerly Vice President of Research for DuPont.

Dr. Stanley Johnson, Iowa State University, U.S.A. Dr. Johnson is a prominent agricultural economist. He is presently Vice Provost for Extension at Iowa State University. He was formerly Director of ISU's Center of Agricultural Research and Development.

Dr. Dmytro Melnychuk, Rector, National Agricultural University, Ukraine. Dr. Melnychuk is Rector of the Agricultural University of Ukraine and an Academician of the Ukrainian Academy of Agricultural Science. He is President of the Global Consortium of Higher Education and Research for Agriculture.

Dr. Susan Offutt, USDA Economic Research Service. Dr. Offut is a prominent agricultural economist. She is presently Administrator of USDA's Economic Research Service.

Dr. Rajendra Singh Paroda, CGIAR. Dr. Paroda was formerly Director-General of the Indian Council of Agricultural Research.

Dr. Per Pinstrip-Andersen, Director-General, International Food Policy Research Institute. Dr. Pinstrip-Andersen is a prominent agricultural economist. Until recently, he was Director-General of the International Food Policy Research Institute. In 2001, he received the World Food Prize.

Dr. Kenneth Quinn, World Food Prize. Dr. Quinn heads The World Food Prize Foundation. He was with the U.S. State Department and was formerly U.S. Ambassador to Cambodia.

Dr. Timothy Reeves, Director-General, CIMMYT. Dr Reeves is a prominent agronomist focusing on sustainable agriculture. He was until recently Director-General of the International Center for maize/corn and wheat breeding (CIMMYT).

Dr. Ismail Seragelden, President, Alexandria Library, Egypt. Dr. Seragelden, an economist, was formerly Vice President of the World Bank and is presently President of the Alexandria Library, Egypt.

I

FRONTIERS IN WORLD FOOD

This section includes a series of essays and reviews on the world food situation. This critical topic is introduced by an overview from the United Nation Food and Agriculture Organization (FAO). The volume is dedicated to Norman E. Borlaug. He is not only a great agriculturist, father of the green revolution, and Nobel Peace Prize winner but also founder of The World Food Prize. We are delighted that Norman Borlaug has contributed a seminal essay on feeding a world of 10 billion people. In addition, Kenneth Quinn, president of the World Food Prize foundation, has contributed a definitive short biography of Dr. Borlaug. The link with The World Food Prize is further emphasized by the publication of the list of past laureates, their statement for the World Food Summit and reviews by laureates, including 2002 laureate Pedro Sanchez and 2001 laureate Per Pinstrup-Andersen.

1

THE STATE OF WORLD FOOD SECURITY
Food and Agriculture Organization of the United Nations

The state of world food security reflects a stark dichotomy between abundance and deprivation. The world produces enough food to provide everyone with an adequate diet; information systems can pinpoint where food is needed; modern transport systems can move food around the globe rapidly; yet nearly one person in seven cannot produce or buy enough food to sustain an active and healthy life. While obesity has become one of the fastest rising health problems in both developed and developing countries, 840 million people do not have enough to eat.

Millions of people, including 6 million children under the age of five, die each year as a result of hunger and malnutrition. Of these millions, relatively few are the victims of famines. Far more die unnoticed, killed by the effects of chronic hunger and malnutrition that leave them weak, underweight, and vulnerable. Health and mortality indicators are closely correlated with the prevalence of hunger. Common childhood diseases are far more likely to be fatal in children who are even mildly undernourished, and the risk increases sharply with the severity of malnutrition. Eliminating hunger and malnutrition could save millions of lives each year (FAO 2002).

Poverty reduction has been the center of the development effort in the last decade, and the lessons learned demonstrate that only purposeful action will alleviate poverty. Poverty is a cause of hunger, but hunger can also be a cause of poverty. Hunger in childhood impairs mental as well as physical growth, impacting negatively on productive capacity in adulthood. People who are chronically undernourished cannot work at their highest potential, and they tend to have lower earnings and shorter working lives. FAO estimates that reducing the number of undernourished people by half by 2015 would yield a value of more than $US120 billion in terms of longer and healthier lives.

Heads of state and government representing 186 countries met at the World Food Summit (WFS) in Rome in 1996 and affirmed their "common and national commitment to achieving food security for all" and

3

agreed to work toward the achievement of the intermediate goal of "reducing the number of undernourished people to half their present number no later than 2015."

They reconvened in June 2002 at the WFS *five years later* to reaffirm this pledge, which was also echoed in the Millennium Development Goals. It is already clear, however, that these targets cannot be met unless purposeful action is taken.

Rapid progress in cutting the incidence of chronic hunger in developing countries is possible if political will is mobilized (see Figure 1.1). FAO believes that an international alliance against hunger is required to launch a coordinated and sustained effort at all levels—international, regional, national, and local—to generate the political will and mobilize the necessary resources for this battle.

A successful anti-hunger program must address the problem on two tracks. The first track stresses action to relieve hunger through direct assistance to the poor. The second track stresses the importance of enhancing access to food by the poor. The two tracks are mutually reinforcing. Programs to enhance direct and immediate access to food

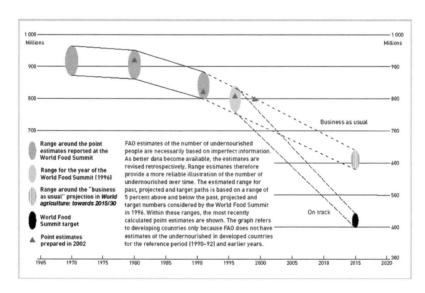

FIGURE 1.1 The number of undernourished in the developing world—observed and projected ranges compared with the World Food Summit target.

offer new outlets for expanded production. Countries that have followed this approach are seeing the benefits.

The purpose of this chapter is to achieve the following:

- Present the status of, and trends in, the food security situation in the world
- Explore the relationship between undernourishment and poverty
- Point to important directions that are necessary for hunger eradication and poverty alleviation

THE CURRENT DIMENSIONS OF FOOD INSECURITY

Food security is the "physical and economic access by all people, at all times, to sufficient, nutritionally adequate, and safe food for an active and healthy life." This definition implies that the mere presence of food does not guarantee that it is accessible to a person, since that person must also have the resources necessary to obtain access to food. How serious is the problem of food insecurity in the world today?

CHRONIC HUNGER AND MALNUTRITION

According to FAO's latest estimates, there were 840 million undernourished people[1] in the world in 1998–2000: 799 million in developing countries, 30 million in countries in transition, and 11 million in developed market economies. More than half of the undernourished (508 million people, 60 percent of the total) live in Asia and the Pacific, while sub-Saharan Africa accounts for almost a quarter (196 million people; 23 percent of the total)(see Figure 1.2).

The prevalence of undernourishment differs markedly by region. Sub-Saharan Africa has the highest prevalence, at 33 percent of the population. The second highest is found in Asia and the Pacific, where 16 percent of the population is undernourished. This regional aggregate disguises important subregional differences; in South Asia, 24 percent of the population is undernourished, and in East and Southeast Asia, the proportion is 10 and 12 percent, respectively. For the Near East and North Africa, the prevalence of undernourishment is 10 percent, and for Latin America and the Caribbean it is 11 percent (see Figure 1.3).

The prevalence of undernourishment in developing countries has fallen from 28 percent of the total population in 1979–81 to 17 percent

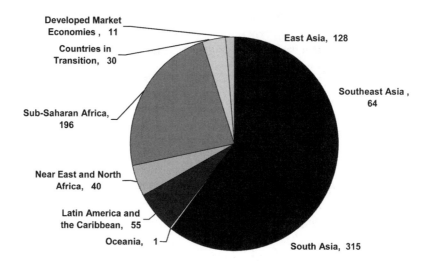

FIGURE 1.2 Undernourished population (in millions) by region, 1998–2000 (FAO).

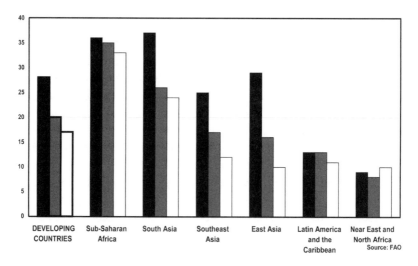

FIGURE 1.3 Percentage of undernourished in total population by region, 1979–81, 1990–92, 1998–2000 (FAO).

in 1998–2000 (see Figure 1.3). This is significant progress. However, it has been uneven and has slowed in recent years. In Asia and the Pacific, the prevalence has been halved since 1979–81, with the most rapid progress occurring in East Asia. In sub-Saharan Africa the incidence of undernourishment has been falling slightly since 1979–81, although with population growth the absolute number of undernourished has continued to rise. In Latin America and the Caribbean, the incidence of undernourishment has fallen marginally in the 1990s, from a low base. In the Near East and North Africa, marginal progress in the 1980s was offset by a slight deterioration during the 1990s, so both the prevalence and the absolute number of undernourished have increased, also from a low base.

Table 1.1 presents three additional indicators of nutritional status: the share of starchy foods (cereals, roots, and tubers) in the overall diet, life expectancy at birth, and under-five mortality rates. The last two indicators measure aspects of a more complex state of human well-being, including nutritional status. To allow a better understanding of trends, Table 1.1 also gives these figures for 1990–92.

The share of the total diet derived from starchy foods is an indicator of variety and quality of the average diet of a population. A satisfactory diet can be had with starchy staples ranging anywhere between 55–75 percent of total dietary energy supply (DES), and diet composition can thus vary considerably from season to season and from culture to culture, without detrimental effect on nutritional status. Nevertheless, starchy staples in excess of 70–75 percent of total DES is cause for concern because there is a positive correlation between this indicator and the number of undernourished. Moreover it is indicative of insufficient nutritional diversity and a deficiency of micronutrients needed for a healthy diet. Note that the figures in Table 1.1 refer to the average diet, so it is likely that the food-insecure people within these populations are more heavily dependent on starchy staples than the average consumer.

In the developing world, in 1998–2000, the share of cereals, roots, and tubers in total energy was around 63 percent, down very slightly from 1990–92. In most of the developing regions, not only was this share high to begin with, but there was no evidence of any decline, indicating that diets had not diversified to any great extent. The only exceptions were East and South East Asia, where this share went down 7 percentage points, and Latin America and the Caribbean where the share was low to begin with, at 45 percent. Everywhere else this share remains above 60 percent and is higher in several individual countries. It is, in fact, 70 percent or more in 15 countries in sub-Saharan Africa, 5 countries in Southeast Asia, and 2 countries in South Asia.

TABLE 1.1. Food security indicators by developing region and time period

Region	East & Southeast Asia	Latin America & Caribbean	Near East & North Africa	South Asia	Sub-Saharan Africa	All Developing Regions
Number undernourished (millions)						
1990–92	292	59	26	276	166	818
1998–2000	193	55	40	315	196	799
Percentage undernourished						
1990–92	16	13	8	26	35	20
1998–2000	10	11	10	24	33	17
DES (Kcal/day/person)						
1990–92	2656	2710	3010	2330	2120	2540
1998–2000	2930	2820	2940	2390	2210	2670
Share of starchy food in total DES (percentage)						
1990–92	74	45	62	68	70	67
1998–2000	67	44	61	65	69	63
Life expectancy at birth in years, female/male						
1990–92	69/66	71/65	66/64	60/59	52/49	65/62[1]
1996–98	71/67	73/67	69/66	63/62	52/49	67/63
Under-five mortality rate, per 1,000						
1990–92	55	49	72	121	155	91[1]
2000	45	37	54	96	162	84

Source: FAO Statistics, World Bank World Development Indicators, 2000 and 2002.
[1] Average for low and middle-income countries (World Bank definition).

Turning now to indicators of general health, we find that except in sub-Saharan Africa, there has been a general improvement in both female and male life expectancy between 1990–92 and 1996–98. By the latter period, the regions with the highest female and male life expectancies were Latin America and the Caribbean and East and South East Asia respectively, which were quite close to the OECD averages of 81 and 75 for that period. Life expectancy is lowest in countries with the highest prevalence of undernourishment, also because undernourishment shortens lives through increased susceptibility to illness. In addition, there are thirty-two countries that have seen life expectancy decline since 1990. Most are countries hit by the AIDS epidemic. Nine countries lost more than three years of life expectancy: Botswana (−10.7), Zambia (−6.6), Kenya (−6.1), Zimbabwe (−5.2), Uganda (−4.3), Kazakhstan (−3.7), Côte d'Ivoire (−3.7), Central African Republic (−3.2), and Namibia (−3.1).

The mortality rate for newborn infants is also an indicator of the nutritional status of their mothers, while those for children under five are suggestive of the nutritional status of the children themselves. It is now recognized that 6 million out of 11 million deaths among children under five—or 55 percent of young child mortality in developing countries—are associated with malnutrition. Though mortality rates in children under five fell in all regions except sub-Saharan Africa, they are declining too slowly to attain the Millennium Development Goal of a two-thirds reduction by 2015: rates should have come down by roughly 30 percent in the 1990s, but they declined by only 8 percent. Between 1990 and 2000, 17 developing countries reduced their under-five mortality rate fast enough to meet the Millennium Development Goal. But over the same period, 14 countries experienced worsening rates, among them the Democratic Peoples Republic of Korea, where the child mortality rate increased from 35 to 90, and Zimbabwe, where it went from 76 to 116 per 1,000. By 2000, even in the region with the lowest under-five mortality rate in the developing world, Latin America and the Caribbean, this rate was still around 37 per thousand. This was enormously higher than the average for the OECD countries, which was about 6 per thousand.

FOOD SHORTAGE EMERGENCIES[2]

As noted above, these stark data on undernourishment and deprivation do not reflect a global scarcity of food. Quite to the contrary, world prices for basic foodstuffs have declined 38 percent since 1996, suggesting there is no shortage of food at the global level. Although that year marked a "spike" in global cereals prices, longer-term trends indicate that world prices have been declining steadily.[3]

In any given year, however, between 5 and 10 percent of the total number of undernourished in the world can be traced to specific events: droughts; floods; armed conflict; and social, political, and economic disruptions. Frequently, these shocks strike countries already suffering from endemic poverty and struggling to recover from earlier natural and manmade disasters. Often, transitory food insecurity transforms into chronic food insecurity as people's assets and savings are wiped out.

Globally, 31 countries were experiencing severe food shortages and required international food assistance as of August 2002. An estimated 67 million people required emergency food aid as a result of these shocks. Both the number of countries and people affected remained almost identical to the figures from a year earlier, as did the causes and locations of these food emergencies. As in previous years, drought and conflict were the most common causes of emergencies and Africa was the most affected region.

In southern Africa, nearly 13 million people need emergency food aid. A combination of droughts, floods, and economic dislocations reduced harvests in several countries to half or less than their normal levels. In Zimbabwe—until recently an exporter of maize—bad weather, political conflict, and economic problems have combined to cripple production. Maize production has fallen sharply in several countries, and prices have risen by as much as 400 percent, seriously undermining access to food for large sections of the population.

In eastern Africa, the food outlook is bleak in several countries due to poor seasonal rains. In Eritrea, Ethiopia, and Kenya a combined figure of 10.3 million people are in need of food assistance due to drought or poor rains. In Somalia, despite a favorable forecast for the main season crops, serious malnutrition rates are reported, reflecting successive droughts and long-term insecurity. The escalation of conflict in northern Uganda has displaced large numbers of people, adding to the more than 1.5 million internally displaced persons, refugees, and other vulnerable people that already depend on food assistance.

In western Africa, dry weather has seriously affected crops, particularly in The Gambia, Guinea-Bissau, Mauritania, and Senegal. In Cape Verde, prospects for the maize crop, normally planted from July, are unfavorable due to delayed onset of rains. By contrast, crop-growing conditions have improved in central and eastern parts of the Sahel. Agricultural activities in Liberia have been disrupted by renewed civil strife, and both Guinea and Sierra Leone remain heavily dependent on international food assistance due to large numbers of internally displaced persons and refugees.

In central Africa, the food supply situation has improved in Burundi and Rwanda following good harvests of the 2002 second season crops. By contrast, the food and nutritional situation in the Democratic Republic of Congo gives cause for serious concern. Persistent civil strife continues to cause massive population displacements, with the number of internally displaced persons currently estimated at 2 million.

In Asia, food assistance has been resumed in the Democratic Peoples Republic of Korea following recent donations from the international community. In Mongolia, another harsh winter and severe spring storms further eroded the food security of nomadic herders. Extreme floods have caused loss of life and damage to infrastructures and crops in Bangladesh, western and central China, and northeastern India. The food supply situation in some Asian countries of the Commonwealth of Independent States is tight due to adverse weather. Emergency food assistance is required in Georgia, Tajikistan, and Uzbekistan. Tajikistan, in addition, has recently experienced a locust infestation, torrential rains, and floods, which have destroyed large areas of crops.

The food situation in Afghanistan is grave, with about 10 million people entirely dependent on food assistance. Even before the events of September 11, 2001, Afghanistan was gripped by a serious food crisis. After a third year of drought, cereal output in 2001 fell to barely half the production level of 1998. Livestock herds, which are critical to the country's economy and food security, had been reduced by an estimated 40 percent. Hunger and malnutrition have increased sharply in a country where stunting of children was already reported as high as 52 percent in 1998, even before the drought set in and food production plummeted. Cereal production has rebounded significantly in 2002, buoyed by increased rainfall and better access to agricultural inputs; however, the worst locust plague in 30 years and floods in some areas have affected crop prospects. Despite this recovery, after years of conflict and drought, large-scale investment is urgently needed to repair rural infrastructure and restore crop and livestock production.

Elsewhere, targeted food assistance continues to be necessary for refugees, the internally displaced, and vulnerable populations in the Federal Republic of Yugoslavia and in Chechnya in the Russian Federation.

In Central America, a severe drought that devastated crops in 2001 combined with the collapse in world coffee prices left families in rural areas in several countries of the region dependent on food aid.

UNDERSTANDING THE RELATIONSHIP BETWEEN HUNGER AND POVERTY

The lack of action in the fight against hunger might have arisen from a belief that success in poverty reduction, resulting from market-driven economic development, would "automatically" take care of the problem of hunger. However, this thinking does not take into account three points:

- Poverty reduction takes time.
- The hungry need immediate relief.
- Hunger is as much a cause as an effect of poverty.

Poverty is at the root of undernourishment and food insecurity. What is often ignored or underestimated is that undernourishment and malnutrition might be a major impediment in peoples' efforts to escape poverty. Table 1.2 lists some important studies and their findings on the effects of hunger and malnutrition on productivity. These and numerous other studies have confirmed that hunger seriously impairs the ability of the poor to develop their skills and reduces the productivity of their labor.

Hunger in childhood impairs both physical and mental growth. There is evidence that poor nutrition is associated with poor school performance in children, thereby damaging their future economic prospects. There are two reasons for this handicap:

- Because of hunger, the child is listless or tired and inattentive and cannot participate in learning activities.
- Cognitive ability might be impaired as a result of prolonged and severe malnutrition.

There is also a risk of intergenerational transmission of poor nutritional status. For example, women who suffer from poor nutrition are more likely to give birth to underweight babies. These babies thus start out with a nutritional handicap. When malnourished children reach adulthood, evidence from household surveys in developing countries shows that people with smaller and slighter body frames caused by undernourishment earn lower wages in jobs involving physical labor. Other studies have shown a strong positive impact of the Body Mass Index (BMI, a measure of weight for a given height) on wages.

TABLE 1.2. Summary of studies on the productivity impact of poor nutrition

Study Authors	Country	Group Studied	Main Findings
Croppenstedt and Muller (2000)	Ethiopia	Rural households, mainly agricultural	Output and wages rise with Body Mass Index (BMI)[1] and Weight-for-Height (WfH). Adult height has positive impact on wages.
Strauss (1986)	Sierra Leone	Rural households, mainly agricultural	Calorie intake has positive impact on productivity.
Satyanarayana et al. (1977)	India	Indian factory workers	WfH is significant determinant of productivity.
Deolalikar (1988)	India	Southern Indian agricultural workers	Significant effect of WfH on farm output & wages.
Alderman et al. (1996)	Pakistan	Rural households, mainly agricultural	Adult height is significant determinant of rural wages.
Haddad and Bouis (1991)	Philippines	Sugarcane growers	Adult height is significant determinant of rural wages.
Thomas and Strauss (1996)	Brazil	Urban population sample	BMI, adult height, have strong, positive impacts on market wages.
Spurr (1990)	Colombia	Sugarcane cutters and loaders	Weight, height are significant determinants of productivity.
Immink et al. (1984)	Guatemala	Coffee and sugarcane growers	Adult height has positive impact on productivity.

[1]The Body Mass Index is weight (kg) / (square of height (cms)).

Micronutrient deficiencies can also reduce work capacity. Surveys suggest that iron deficiency anemia reduces the productivity of manual laborers by as much as 17 percent. As a result, hungry and malnourished adults earn lower wages when they work. Hunger also affects the productivity of workers through its impact on health because poorer nutritional status leaves people more susceptible to illness. Workers who are malnourished are frequently unable to work as many hours and years as those who are well-nourished, because they fall sick more often and have shorter life spans.

People who live on the edge of starvation can be expected to follow a policy of safety first with respect to investments—that is, they will avoid taking risks since the consequences for short-term survival of a downward fluctuation in income will be catastrophic. But less risky investments also tend to have lower rewards. Again, the tendency is for poor nutrition to be associated with lower earnings.

The cumulative impact of all these factors is staggering. Recent research on some Asian and African countries with high levels of hunger and malnutrition concludes that in Asia as many as 2.8 million children and close to 300,000 women die needless deaths every year because of malnutrition. In some Asian countries, estimates of adult productivity losses arising from the combined effect of stunting, iodine deficiency, and iron deficiency are equivalent to about 3 percent of GDP every year in these countries. A study on Africa came to broadly similar conclusions; undernourishment may have cost the countries of sub-Saharan Africa an average of 1 percentage point in GDP growth per year over the period from 1960 to 1990. Similarly the work of Nobel Laureate Robert Fogel shows, based on historical longitudinal studies of single countries, that improvements in nutrition and health explained half of British and French economic growth in the 18th and 19th centuries. Table 1.3 shows some of the estimated impacts of malnutrition.

These studies clearly show the vicious cycle of hunger and poverty from which it is difficult for the poor and the hungry to escape without external help.

However, if the cycle were broken, the benefits would be enormous. FAO estimates that reducing the number of undernourished people by half by 2015 would yield a value of more than $US120 billion in terms of longer and healthier lives. Similar calculations in the report of the World Health Organization Commission on Macroeconomics and Health suggest gains from improved nutrition and health of hundreds of

TABLE 1.3. Estimates of productivity costs of malnutrition, selected Asian countries (annual, as percent of GDP)

| Country | Losses of Adult Productivity | | |
	Stunting	Iodine Deficiency	Iron Deficiency
India	1.4	0.3	1.25
Pakistan	0.15	3.3	0.6
Vietnam	0.3	1.0	1.1
	Losses Including Childhood Cognitive Impairment Associated with Iron Deficiency	
Country	Cognitive Only	Cognitive Plus Manual Work
Bangladesh	1.1	1.9
India	0.8	0.9
Pakistan	1.1	1.3

Source: Horton, 1999, p. 251.

billions of dollars per year if the goals can be met. Thus fighting hunger is not only a moral imperative; it also brings large economic benefits.

Success in reducing hunger is also likely to produce large benefits in terms of sustainable development. The economic prosperity resulting from hunger reduction should create demand for sustainable use of the environment and of common property resources.

Finally, better nourishment may also reduce the likelihood of conflict and thus contribute to global stability and lower expenditure on conflict prevention and rehabilitation of war-torn areas. A study by the United States Agency for International Development found that meeting the WFS target would lower the cost of peacekeeping and humanitarian operations by about $US2.5 billion per year.

Halving hunger is not only a valid goal in itself, but is also closely linked to the achievement of other key goals set by the international community, most of which are reflected in the Millennium Declaration.

ALLEVIATING POVERTY AND HUNGER

A successful strategy for alleviating poverty and hunger in developing countries must include two essential elements. First it should aim to promote agriculture and rural development. But, agricultural development cannot contribute fully unless the productivity gap caused by hunger is dealt with directly. Therefore, a strategy of promoting agricultural development when deployed in conjunction with policies to enhance the capabilities of the poor and to ensure food security offers the best hope of swiftly reducing mass poverty and hunger. Although those are but two of the many measures needed to effectively fight hunger, they are necessary in most societies where hunger is widespread.

Why focus on agricultural development? First, there is the seemingly obvious but often ignored fact that more than 70 percent of the poor and undernourished people live in rural areas and derive the basis for their livelihoods directly from agriculture as producers or laborers or from rural non-farm (RNF) activities dependent on agriculture.

Although hunger and poverty are multifaceted phenomena, sustainable hunger and poverty reduction require enhancement of the sources from which the poor derive their livelihoods, especially their income sources. Hence, pro-poor income growth needs to be encouraged. The question then becomes, under what circumstances is income growth pro-poor? The short answer is that income growth originating in agricultural development, when coupled with growth in the RNF sector is likely to be strongly poverty reducing provided that it does not occur against a backdrop of high inequality in asset ownership, especially of land. The argument is fleshed out in the next sections.

THE IMPORTANCE OF AGRICULTURAL DEVELOPMENT

As noted above, poverty in developing countries is concentrated in rural areas. This is particularly true in countries with high undernourishment. These countries depend on agriculture, directly or through related activities, for the largest part of the employment of the labor force, and for a high proportion of their economic output and export earnings. The percentage of the labor force employed in agriculture in 1999 was 56 percent on the average in developing countries. In Africa south of the Sahara, where 33 percent of the population was undernourished in 1998–2000, employment in agriculture is as high as 67 percent of the total labor force (see Table 1.4).

Regional data mask important differences between countries. Table 1.5 reports data for countries classified according to the prevalence of hunger. The five categories in the table correspond to the share of population estimated to be undernourished.[4] The table shows that for coun-

TABLE 1.4. Dimensions of agriculture in developing countries

	Share of Rural to Total Population (Percent)		Share of Agricultural to Total Labor Force (Percent)		Share of Agriculture in Total GDP (Percent)	
	1990	1999	1990	1999	1990	2000
Developing countries	66	60	61	56	16	12
Latin America and Caribbean	29	25	25	20	9	7
Near East and North Africa	46	40	39	34	15	14
Sub-Saharan Africa	74	68	72	67	18	17
East and Southeast Asia	71	64	68	62	20	13
South Asia	75	72	63	59	31	25

Source: FAOSTAT and World Bank, World Development Indicators 2002.

TABLE 1.5. The importance of agricultural sector in developing countries by prevalence of undernourishment category (1990–93 and 1995–98)

Indicator	Undernourishment Prevalence Categories									
	1 <2.5%		2 2.5–4%		3 5–19%		4 20–34%		5 >35%	
	90/93	95/98	90/93	95/98	90/93	95/98	90/93	95/98	90/93	95/98
Share of agriculture in GDP (%)	10.9	9.7	11.6	12.7	16.4	18.2	23.1	22.3	31.6	21
Share of agricultural exports in total exports (%)	10.7	9.0	11.2	8.7	11.7	10.7	18.2	15.0	42.2	43.5
Share of rural population in total population (%)	31.8	27.3	52.0	51.6	63.5	59.9	72.8	71.2	74.8	75.8

Note: Prevalence categories represent countries grouped by the percentage of the population undernourished.

tries in which undernourishment is high, agriculture is of paramount importance for overall GDP, exports, and especially employment. It is therefore hard to imagine how overall growth (an essential requirement for sustainable poverty alleviation) could occur without growth in the agricultural sector, or how employment opportunities for the poor could be created in very poor countries without growth in the rural sector, which accounts for as much as three-quarters of overall employment.

The poor in rural areas depend on agriculture to produce the food they eat or to generate income. Rural households generate income from agricultural activities and/or from employment in rural non-farm activities, which are in most cases linked to the agricultural sector.

Agricultural development, by raising incomes in agriculture, creates demand for the products of the rural non-farm sector, the output of which is demand-constrained and makes it possible for that sector to grow. Since the goods and services produced by the rural non-farm sector require few skills and little capital, the barriers to entry in those sectors and activities (either as laborers or entrepreneurs) are low and, as such, those activities are easily accessible by the poor, raising and diversifying their income sources, thus reducing their exposure to risk. Thus, productivity-induced growth in agricultural output, in addition to increasing income for those directly involved in farming, creates secondary or multiplier effects in the local economy and to the incomes of those involved in it. A recent study in five countries in sub-Saharan Africa shows that adding $US1.00 of new farm income potentially increases total income in the local economy—beyond the initial $1.00—by an additional $1.88 in Burkina Faso, by $1.48 in Zambia, by $1.24–$1.48 in two locations in Senegal, and by $0.96 in Niger.   This process cannot work, however, if there are marked initial inequalities in access to agricultural and other assets. Large holdings are likely to use capital-intensive technologies, so additional needed inputs are more likely to be imported either from abroad or from large urban centers. At least part of the additional income derived from increased agricultural productivity is likely to be spent on more luxury goods, also imported from outside the local area. Such a process is thus characterized by "leakages," which are likely to break the virtuous circle described above.[5]

CHALLENGES TO AGRICULTURAL DEVELOPMENT IN DEVELOPING COUNTRIES

The effort to promote productivity-based agricultural growth faces some critical challenges. Some of the indicators of the difficulties facing developing countries in developing their agriculture can be seen in Table 1.6. The table shows that in various degrees, developing countries are handicapped in their effort to compete with higher-income ones in an increasingly global market. Particularly serious, in almost all indicators of modern input use and agricultural performance is the handicap of sub-Saharan Africa, although agriculture is the mainstay of most countries in the region.

The ability to rapidly adjust to the changing context in which agricultural development takes place is a challenge to all developing countries. And while reductions in barriers to trade expand the opportunities for raising sectoral output, globalization raises risks of marginalization for countries which—because of their resource endowment, location, size, or lack of skills and infrastructure—remain uncompetitive in world markets and unable to attract investment.

The extent to which developing countries are able to take advantage of new market opportunities emerging from globalization ultimately depends on their competitiveness and their capacity to increase the production of goods that are in demand. This may require substantial investments in infrastructure, technology, and communications aimed at reducing costs and speeding up transport. But it also calls for developing institutional capacities to set and enforce high standards for training farmers in production of marketable products.

A closely related challenge is posed by new production technologies. Improved crop varieties and animal breeds, greater use of fertilizers and pesticides, better farm equipment, and improvements in livestock care and health have all contributed to the growth in agricultural output, which has responded not only to the demands of a population that has doubled from 3 billion in 1960 to 6 billion today but also enabled average daily food intake to rise from less than 2000 kcal per head in 1960 to almost 2,700 kcal per head per day in the developing world in 1997–99.

But can this trend continue—and if so, how? It seems clear that the majority of yield increases will continue to result from improvements of classical or conventional technology—in particular, improvement of water use efficiency and improved nutrient uptake by plants and animals. Access to conventional technologies is still beyond the means of many farmers, as shown by the very low levels of fertilizer utilization and irrigation in some developing regions (Table 1.6). Regarding the use of modern varieties, Table 1.7 shows that some regions are better prepared

TABLE 1.6. Agricultural indicators: developing, middle income and high income countries (1999)

Indicator	Africa	Sub-Saharan Africa	Near East/ North Africa	South Asia	East and Southeast Asia	Latin America/ Caribbean	Middle Income	High Income	World
Ag. capital stock/Ag. worker[1]	2,102	1,442	9,110	1,866	1,592	10,441	1,764	107,156	3,920
Irrigation, % of arable land	7.0	3.8	28.7	39.3	31.9	11.6	19.9	11.9	20.0
Ag. value added/ worker ($/yr)	416	285	1,859	412	461	3,028	335	17,956	645
Cereal production/ caput (kg/yr)	147	128	128	224	336	259	339	746	349
Cereal yields (kg/ha)[2]	1,225	986	1,963	2,308	4,278	2,795	2,390	4,002	2,067
Livestock productivity[3]	164	128	147	121	150	198	191	248	193
Fertilizer use (kg/ha)[4]	22	9	69	109	241	85	111	125	100

[1]In 1995 U.S. dollars, 1999 figures.
[2]Average 1997–99.
[3]Carcass weight (kg)/animal, cattle and buffalo, 1997–99 average.
[4]Per ha of arable land.

TABLE 1.7. Modern variety diffusion 1980, 1990, and 1998—percent area planted to modern varieties

	Latin America			Asia (Including China)			Middle East/ North Africa			Sub-Saharan Africa		
	1980	1990	1998	1980	1990	1998	1980	1990	1998	1980	1990	1998
Wheat	46	82	90	49	74	86	18	38	66	22	32	52
Rice	22	52	65	35	55	65				2	15	40
Maize	20	30	46	25	45	70				4	15	17
Sorghum				20	54	70				8	15	26
Millets				30	50	78				0	5	14
Barley							7	17	49			
Lentils							0	5	23			
Beans	2	15	20									
Groundnut				15	20	50				0	20	40
Cassava	1	2	7	0	2	12				0	2	18
Potatoes	54	69	84	50	70	90				25	50	78
All crops	23	39	52	43	63	82	13	29	58	4	13	27

than others to face the challenge posed by the need to increase productivity and competitiveness.

There are undoubtedly benefits associated with the new generation of biotechnologies, although there are also risks and uncertainties. Assuming that satisfactory safeguards can be developed and applied to limit risks to plant, animal, and human health (biosecurity risks), a major issue is the extent to which new biotechnology applications will benefit farmers, especially small farmers, in developing countries. Given that most current biotechnology research is by the private sector, it is strongly market-driven, and therefore low priority is given to applications of relevance to developing countries, where the purchasing power of farmers is relatively limited and the possibility of enforcing breeder's rights may be in doubt. One possibility to overcome these obstacles is to sponsor biotechnology development for developing countries by the public sector, including the International Agricultural Research Centres, which operates within the ambit of the Consultative Group for International Agricultural Research (CGIAR).

Public awareness of food safety issues has increased dramatically, especially in developed countries, in the five years since the WFS. The common attribute in many of these crises has been the public opinion that the measures in place are ineffective or inefficiently applied, or applied only in the interest of increased trade or benefits to producers or to industry, and not necessarily in the interest of consumers.

A science-based approach to international standard setting should be followed, particularly within the context of WTO rules for sanitary and phyto-sanitary measures. The Codex Alimentarius Commission is the only international standard setting mechanism recognized by the WTO for food safety purpose. Four major challenges emerge in the effort to address food safety:

- Taking the concerns of consumers very seriously and communicating better with them.
- Placing more emphasis on science-based risk assessments of emerging food safety issues.
- Promoting an exchange of information and experience between food safety regulators in developed and developing countries.
- Assisting developing countries to build the capacity of their food safety regulatory systems, and to participate more actively in international standard setting fora. Building such capacity in developing

countries will not only benefit consumers in those countries, it will also enhance their access to developed countries' markets and contribute to the protection of the health of consumers in these countries.

Resources for Agricultural Development

The preceding analysis points to the need for substantial and swift action to foster agricultural productivity, especially in countries with high levels of poverty and undernourishment. Action has to be taken across multiple fronts: institutional, policy, and resource mobilization geared toward agricultural development.[6]

Public investments are required in supporting infrastructure and in public goods, such as basic agricultural research and institution building. The analysis that follows shows that the trends so far regarding resource mobilization for agricultural development is not commensurate with agriculture's role in developing countries.

Although most investments will come from private sources, notably the farmers themselves, the public sector has a critical role to play through its expenditures on agriculture, forestry, and fisheries sectors in creating an environment conducive to private investment (economic incentives) and to ensuring sufficient availability of public goods (basic infrastructure and rules of law, peace, and security). But countries with high levels of undernourishment do not have the ability to mobilize public resources to the extent required. In most developing countries, especially the poorest ones, resources for agricultural development have to be provided by international sources in the forms of grants or lending.

When agricultural investment data are examined by groups of countries experiencing similar prevalence of undernourishment levels, it appears that the capital stock in agriculture per agricultural worker in the group of countries belonging to high undernourishment prevalence categories is low when compared to the group with low levels of undernourishment (see Figure 1.4). The wide divergence among capital-labor ratios in agriculture is partly reflected in labor productivity differences across prevalence categories (see Figure 1.5). Thus, the value added per worker in agriculture in the lowest prevalence category was 14–17 times that of the highest prevalence category in the period between 1990 and 1998, which is much higher than the difference in capital-labor ratios. This result points not only to higher levels of capital resources but also

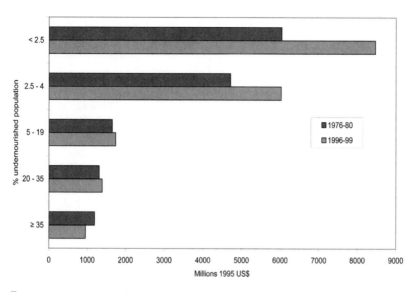

FIGURE 1.4 Capital stock per agricultural worker by undernourishment prevalence category (FAO).

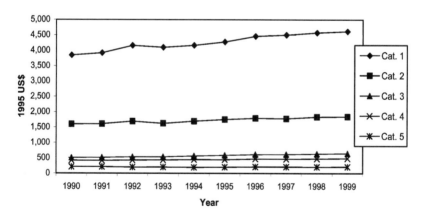

FIGURE 1.5 Agricultural value added per agricultural worker by undernourishment prevalence category.

higher overall productivity of resources in countries with low levels of undernourishment relative to countries in the higher categories.

External assistance to agriculture counts for as much as 86 percent of gross domestic investment and 51 percent of government expenditures in the countries with the highest prevalence of undernourishment. Developing countries face a decline in the overall official development assistance (ODA) by the major bilateral and multilateral donors, and in the share of ODA directed toward the agricultural sector (see Figure 1.6). Furthermore, private foreign direct investment (FDI) has so far bypassed most of the poor countries, and, of the overall private investment going to the poorer countries, relatively little goes to the food and agriculture sectors of the poorest of them.[7]

The Way Forward: An Anti-Hunger Program

To conclude this chapter, this section presents an anti-hunger program aimed at meeting the WFS goals. The anti-hunger program includes five priorities for action. The first four priorities relate to the agriculture and rural development track. The fifth relates to measures to enhance access to food.

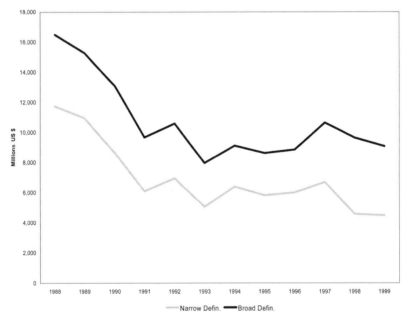

FIGURE 1.6 Official development assistance toward agriculture and rural development —narrow and broad definitions (millions of 1995 U.S. dollars).

1. IMPROVE AGRICULTURAL PRODUCTIVITY AND ENHANCE LIVELIHOODS AND FOOD SECURITY IN POOR RURAL COMMUNITIES

Improving the performance of small farms in poor rural and peri-urban communities offers one of the best and most sustainable avenues for reducing hunger by increasing the quantity of locally available food and improving its quality. But it also creates a base for expanding and diversifying farm output into tradable products, opens employment opportunities, and slows rural-urban migration. The scale of the program must be massive if it is to have a meaningful impact on reducing hunger and poverty.

2. DEVELOP AND CONSERVE NATURAL RESOURCES

If food demand is to be met in the future, increased outputs will have to come mainly from intensified and more efficient use of the limited land, water, and plant and animal genetic resources available. At the same time, action must be taken to arrest the destruction and degradation of the natural resource base. Achieving these apparently conflicting tasks requires investments to manage the resource base, improve technical production efficiency (yields), and develop practices that foster sustainable and intensified food production. International agreements, such as the International Treaty on Plant Genetic Resources for Food and Agriculture adopted at the 2001 FAO Conference, can provide agreed frameworks for the conservation and sustainable utilization of key agricultural resources and the fair and equitable sharing of the benefits.

3. EXPAND RURAL INFRASTRUCTURE (INCLUDING CAPACITY FOR FOOD SAFETY AND PLANT AND ANIMAL HEALTH) AND BROADEN MARKET ACCESS

Throughout the 1990s, many developing countries have invested substantially in infrastructure. While such investments have done much to improve living standards and increase productivity, the rural areas of most developing countries still face inadequate levels of services and often a deteriorating stock of rural infrastructure. This infrastructural handicap has resulted in, inter alia, reduced competitiveness of the agriculture of developing countries in domestic and international markets, and has increased the costs of supplying growing urban markets from national farm production. Highest priority must go to improving the rural road network. Substantial investments are also required in monitoring and surveillance systems and in building the capacity of institutions responsible for plant and animal health. Finally, action is urgently needed to develop food handling, processing, distribution, and marketing enterprises by promoting the emergence of small farmers' input supply, processing, and marketing cooperatives and associations.

4. Strengthen capacity for knowledge generation and dissemination (research, extension, education, and communication)

As noted, success in promoting rapid improvements in livelihoods and food security through on-farm investments depends on small farmers having good access to relevant knowledge. Increasingly, agricultural research and technology development will be dominated by the private sector, especially suppliers of inputs and companies purchasing farm products. There remain, however, many areas of research and extension that yield few benefits for the private sector, but which are still vital for agricultural development and the sustainable management of natural resources. These include most forms of pro-poor technology development and most approaches to farm development that do not depend on the increased use of purchased inputs—such as integrated pest management, measures to raise the organic matter content of soils, or to improve fertilizer use efficiency (e.g., through biological nitrogen fixation), or to conserve genetic resources. Research in all such topics tends to be seriously underfunded.

To improve the chances of success, a strategy for agricultural and rural development should adopt an approach in which research, extension, education, and communication components are integrated.

5. Ensure access to food for the most needy through safety nets and other direct assistance

All governments committed to achieving the WFS goal need to put programs in place that ensure that their citizens have access to adequate food where this goal is not being met through traditional extended family and community coping arrangements, market mechanisms, and the process of economic growth. Options include targeted direct feeding programs, food-for-work programs, and income-transfer programs.

FAO estimates that the combined public cost of investment requirements indicated for financing all five of these priority actions is approximately $US24 billion at 2002 prices. Of this, some $US5 billion will be for addressing the hunger problem through direct transfers to the undernourished. Another $US19 billion will be required for addressing the problem of undernourishment and rural poverty through agricultural growth and productivity enhancement in rural areas.

CONCLUSION

FAO considers that the slow progress in hunger reduction points to the need for increasing resources to meet the five priorities. An overall strategy involving these direct measures is needed to enhance access to food for those in extreme poverty, while at the same time promoting board-based agriculture and rural development to create the opportunities.

REFERENCES

FAO, 2002. *The State of Food Insecurity in the World 2002*. United Nations: Food and Agriculture Organization.

NOTES

1. FAO's primary indicator of food security is the number of people whose diet does not allow them to consume a sufficient number of calories for a healthy diet. This indicator is based on country-level estimates of the average per person dietary energy supply (DES) from local food production, trade, and stocks; the number of calories needed by different age and gender groups; the proportion of the population represented by each group; and country-specific coefficients of income/expenditure distribution to take account of inequality of access to food.
2. This report is based on information available as of August 2002. Up-to-date information can be found in FAO's *Foodcrops and Shortages*, issued every two months.
3. FAO, Food Stocks and Price Monitor.
4. Thus, countries in category 1 have less than 2.5 percent of the population undernourished; for category 2, the percentage is between 2.5 and 4 percent; etc.
5. The argument can be generalized for the case in which the intra-national distribution of income and access to assets is very skewed.
6. An extensive review of issues related to resource mobilization especially for agricultural growth can be found in "Mobilising Resources to Fight Hunger," Information Paper for the 27th session of the Committee on World Food Security, Rome May 28, June 1 2001 (http://www.fao.org/docrep/meeting003/Y0006E/Y0006E00.htm).
7. A rough estimate shows that for developing countries as a whole, foreign direct investment accounts for less than 1 percent of the total capital stock in agriculture.

2

FEEDING A WORLD OF 10 BILLION PEOPLE: OUR 21ST CENTURY CHALLENGE[1]

Norman E. Borlaug

It is a great pleasure to give the inaugural lecture of the newly created Borlaug lecture series at Iowa State University. My purpose here today is to explore the role of science and technology in the coming decades to improve the quantity, quality, and availability of food for the world's people.

Ten days ago, on October 5th, I began my 59th year of continuous involvement in agricultural research and production in the low-income, food-deficit developing countries. I have worked with many colleagues, political leaders, and farmers to transform food production systems. During the past 40 years, thanks to a continuing stream of high-yielding varieties that have been combined with improved crop management practices, food production has more than kept pace with global population growth. Globally, per capita world cereal supplies are 23 percent higher and real prices are 65 percent lower than in 1961. Because much of these gains has come through the adoption of productivity-enhancing technology, both producers and consumers—especially poor consumers—have gained.

Despite this tremendous expansion in food production, 2 billion people still lack reliable access to safe, nutritious food, and 800 million of them—including 150 million children—are chronically malnourished. Thus, there is no room for complacency on the food production and poverty-alleviation fronts.

AGRICULTURE AND POPULATION

In geologic terms, the domestication of plant and animal species is a recent event. Archaeological evidence indicates that all the major cereals, economically important legumes, root crops, and animal species that are our principal sources of food were domesticated over a period of only 2,000–3,000 years. The process may well have begun when Neolithic women, faced with food shortages when their menfolk failed to bring home enough food from hunting forays, decided that something had to be done and began searching for a means to assure a more permanent and reliable supply. This was achieved by sowing seed of the same wild grain species they had been collecting for untold millennia to supplement their meat diet. Thus, agriculture was born, some 10,000–12,000 years ago. With the development of agriculture, the condition of humankind began to improve markedly, and human numbers, estimated to have been 15 million at that time, began to increase at an accelerated rate. A more stable food supply resulted in better nutrition and the development of a settled way of life, leading to higher survival rates and yet more rapid population growth.

World population presumably doubled four times—to about 250 million—from the beginning of agriculture to the start of the Christian era. It doubled again, to 500 million, by about 1650. The next doubling required only 200 years, producing a population of one billion by 1850. At about that time the discovery of the nature and cause of infectious diseases—the dawn of modern medicine—began to lower death rates. It took only 80 years for the next doubling—to two billion people—which occurred about 1930. Shortly thereafter, the development of sulfa drugs, antibiotics, and improved vaccines led to a further substantial reduction in death rates, especially among infants and children. The next doubling of population took only 45 years—to about 1975, when global population reached four billion. The next doubling is projected by 2020, again only 45 years, representing a 533-fold increase since the discovery of agriculture.

While growth of world population overall is now slowing, the current rate in much of the developing world is still frighteningly high. Over the next 50 years, world population is likely to swell to at least 10 billion people, with 90–95 percent born in low-income developing countries, and very likely into conditions of poverty. Hopefully, by the end of the 21st century, world population will stabilize at 11–12 billion people, and much of the poverty that still haunts the world will have been abated.

There are two aspects to the problem of feeding the world's people. The first is the complex task of producing sufficient quantities of the desired foods to satisfy people's needs. The second task, equally or even more complex, is to distribute the food equitably. The chief impediment to equitable food distribution is poverty—lack of purchasing power.

About 42 percent (2.6 billion) of the world's people still live on the land and rely on their own efforts to feed themselves. Millions of these rural poor remain food-insecure. Thus, only by increasing agricultural productivity in food-deficit areas can both aspects of the world food problem be ameliorated.

FOOD PRODUCTION AND THE ROLE OF SCIENCE

In 2000, global food production of all types stood at 5.2 billion metric tons, representing some 2.7 billion tons of edible dry matter (Table 2.1).

Of this total, 99 percent was produced on the land; only 1 percent came from the oceans and the inland waters. Plant products constitute

TABLE 2.1. World food supply, 2000

	Production, Million Metric Tons		
Commodity	Gross Tonnage	Edible Matter[1]	Dry Protein[1]
Cereals	2,064	1,718	171
Maize	593	524	54
Wheat	585	516	61
Rice	601	407	34
Barley	135	118	12
Sorghum/millet	84	76	7
Roots and Tubers	698	186	12
Potato	328	71	8
Sweet potato	139	42	2
Cassava	177	66	1
Legumes, oilseeds/nuts	165	112	39
Sugarcane and sugar beet[2]	150	150	0
Vegetables and melons	692	81	7
Fruits	466	64	3
Animal products	992	196	87
Milk, meat, eggs	866	164	66
Fish	126	32	23
All food	5,227	2,507	319

Source: FAOSTAT, July 2002.

[1]At zero moisture content, excluding inedible hulls and shells.

[2]Sugar content only.

92 percent of the human diet, with about 30 crop species providing most of the world's calories and protein. These included eight species of cereals, which collectively accounted for 70 percent of the world food supply. Animal products, constituting 7 percent of the world's diets, come indirectly from plants. A third source of food, microbial fermentation, is used primarily to produce certain vitamins and amino acids. These products are important nutritionally, but the quantities are relatively small and they are not included in the survey.

Until the 19th century, crop improvement was in the hands of farmers, and food production grew largely by expanding the cultivated land area. Improvements in farm machinery expanded the area that could be cultivated by one family, especially in the U.S.A. Machinery made possible better seedbed preparation, moisture utilization, and improved planting practices and weed control, resulting in modest increases in yield per hectare.

By the mid-1800s, German scientist Justus von Leibig and French scientist Jean-Baptiste Boussingault had laid down important theoretical foundations in soil chemistry and crop agronomy. Sir John Bennett Lawes produced super phosphate in England in 1842, and shipments of Chilean nitrates (nitrogen) began arriving in quantities in European and North American ports in the 1840s. However, the use of organic fertilizers (animal manure, crop residues, green manure crops) remained dominant into the early 1900s. Of course, the most skillful and dedicated users of organic fertilizers (which also included human waste) were the Chinese, Japanese, and Koreans.

The groundwork for more sophisticated genetic improvement of crop plant species was laid by Charles Darwin, in his writings on the variation of life species (published in 1859), and by Gregor Mendel through his discovery of the laws of genetic inheritance (reported in 1865). Darwin's book immediately generated a great deal of interest, discussion, and controversy. Mendel's work was largely ignored for 35 years. The rediscovery of Mendel's work in 1900 provoked tremendous scientific interest and research in plant genetics.

The first decade of the 20th century brought a fundamental scientific breakthrough that was followed by the rapid commercialization of the breakthrough. In 1909, Fritz Haber, Nobel Laureate in Chemistry (1918), demonstrated the synthesis of ammonia from its elements. Four years later—in 1913—the company BASF, thanks to the innovative solutions of Karl Bosch, began operation of the world's first ammonia plant. The expansion of the fertilizer industry was soon arrested by WWI (ammonia used to produce nitrate for explosives), then by the great economic depression of the 1930s, and then again by the demand for explosives during WWII. However, after the war, rapidly increasing amounts of nitrogen became available and contributed greatly to boosting crop yields and production.

By the 1930s, much of the scientific knowledge needed for high-yield agricultural production was available in the United States. However, widespread adoption was delayed by the great economic depression of the 1930s, which paralyzed the world agricultural economy. It was not until WWII brought a much greater demand for food supplies that the new research findings began to be applied widely (excluding nitrogen fertilizer), first in the United States and later in many other countries.

It is only since WWII that fertilizer use, and especially the application of low-cost nitrogen derived from synthetic ammonia, has become an indispensable component of modern agricultural production (nearly 82 million nutrient tonnes of nitrogen consumed annually). It is estimated that 40 percent of today's 6.2 billion people are alive, thanks to the Haber-Bosch process of synthesizing ammonia (Vaclav Smil, University Distinguished Professor, University of Manitoba).

Hybrid U.S. corn (maize) cultivation led the modernization process. In 1940, U.S. farmers produced 56 million metric tons of corn on roughly 31 million hectares (77 million acres), with an average yield of 1.8 t/ha. In 2000, U.S. farmers produced 252 million tons of maize on roughly 29 million hectares, with an average yield of 8.6 t/ha. This 4.5-fold yield increase is the impact of modern hybrid seed-fertilizer-weed control technology!

International agricultural development assistance began in the early 1940s, thanks in no small measure to the efforts of Henry A. Wallace, whom I had the pleasure of knowing. Wallace was a paradoxical figure —a far-sighted politician, brilliant geneticist, self-taught biometrician and economist, prolific author, successful businessman, and great humanitarian. He was a leader of the hybrid maize revolution, developing some of the first commercial hybrids and founding what today is Pioneer Hi-bred International, the largest seed company in the world. In addition, the "Hy-line" chicken breeds he developed at one point accounted for three-quarters of all the egg-laying poultry sold worldwide.

His career in politics was controversial. The son of a prominent Iowan Republican family, he took over the family publishing business, which produced the Wallace Farmer, one of the most influential agricultural weekly newspapers of the day. He then became a Democrat and held two cabinet posts, served four years as Vice President to Franklin Roosevelt, and ran for President himself in 1948, as a candidate for the Progressive Party. Wallace was seen by many, as a dangerous leftist, by others as the "Prophet of the Common Man." The author of many of the New Deal programs that lifted agriculture and rural America out of the great economic depression of the 1930s, many consider him the most influential Secretary of Agriculture in U.S. history.

Wallace was a great internationalist. He was instrumental in the establishment in 1943 of the Inter-American Institute of Agricultural Sciences in Turrialba, Costa Rica, to promote better food crops and as a symbol of "effective Pan Americanism." He also had a significant influence over the direction that my life took. It was he who helped convince the Rockefeller Foundation to establish the first U.S. foreign agricultural assistance program in 1943—the Mexican Government-Rockefeller Foundation Cooperative Agricultural Program—which I joined in 1944 as a young scientist assigned to wheat research and development. Wallace would visit our research program in Mexico from time to time. He liked to tease me, asking, "what's a good Iowa farm boy like you, Norman, messing around with a crop like wheat? You should be working on corn." The last time I saw him was in 1963 at the 50th Anniversary celebration of the Rockefeller Foundation. He was very interested in the new semi-dwarf wheats and their potential to revolutionize production in Asia. "Are your new wheats going to make a difference in Asia?" he asked. I responded, "Give us five years and South Asia will be self-sufficient in wheat production." As it turns out, this milestone was reached in 1968 in Pakistan and in 1972 in India. Sadly, Wallace did not live to see this happen.

THE GREEN REVOLUTION

Over the past four decades, sweeping changes have occurred in the factors of production used by farmers in many parts of the developing world, but nowhere more dramatically than in India, Pakistan, China, and other developing countries of Asia (Table 2.2):

TABLE 2.2. Changes in factors of production in developing Asia

Year	Adoption of Modern Varieties		Irrigation	Fertilizer/Nutrient Consumption	Tractors
	Wheat	Rice			
	M ba	*(% Area)*	*Million ba*	*Million metric tons*	*Million*
1961	0 (0%)	0 (0%)	87	2	0.2
1970	14 (20%)	15 (20%)	106	10	0.5
1980	39 (49%)	55 (43%)	129	29	2.0
1990	60 (70%)	85 (65%)	158	54	3.4
2000	70 (84%)	100 (74%)	175	70	4.8

Source: FAOSTAT, July 2002, and author's estimates on modern variety adoption, based on CIMMYT and IRRI data.

- High-yielding semi-dwarf varieties are now used on 84 and 74 percent of the wheat and rice areas, respectively.
- Irrigation has more than doubled to 175 million hectares; fertilizer consumption has increased more than thirtyfold, and now stands at about 70 million tons of total nutrients.
- Tractor use has increased from 200,000 to 4.8 million units.

As a result, rice and wheat production has increased from 127 million tons to 762 million tons over this 40-year period (FAOSTAT 2002).

Green revolution critics have tended to focus too much on the wheat and rice varieties, per se, as if they alone can produce miraculous results. Certainly, modern varieties can shift yield curves higher due to more efficient plant architecture and the incorporation of genetic sources of disease and insect resistance. However, modern, disease-resistant varieties can achieve their genetic yield potential only if systematic changes are also made in crop management, such as in dates and rates of planting, fertilization, water management, and weed and pest control. Moreover, many of these crop management changes must be applied simultaneously if the genetic yield potential of modern varieties is to be fully realized. For example, higher soil fertility and greater moisture availability for growing food crops also improves the ecology for weed, pest, and disease development. Thus, complementary improvements in weed, disease, and insect control are also required to achieve maximum benefits.

Despite the successes of small-holder Asian farmers in applying green revolution technologies to triple cereal production since 1961, millions of miserably poor people remain, especially in South Asia. A comparison of China and India—the world's two most populous countries—is illustrative of the point that increased food production alone is not sufficient to achieve food security (Table 2.3). Over the past 25–35 years, both countries have achieved remarkable progress in food production. Huge stocks of grain have accumulated in India over the past several years, while tens of millions need more food but do not have the purchasing power to buy it.

China has been more successful in achieving broad-based economic growth and poverty reduction than India. Nobel Economics Laureate, Professor Amartya Sen, attributes the difference to the greater priority that the Chinese government has given to investments in rural education and health care services. Nearly 80 percent of the Chinese population is

Table 2.3. Social development indicators in China and India

	China	India
1961 population, millions	669	452
2000 population, millions	1,290	1,016
Population growth, 1985–95, %/year	1.3	1.9
GDP per capita, U.S. dollars, 1995	780	440
Percent in agriculture, 1990	74	64
Poverty, % pop. below $1/day, 1999	29	53
Child malnutrition, % underweight, 1993–99	9	45
% Illiterate population (over 15), 1995	22	50

Sources: 2001 World Bank Atlas; FAOSTAT, 2001.

literate, while only 50 percent of the Indian population can read and write. Only 9 percent of Chinese children are malnourished compared to 45 percent in India. With a healthier and better-educated rural population, China's economy has been able to grow about twice as fast as the Indian economy over the past two decades and today China has a per capita income nearly twice that of India.

Africa Is the Greatest Worry

More than in any other region of the world, food production south of the Sahara is in crisis. High rates of population growth and little application of improved production technology have resulted in declining per capita food production, escalating food deficits, and deteriorating nutritional levels, especially among the rural poor. While there are some signs during the 1990s that small-holder food production is beginning to turn around, this recovery is still very fragile.

Sub-Saharan Africa's extreme poverty, poor soils, uncertain rainfall, increasing population pressures, lack of secure land titles, political and social turmoil, shortages of trained agriculturalists, and weaknesses in research and technology delivery systems all make the task of agricultural development more difficult. But we should also realize that to a considerable extent, the present food crisis is the result of the long-time neglect of agriculture by political leaders. Even though agriculture provides the livelihood to 60–80 percent of the people in most countries, agricultural and rural development has been given low priority. Investments in distribution and marketing systems and in agricultural research and education are woefully inadequate. Furthermore, many governments pursued and continue to pursue a policy of providing cheap food for the politically volatile urban dwellers at the expense of production incentives for farmers.

In 1986 I became involved in food crop production technology transfer projects in sub-Saharan Africa, sponsored by the Nippon Foundation and its Chairman, the late Ryoichi Sasakawa, and enthusiastically supported by former U.S. President Jimmy Carter. Our joint program is known as Sasakawa-Global 2000 (SG 2000), and currently operates in 10 sub-Saharan African countries, and previously, in four other countries.

Working hand-in-hand with national extension services during the past 16 years, SG 2000 has helped small-scale farmers grow more than one million demonstration plots, ranging in size from 1,000–5,000 square meters. These demonstration plots have been concerned with demonstrating improved technology for basic food crops: maize, sorghum, wheat, rice, cassava, and grain legumes.

The packages of recommended production technology include

- The use of the best available commercial varieties or hybrids
- Proper land preparation and seeding to achieve good stand establishment
- Proper application of the appropriate fertilizers and, when needed, crop protection chemicals
- Timely weed control
- Moisture conservation and/or better water utilization, if under irrigation

SG 2000 also has helped small-holder farmers to improve on-farm grain storage, both to reduce losses due to spoilage and infestation and to allow farmers to hold stocks longer, so that they can sell when market prices are more favorable.

Virtually without exception, demonstration plot yields are two to three times higher than the control plots employing the farmer's traditional methods. Hundreds of field days, attended by thousands of farmers, have been organized to demonstrate and explain the components of the production package. In areas where the projects are operating, farmers' enthusiasm is high and political leaders are taking much interest in the program.

While the program clearly has demonstrated the availability of technology to greatly improve yields, sustained adoption by farmers of the recommended technologies has been disappointing. High prices and undeveloped input supply systems keep most small-holder farmers from purchasing improved seeds, fertilizers, crop protection chemicals, and farm machinery. Similarly, poorly functioning rural finance systems and high interest rates conspire to deprive most farmers of access to production credit.

Fertilizer use in sub-Saharan Africa is the lowest in the world (Table 2.4). To make matters worse, there was no improvement in total fertilizer consumption between 1985 and 2000. In consequence, it is estimated that 600 kg/ha of nitrogen, 200 kg/ha of phosphorus, and 450 kg/ha of potassium have been "mined" (through cropping and erosion) on each of 100 million hectares of cultivated land over the past 30 years. In marked contrast, soil nutrient levels on farms in North American and Europe have increased over this same time period (sometimes resulting in groundwater and stream pollution). The soil nutrient losses in sub-Saharan Africa are an environmental time bomb. Unless we wake up soon and reverse these disastrous trends, the future viability of African food systems will indeed be imperiled.

To date, while all governments provide lip service about the importance of agriculture, few have given priority to agriculture and rural development. Indeed, most governments have reduced their funding for agriculture. Over the past 10–15 years, the hope had been that the private sector would fill the void of a retreating public sector. This has not happened, nor will it while political and economic instability continues.

The lack of infrastructure—especially roads and transport, but also potable water and electricity—is also a serious obstacle to progress. Improved transport systems would greatly accelerate agricultural production, break down tribal animosities, and help establish rural schools and clinics in areas where teachers and health practitioners are heretofore unwilling to venture.

So far, the forces of globalization have brought few, if any, benefits to sub-Saharan Africa. Traditional exports—such as cocoa, coffee, and palm oil—have been hobbled by depressed world prices and increasing competition from Asian countries. Some progress has been made in expanding nontraditional exports, especially in fruits and vegetables.

Table 2.4. Fertilizer nutrient consumption (in kg of nutrients) per hectare of arable land in selected countries, 2000

Country	Kg
Uganda	1
Ghana	3
Guinea	4
Mozambique	4
Tanzania	6
Nigeria	7
Burkina	9
Mali	11
Ethiopia	16
Malawi	16
Cuba	37
South Africa	51
India	103
U.S.A.	105
Brazil	140
France	225
China	279
U.K.	288
Japan	325
Vietnam	365
Netherlands	578

Source: FAOSTAT, July 2002.

FUTURE SOURCES OF INCREASED FOOD SUPPLY

The potential for further expansion in the global arable land area is limited for most regions of the world. This is certainly true for densely populated Asia and Europe. Only in sub-Saharan Africa and South America do large unexploited tracts exist, and only some of this land should eventually come into agricultural production. Much has acid soils, such as the Cerrados of Brazil and in parts of central and southern Africa, and researchers and farmers have developed varieties and soil management practices that can make these former "wastelands" productive for agriculture. In populous Asia, home to more than half of the world's people, there is very little uncultivated land left. Indeed, some of the land, especially in South Asia, currently in production should be taken out of cultivation, because of high susceptibility to soil erosion.

The International Food Policy Research Institute (IFPRI) estimates that more than 85 percent of future growth in cereal production must come from increasing yields on lands already in production. Such productivity improvements will require varieties with higher genetic yield potential and greater tolerance of drought, insects, and diseases. To achieve these genetic gains, advances in both conventional and biotechnology and research will be needed. In crop management, we can expect productivity improvements in soil and water conservation, reduced tillage, fertilization, weed and pest control, and postharvest handling.

Irrigated agriculture—which accounts for 70 percent of global water withdrawals—covers some 17 percent of cultivated land (about 275 million ha) yet accounts for nearly 40 percent of world food production. The rapid expansion in world irrigation and in urban and industrial water uses has led to growing shortages. Indeed, the U.N.'s 1997 Comprehensive Assessment of the Freshwater Resources of the World estimates that, by the year 2025, as much as two-thirds of the world's population could be under stress conditions.

Clearly, we need to rethink our attitudes about water, and move away from thinking of it as nearly a free good, and a God-given right. Pricing water delivery closer to its real costs is a necessary step to improving use efficiency. Farmers and irrigation officials (and urban and industrial consumers) will need incentives to save water. Moreover, management of water distribution networks, except for the primary canals, should be decentralized and turned over to farmers.

There are many technologies for improving the water use efficiency in agriculture. Wastewater can be treated and used for irrigation, especially important for peri-urban agriculture, which is growing rapidly around many of the world's mega-cities. New crops requiring less water (and/or new improved varieties), together with more efficient crop sequencing and timely planting, also can achieve significant savings in water use.

Proven technologies are also available that save water, reduce soil salinity, and increase water productivity (yield per unit of water used). Various new precision irrigation systems—like drip and sprinkler systems—are available that will supply water to plants only when they need it. Technologies like planting on raised beds or conservation (zero) tillage also use water more efficiently, especially in irrigation. Improved small-scale and supplemental irrigation systems also are now available to increase the productivity of rainfed areas, which offers much promise for small-holder farmers.

In order to expand food production for a growing world population within the parameters of likely water availability, the inevitable conclusion is that humankind in the 21st century will need to bring about a "blue revolution" to complement the so-called green revolution of the 20th century. In the new blue revolution, water-use productivity must be wedded to land-use productivity. New science and technology must lead the way.

What to Expect from Biotechnology?

In the last 20 years, biotechnology based upon recombinant DNA has developed invaluable new scientific methodologies and products in food and agriculture. This journey deeper into the genome—to the molecular level—is the continuation of our progressive understanding of the workings of nature. Recombinant DNA methods have enabled breeders to select and transfer single genes, which has not only reduced the time needed in conventional breeding to eliminate undesirable genes, but also allowed breeders to access useful genes from other distant taxonomic groups. So far, these gene alterations have conferred producer-oriented benefits, such as resistance to pests, diseases, and herbicides. Other benefits likely to come (through biotechnology and conventional plant breeding) are varieties with greater tolerance of drought, waterlogging, heat and cold—important traits given current predictions of climate change. In addition, many consumer-oriented benefits, such as improved nutritional and other health-related characteristics, are likely to be realized over the next 10–20 years.

The following are two of my own dreams for biotechnology. Among all the cereals, rice is unique in its immunity to the rusts (*Puccinia* spp). All the other cereals—wheat, maize, sorghum, barley, oats, and rye—are attacked by two to three species of rusts, often resulting in disastrous epidemics and crop failures. Enormous scientific effort over the past 80 years has been devoted to breeding wheat varieties for resistance to stem, leaf, and yellow rust species. After many years of intense crossing and selecting, and multi-location international testing, a good, stable, but poorly understood, type of resistance to stem rust was identified in 1952 that remains effective worldwide to the present. However, no such success has been obtained with resistance to leaf or yellow rust, where genetic resistance in any particular variety has been short-lived (3–7 years). Imagine the benefits to humankind if the genes for rust immunity in rice could be transferred into wheat, barley, oats, maize, millet, and sorghum. Finally, the world could be free of the scourge of the rusts, which have led to so many famines over human history.

On another front, bread wheat has superior dough for making leavened bread and other bakery products due to the presence of two proteins—gliadin and glutenen. No other cereals have this combination. Imagine if the genes for these proteins could be identified and transferred to the other cereals, especially rice and maize, so that they, too, could make good-quality "leavened" bread. This would help many countries—and especially the developing countries in the tropics, where bread wheat flour is often the single largest food import—to save valuable foreign exchange.

Despite the formidable opposition in certain circles to transgenic crops, commercial adoption by farmers of the new varieties has been one of the most rapid cases of technology diffusion in the history of agriculture. Between 1996 and 2001, the area planted commercially to transgenic crops has increased from 1.7 to 52.4 million hectares (Clive James 2002).

Although there have always been those in society who resist change, the intensity of the attacks against GMOs by certain groups is unprecedented, and in certain cases, even surprising, given the potential environmental benefits that such technology can bring in reducing the use of crop protection chemicals. It appears that many of the most rabid crop biotech opponents are driven more by a hate of capitalism and globalization than by the actual safety of transgenic plants. However, the fear they have been able to generate about biotech products among the public is due in significant measure to the failure of our schools and colleges to teach even rudimentary courses on agriculture. This educational gap has resulted in an enormous majority, even among well-educated people, who seem totally ignorant of an area of knowledge so basic to their daily lives and indeed, to their future survival. We must begin to address this ignorance without delay—especially in the wealthy urban nations—by making it compulsory for students to study more biology and to understand the workings of agricultural and food systems.

Much of the current debate about transgenic crops in agriculture has centered around two major issues—safety and concerns of access and ownership. Part of the criticism about GMO safety holds to the position that introducing "foreign DNA" into our food crop species is unnatural and thus an inherent health risk. Since all living things—including food plants, animals, and microbes—contain DNA, how can we consider recombinant DNA to be unnatural? Even defining what constitutes a "foreign gene" is also problematic, since many genes are common across many organisms.

Almost all of our traditional foods are products of natural mutation and genetic recombination, which are the drivers of evolution. Without this ongoing process, we would probably all still be slime on the bottom of some primeval sea. In some cases, Mother Nature has done the genetic modification, and often in a big way. For example, the wheat groups we rely on for much of our food supply are the result of unusual (but natural) crosses between different species of grasses. Today's bread wheat is the result of the hybridization of three different grasses, each containing a set of seven chromosomes, and thus could easily be classified as transgenic. Corn is another crop that is the product of transgenic hybridization (probably of *Teosinte* and *Tripsacum*).

Several hundred generations of farmers have accelerated genetic modification through recurrent selection of the most prolific and hardiest plants and animals. To see how far the evolutionary changes have come, one only needs to look at the 5,000-year-old fossilized maize cobs found in the caves of Tehuacan in Mexico, which are about 1/10 the size of modern maize varieties. Over the past 100 years or so, scientists have been able to apply a growing understanding of genetics, agronomy, plant physiology, pathology, and entomology to accelerate the process of combining high genetic yield potential with greater yield dependability under a broad range of biotic and abiotic stresses.

Obviously, it does make sense for GMO foods to carry a label if the food is substantially different from similar conventional foods. This would be the case if there is a nutritional difference, or if there is a known allergen or toxic substance in the food. If the food is essentially identical to regular versions of the same food, what would be the utility? To us, this would undermine the central purpose of labeling, which is to provide useful nutritional or health-related information to allow consumers to make informed choices.

On the environmental side, I find the opposition to the transgenic crops carrying the *Bacillus thuringiensis* (Bt) gene to be especially ironic. Rachel Carson, in her provocative 1962 book, *Silent Spring*, was especially effusive in extolling the virtues of Bt as a "natural" insecticide to control caterpillars. But anti-GMO activists have decried the incorporation of the Bt gene into the seed of different crops, even though this can reduce the use of insecticides and is harmless to other animals, including humans. Part of their opposition is based upon the prospect that widespread use of Bt crops may lead to mutations in the insects that eventually will render the bacterium ineffective. This seems incredibly naïve. We can be quite sure that the ability of a particular strain of *Bacillus thuringiensis* to confer insect resistance inevitably will break down, and this is why dynamic breeding programs—using both conventional and recombinant DNA techniques—are needed to develop varieties with new gene combinations to keep ahead of mutating pathogens. This has been the essence of plant breeding programs for more than 70 years.

Of course, scientists and researchers employing recombinant DNA must pay attention to public values and concerns, and must explore all legitimate and reasonable questions about the potential impacts of their activities. However, today we are seeing too many opponents of biotechnology dismiss the many safety and regulatory checks that govern whether a new product is brought to the marketplace. Unfortunately, they willfully choose to emphasize highly unlikely potential risks.

In the United States, at least three federal agencies provide scrutiny over the safety of GMOs:

- The U.S. Department of Agriculture (USDA), which is responsible for seeing that the plant variety is safe to grow
- The Environmental Protection Agency (EPA), which has special review responsibilities for plants that contain genes that confer resistance to insects, diseases, and herbicides to protect against adverse environmental effects
- The Food and Drug Administration (FDA), which is responsible for food safety

The data requirements imposed upon biotechnology products are far greater than they are for products from conventional plant breeding, and even from mutation breeding, which uses radiation and chemicals to induce mutations. But we must also understand that there is no such thing as "zero biological risk." It simply doesn't exist, which makes, in my opinion, the enshrinement of precautionary principle just another ruse by anti-biotech zealots to stop the advance of science and technology.

There is no reliable scientific information to date to substantiate that GMOs are inherently hazardous. Recombinant DNA has been used for 25 years in pharmaceuticals, with no documented cases of harm attributed to the genetic modification process. So far, this is also the case in GMO foods. This is not to say that there are no risks associated with particular products. There certainly could be. But we need to separate the methods by which GMOs are developed—which are not inherently unsafe—from the products, which could be if certain toxins or allergens are introduced.

There certainly have been errors in the GMO certification process. A recent example was the "restricted" approval in the United States by the EPA of a Bt corn hybrid, Starlink, for use only as an animal feed because of possible allergenic reactions that this strain of Bt might have in humans. EPA granted this approval knowing full well that marketing channels did not exist to segregate maize destined for animal feed from that destined for human consumption. As a result, Starlink corn got into various corn chips and taco shells, and undermined public confidence. Lost in the furor, however, was the fact that there was little reason to believe that the maize was actually unsafe for human consumption—only an unsubstantiated fear that it might cause allergic reactions. Subsequently, a blue-ribbon scientific panel confirmed that Starlink corn was safe for human consumption. Still, it has now become policy that no variety will be released without approval for both food and feed uses.

A second controversial aspect of transgenic varieties involves issues of ownership and access to the new products and processes. Since most of GMO research is being carried out by the private sector, which aggressively seeks to patent its inventions, the intellectual property rights issues related to life forms and to farmer access to GMO varieties must be seriously addressed. Traditionally, patents have been granted for "inventions" rather than the "discovery" of a function or characteristic. How should these distinctions be handled in the case of life forms? Moreover, how long, and under what terms, should patents be granted for bioengineered products?

The high cost of biotechnology research also appears to be leading to a rapid consolidation in the ownership of agricultural life science companies. Is this desirable? I must confess to uneasiness on this score, and believe that the best way to deal with this potential problem is for governments to ensure that public sector research programs are adequately funded to produce public goods. By this I mean, true public sector funding, which is quite different to the "commissioned" research that public institutions are now doing for private companies.

Unfortunately, during the past two decades, support to public national research systems in the industrialized countries has seriously declined, while support for international agricultural research has dropped so precipitously to border on the disastrous. If these trends continue, we risk losing the broad continuum of agricultural research organizations—both public and private, and from the more-basic to the more-applied—which are needed to keep agriculture moving forward. We need to ensure that farmers and consumers never become hostages to possible private sector monopolies. So, yes, I am all for private sector research and believe that private companies need to be fairly compensated for their research investments and have their intellectual property protected. But the public sector must always retain a moderating hand, in order to ensure that the public good continues to be served and also to educate and train future generations of scientists.

Growing Restrictions in International Scientific Exchange

I am concerned about growing restrictions in the sharing of improved germplasm and scientific information around the world. In particular, plant quarantine regulations and restrictions related to plant quarantine regulations seem to be especially problematic. The past benefits of relatively unfettered international germplasm exchange have been enormous. Organized international germplasm exchange and testing only began in the early 1950s, in response to a devastating stem rust epidemic in wheat in North America, caused by race 15 B, to which all commercial varieties were susceptible. Faced with this crisis of epidemic proportions, the departments of agriculture in the United States and Canada appealed to other research programs in the Americas, to exchange a broad range of their best early- and advanced-generation breeding materials, and to test these materials at many locations simultaneously. The Mexican Government-Rockefeller Cooperative Agricultural Program with which I was associated, and several national agricultural research programs in South America, responded rapidly. Out of this initial effort, new sources of stem rust resistance were identified that have held up to this day. Indeed, no stem rust epidemics have occurred in the Americas in nearly 50 years.

International germplasm networks became a hallmark of the international centers supported by the Consultative Group for International Agricultural Research (CGIAR). International sharing of germplasm and information broke down the psychological barriers that previously had isolated individual breeders from each other, and led to the introduction of enormous new quantities of useful genetic diversity. It became accepted policy that individual breeders could use any material from these international nurseries, either for further crossing or for direct commercial release, as long as the original source was recognized. This led to the accelerated development of new high-yielding, disease- and insect-resistant varieties, and ushered in a golden era in plant breeding around the world.

For more than two decades, I also have become increasingly concerned about various types of restrictive plant quarantine regulations that jeopardize cooperative international agricultural research. After the terrorist attacks on the United States on September 11, 2001, such regulations are likely to become even more restrictive. I certainly agree that plant quarantine regulations have, in the past, played an important role in excluding and/or delaying the spread of many pathogens into new areas. And quarantine regulations will no doubt continue to play an important role in protecting crops against foreign pests and protecting exports of agricultural commodities from embargoes. However, it is obvious that the effectiveness of quarantines, since the end of World War II, has been complicated by both great expansion in volumes of agricultural commodities moving in international trade, and by the method and speed of transport, in which millions of people cross the oceans in a few hours. It seems to me that the effectiveness of quarantines against the different disease, insect, and weed pests needs review and adjustments to be pertinent to the new conditions. In the case of cereal diseases, I believe we need to combine effective disease surveillance programs with aggressive plant breeding programs to develop resistant varieties for areas likely to be affected. In the world of increasing globalization—in trade and travel—the concept of "exclusion" seems woefully outdated.

Intellectual property rights are also restricting the flow of useful germplasm between nations. Scientists increasingly are reluctant to share their early-generation breeding materials, for fear that this will preclude the possibility of patenting. The biotech opponents are now trying to convince Third World nations that their plant species are at risk of being stolen by the private-sector gene prospectors/bio-pirates, and are recommending legal barriers to stop the flow of germplasm. This is unfortunate, especially since the concept of what constitutes "indigenous" germplasm has been greatly blurred over the past 500–600 years. Maize, beans, groundnuts, cassava, potatoes, cocoa, and peppers—to name only a few —were originally domesticated in the Americas and spread by explorers and traders throughout Europe, Asia, and Africa. Rice, wheat, barley, oats, rye, soybeans, and peas spread from Asia to other continents, and sorghum, millet, and coffee spread from Africa around the world. Thus, historically speaking, all nations are "bio-pirates" in one way or another.

Hopefully, sufficient goodwill and humanity will exist in current and future generations so that new forms of public-private collaboration come into being to ensure that all farmers and consumers worldwide will have the opportunity to benefit from the new genetic revolution. In this quest, we must take care not to confuse science with politics. So when scientists lend their names and credibility to unscientific propositions, what are we to think? Is it any wonder that science is losing its constituency? We must maintain our guard against politically opportunistic researchers, like the late T.D. Lysenko, whose pseudo-science in agriculture and vicious persecution of anyone who disagreed with him, contributed greatly to the collapse of the former U.S.S.R.

Agriculture and the Environment

The current backlash against agricultural science and technology evident in some industrialized countries is hard for me to comprehend. Thanks to science and technology that has permitted increasing yields on the lands best suited to agriculture, world farmers have been able to leave untouched vast areas of land for other purposes.

Had the U.S. agricultural technology of 1940 (with relatively little chemical fertilizer and agricultural chemicals used) still persisted today, we would have needed an additional 575 million acres of agricultural lands. This would need to have been of the same quality—to match current production of 700 million tons for the 17 main food and fiber crops produced in the United States (Figure 2.1). The area spared for other land uses is slightly greater than all the land in 25 states east of the Mississippi River.

Worldwide, had 1950 average global cereal grain yields still prevailed in 1998, instead of the 600 million hectares that were used for production, we would have needed nearly 1.8 billion ha of land of the same quality to produce the current global harvest (Figure 2.2), land that generally was not available, especially in highly populated Asia. Moreover, had more environmentally fragile land been brought into agricultural production, the impact on soil erosion, loss of forests, grasslands, and biodiversity, and extinction of wildlife species would have been enormous.

The attacks against chemical fertilizers are also hard to swallow. Biochemically, it makes no difference to the plant whether the nitrate ion it "eats" comes from a bag of fertilizer or decomposing organic matter. Yet, to hear many uninformed people, chemical fertilizer is seen more as

Million of hectares

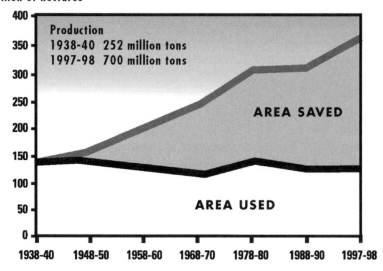

FIGURE 2.1 U.S.A. total crop area spared by application of improved technology on 17 food, feed, and fiber crops in the period of 1938–40 to 1997–98.

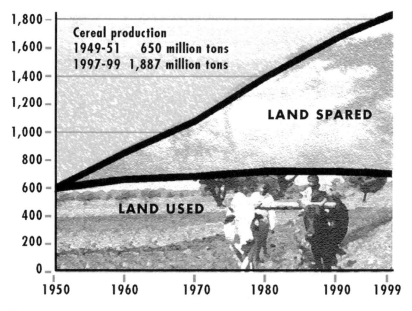

FIGURE 2.2 World cereal production areas saved through improved technology, 1950–99 (FAO Production Yearbooks and AGROSTAT).

a poison than the plant food that it really is. Equally misinformed is the notion that "organically" produced food has higher nutritive value. This is not so. While the affluent nations can certainly afford to pay more for food produced by the so-called "organic" methods, the one billion chronically undernourished people of the low-income, food-deficit nations cannot. Indeed, it would be impossible for organic sources to replace the 80 million metric tons of nitrogen contained in chemical fertilizer. If we tried to do it with cattle, the world beef population would have to increase from about one billion to six or seven billion head, with all of the resulting overgrazing, erosion, and destruction of wildlife habitat this would result. It would produce quite a heap of animal dung, too, and quite an aroma!

Agriculture and Peace

Almost certainly, the first essential component of social justice is adequate food. And yet there are upwards of 1 billion people who go to bed every night hungry. Particularly disheartening are the 150 million young children who go hungry each day, with this undernourishment often leading to irreversible damage to their bodies and minds.

Of the developing countries with the lowest undernourishment, only 8 percent were mired in conflict. In contrast, of those countries where more than half of the population was underfed, 56 percent were experiencing civil conflict (*Assessment of the World Food Security Situation*, FAO, 1999). Since agriculture provides employment for the majority of people in low-income developing countries, it is not surprising that when this sector is allowed to falter, armed conflict often ensues.

It is indeed troubling to see the persistence of large military budgets around the world, including in the United States. In total, something on the order of $US850 billion is spent annually on the military. The U.S.A. accounts for 47 percent of this total (about $US400 billion), and spends 40 times more on the military than it does on overseas development assistance. Indeed, trends in foreign assistance for agricultural and rural development have been declining, not only in the United States, but also in many other donor countries and institutions as well. In 2000, the World Bank reported its lowest level of support to agriculture in its history.

One of my fondest dreams would be to see the achievement of primary education for all and the elimination of gender inequalities in secondary education. Still today, an estimated 120 million primary age children do not go to school, and 870 million adults—nearly two-thirds of them women—cannot read and write. What a waste! If we are to reduce poverty, control population growth, and build a more equitable global society, such disparities must be narrowed.

Closing Comments

Thirty-two years ago, in my acceptance speech for the Nobel Peace Prize, I said that the green revolution had won a temporary success in man's war against hunger, which if fully implemented, could provide sufficient food for humankind through the end of the 20th century. This has happened. But I also warned that unless the frightening power of human reproduction was curbed, the success of the green revolution would only be ephemeral.

I now say that the world has the technology—either available or well advanced in the research pipeline—to feed a population of 10 billion people. Improvements in crop productivity can be made all along the line—in tillage, water use, fertilization, weed and pest control, and harvesting. Both conventional breeding and biotechnology research will be needed to ensure that the genetic improvement of food crops continues at a pace sufficient to meet growing world populations.

The more pertinent question today is whether farmers and ranchers will be permitted to use this new technology. Extremists in the environmental movement from the rich nations seem to be doing everything they can to stop scientific progress in its tracks. Small, but vociferous and highly effective and well-funded, anti-science and technology groups are slowing the application of new technology, whether it be developed from biotechnology or more conventional methods of agricultural science.

Only around four percent of the population in industrialized countries (less than two percent in the U.S.A.) is directly engaged in agriculture. With low-cost food supplies and urban bias, is it any wonder that consumers don't understand the complexities of reproducing the world food supply each year in its entirety, and expanding it further for the 80 million additional people that are added annually.

I believe we must seek to redress this educational gap in industrialized urban nations by making it compulsory in secondary schools and universities for students to take courses on biology and science and technology policy. In this regard, I wish to pay homage to Iowa's own John Ruan, not only for endowing the World Food Prize, which recognizes individual achievements to improve the quantity, quality, and availability of global food supplies, but also for establishing The World Food Prize Foundation Youth Institute Program, which involves more than 100 high schools, mainly in Iowa, and is helping to build the understanding of new generations of youth about international agriculture. Last summer, 13 outstanding graduating seniors who have participated previously in this program were awarded internships to work in research institutions in Asia, Africa, and Latin America. The experiences of these young Iowans have been tremendous and, hopefully, this will encourage some of them to pursue careers in research and development.

In conclusion, permit me to leave you with this thought, so eloquently expressed by Andre and Jean Mayer, two American nutritionists, in an article, "Agriculture—The Island Empire," published in 1974 in the journal *Daedulus* of the American Academy of Arts and Sciences.

> Few scientists think of agriculture as the chief, or the model science. Many, indeed, do not consider it a science at all. Yet it was the first science —the mother of all sciences; it remains the science that makes human life possible; and it may well be that, before the century is over, the success or failure of Science as a whole will be judged by the success or failure of agriculture.

NOTES

1. Inaugural Lecture, Norman Borlaug Lecture Series; Iowa State University; Ames, IA; October 15, 2002.

3

DR. NORMAN E. BORLAUG: 20TH CENTURY LESSONS FOR THE 21ST CENTURY WORLD
Kenneth M. Quinn[1]

Norman Borlaug's approach to increasing global food production resulted in the saving of as many as one billion people worldwide from famine, starvation, and death and earned him the title "Father of the Green Revolution."

Indeed, writing in *The Atlantic Monthly* in 1997, Gregg Easterbrook wrote that "Norman Borlaug has already saved more lives than anyone who has ever lived."

Over the past three years, it has been my privilege to work directly with Dr. Norman Borlaug—comparing notes on our experiences in rural development, planning World Food Prize International Symposia, selecting young students for International Internships and accompanying him to the 100th Anniversary of the Nobel Prize in Oslo. In the course of doing this, I have spent many fascinating hours talking to him about his life and career. It is a great story.

THE FORMATIVE YEARS

Named by *Time Magazine* as one of the 100 most influential minds of the twentieth century, Norman Borlaug is a quintessential American success story. Norm, as he is known to all who work with him, was born in 1914 to Norwegian-American parents outside Cresco in the northeastern part of the American state of Iowa. His boyhood was spent on a Norman Rockwell–esque farm, where he had indelibly etched on his psyche the value of hard work, first inculcated by his family and later by his teachers and mentors.

His formal education began in a one-room schoolhouse. It was there that a young Norm Borlaug first learned the lesson that confronting the harsh realities of prairie farm life could bring disparate people together and impel them to cooperate. Immigrant children from Norway and Bohemia discovered in that small Iowa school that they had much in common, just as their parents found that working together to ensure sufficient food for all was more important than any ethnic or linguistic differences that might initially divide them. It was an insight that would remain with Borlaug throughout his life and would come to permeate his work.

Norm developed a dogged tenacity—another quality that would play a crucial role in some of his greatest achievements—from participating in his high school wrestling program. His coach, Dave Bartelma, taught him never to give up, no matter how formidable his adversary. This attitude propelled Borlaug to the top of the intercollegiate wrestling world and would later earn him induction in the NCAA Wrestling Hall of Fame.

Still another lesson Norm Borlaug absorbed was the critical importance of rural roads to spreading the word about the latest agricultural innovation and helping farmers get crops to market. Iowa was dramatically affected by the Great Depression, with foreclosures on family farms bringing displaced families close to insurrection. The network of farm-to-market roads being built all over the state not only facilitated agricultural production, but also the transport of children to school and access to medical care. The roads uplifted an entire generation of rural Iowans in a way almost nothing else could. Life improved and the specter of political unrest receded.

All these factors came together in a way that steered Norman Borlaug to seek a university education, the first person in his family to do so. This was a particularly arduous undertaking in the heart of the Great Depression. I recall that as we drove through Oslo in December 2001 during the Nobel Anniversary, Norm recounted to me how, after graduating from high school, he labored for 50 cents a day as a hired farmhand to save enough to pay for a year of college. Eventually Norm had enough money to make his way to the University of Minnesota where he would major in agricultural science, become an accomplished wrestler, meet his wife Margaret, and earn a Ph.D. in Plant Pathology.

IN TOUCH WITH THE ENVIRONMENT

To help pay his way through college, Borlaug worked in a coffee shop, served meals in a sorority house, and parked cars. In the summer, thanks to his major in forestry, he obtained a job as a U.S. forest ranger along the Salmon River in a remote part of the western state of Idaho. He came to embrace the solitude of the forest and cared deeply about the plants and wildlife that was sustained in this habitat. His expectation was that upon graduation, he would become a full-time employee of the Forest Service.

However, fate intervened to redirect his life and to impact human history. As Norm tells the story, just a few weeks before graduation, he received a letter from his supervisor in the U.S. Forest Service informing him that a tight budget situation meant that he could not start his new full-time forest ranger position for another six months. A disappointed Borlaug agreed to delay his arrival and decided to take some additional courses on the Minnesota campus. One day, he saw a notice on a bulletin board for a lecture by Dr. Elvin Stakman, the head of the university's plant pathology department. Borlaug decided to attend.

Norm was riveted by Stakman's lecture on rust, the parasitic fungus that attacked a wide variety of plants and trees. As Lennard Bickel wrote in his biography of Borlaug, "that night . . . Stakman was a magnetic and compelling teacher. His style, his sincerity, the intensity of his delivery made his words ring in Borlaug's ears."

Stakman ended his discourse with a moving charge that it was science which would " . . . go further than has ever been possible to eradicate the miseries of hunger and starvation from this earth." Norman Borlaug was hooked. He went to see Stakman to ask to be admitted to the Ph.D. program in plant pathology and gave up the possibility of a career in the Forest Service. It was a decision that would change his life, and save one billion people.

CONFRONTING POVERTY—MEXICO

Graduating in the middle of World War II, Dr. Borlaug went to work for the DuPont Corporation. But he was soon approached about joining a fledgling research project being initiated by the Rockefeller Foundation in rural Mexico. There, he first saw the plight of poverty-stricken wheat farmers barely able to sustain themselves due to repeatedly poor harvests. Once again, Borlaug found a wide chasm to be bridged. There was an instinctive hesitation to adopt untried new technologies on the part of most subsistence farmers. And, there was an understandable reluctance to trust the word of an expatriate American college boy who didn't even speak their language.

Borlaug admits to being extremely discouraged in this initial venture into the developing world. But his commitment to learn the language, a healthy dose of the determination he learned in high school sports, and his willingness to get his hands dirty working in the fields eventually enabled him to connect with some farmers who tried his new approach to wheat production. As Professor R. Douglas Hurt observed:

> In 1944, when Borlaug arrived in Mexico, its farmers raised less than half of the wheat necessary to meet the demands of the population. Rust perennially ruined or diminished the harvest . . . . Borlaug labored for 13 years before he and his team of agricultural scientists developed a disease resistant wheat. (But) still problems remained."

While the new wheat variety he had developed increased yields and resisted rust, it did not have stems strong enough to hold the now heavy heads of grain. Plants would topple over in the wind and rain. Dr. Borlaug then turned to Japanese dwarf strains which he crossbred with the varieties being raised in the hot, dry climate of northern Mexico. Then, using what a *Wall Street Journal* article called "shuttle breeding," Borlaug and his team would rush the plants to southern Mexico where it was possible to carry out two growing seasons in one year.

The results were as astonishing as they had been difficult to attain.

Aided by the use of fertilizer and irrigation, Borlaug's new spring wheat enabled Mexico to achieve self-sufficiency in 1956. His belief in scientific research and a hands-on connection to the farmers paid off in what was considered an agricultural miracle.

THE GREEN REVOLUTION,
FROM CIMMYT TO SOUTH ASIA

Inspired by his breakthrough achievements, the U.N. Food and Agriculture Organization and the Rockefeller Foundation asked Borlaug to turn his attention to the Middle East and South Asia. The problem of extreme poverty and failing harvests afflicted not only Mexico, but also much of the developing world, and was exacerbated by the post-war population explosion. As Indian agribusiness pioneer and World Food Prize Laureate Dr. B. R. Barwale noted, in the immediate aftermath of World War II, famine and the prospect of mass starvation haunted the

Indian subcontinent and other parts of the globe. The great Bangladesh famine in the late 1940s seemed an ominous harbinger of pandemic starvation, which would extract a devastating toll, adding to the more than 160 million people worldwide who had already died of famine or starvation during the previous 100 years.

But much of the developing world was pulled back from the precipice of enormous human tragedy by the scientific pioneers who promulgated the Green Revolution. Leading them was Dr. Norman Borlaug and the young agricultural scientists he had trained at the Centro Internacional de Mejoramiento de Maiz y Trigo—the International Maize and Wheat Improvement Center (CIMMYT) located outside Mexico City. Having overcome great resistance by farmers in Mexico, Borlaug and his compatriots faced the seemingly impossible task of convincing the leaders of both India and Pakistan—two countries bitterly divided—to embrace an entirely new approach to agriculture. Borlaug recalled that going in to speak to these two most powerful political leaders required summoning the same amount of courage as when he stepped on the wrestling mat. But he went forward and presented the options available. Borlaug's breakthrough technology arrived just in time to prevent a human catastrophe. By increasing crop yields in Pakistan and India fourfold, those traditionally food deficit countries became self-sufficient in an amazingly short time, saving hundreds of millions of lives.

Borlaug's achievements in wheat spread throughout the Middle East and North Africa with similar lifesaving results. Beginning in the early 1960s, his approach to wheat breeding was introduced in Egypt, Tunisia, Syria, Iran, Libya, Jordan, Lebanon, Turkey, Iraq, Afghanistan, Algeria, and Saudi Arabia, in many cases, through young scientists who had studied with him at CIMMYT. Just how significant an impact he had was brought home to me four decades later, while I was visiting with the Egyptian Minister of Agriculture in Cairo. When I mentioned Norman Borlaug's name, the Minister immediately stopped the meeting and sent several aides rushing from the room. They returned a few minutes later with displays of robust wheat plants which the Minister proudly showed me. "We know Norman Borlaug very well," the Minister declared, going on to point out how Borlaug's innovations had helped transform agriculture in his country and throughout the region, to the benefit of millions upon millions of the citizens of all these countries.

THE IMPACT IN ASIA

Borlaug's successes in wheat were quickly replicated in other grains, most notably rice, by scientists such as M. S. Swaminathan in India, and Robert Chandler, Henry "Hank" Beachell, and Gurdev Khush at the International Rice Research Institute in the Philippines. Together, with countless others, they helped avert famine and starvation in much of the developing world in the second half of the 20th century.

I was a young development worker in the Mekong Delta in 1968 when the new "miracle rice" from the Philippines arrived. Its impact in the eight villages in which I worked was as stunning as it was immediate. The four villages that were accessible by road experienced dramatic improvements, both in terms of nutrition and the well being of the people. New IR-8 rice spread rapidly as peasant farmers with small plots were suddenly able to experience both increased yields and double crops. This in turn led to tangible improvements in the quality of life: child mortality dropped; malnutrition abated; and children, especially girls, stayed in school longer.

At the same time, there was a rapid corresponding decrease in the level of armed conflict and military hostilities. It was as though the combination of new roads and new rice seed caused the roots of violent extremism to wither and disappear in a way that military action alone could not. By contrast, the four other villages, with no bridges and no road access, remained mired in poverty: the new "miracle seeds" were not put to use; children remained stunted; and warfare and political dissidence continued there unabated.

This experience seemed to confirm one of the central lessons of Norman Borlaug's boyhood—the ability of agricultural innovation and rural roads to dramatically change social conditions.

THE NOBEL PRIZE

Dr. Norman Borlaug was presented the Nobel Peace Prize in 1970 for his accomplishments in India and Pakistan and for his role as Father of the "Green Revolution." It is indicative of the kind of person he is that when, on October 20, 1970, the phone call came to advise him of his selection as the Laureate, Borlaug was in a remote farm field in Mexico. Margaret had to drive for over an hour to tell him the news and ask him to return home to respond to the calls, and the concomitant press requests for interviews. Lennard Bickel, in his 1974 biography of Borlaug, *Facing Starvation*, describes Norm's reaction: He told Margaret that he didn't see how he could possibly come to speak on the phone since he and his assistants still had much more work to do. He then went back to recording data on his test plots. It was there that the TV camera crews found him two hours later.

In a sense, when Borlaug received the Peace Prize on December 10, 1970, his life had come full circle. Here he was, the son of immigrants who had left Norway due to extreme food shortages, now, back in their country of origin to receive one of the world's highest honors for his role in increasing the world's food supply. As he stood in the great hall of the University of Oslo, Borlaug was lauded as an "indomitable man who fought rust and red tape . . . (and) who more than any other single man of our age, has provided bread for the hungry world."

Borlaug remains the only agricultural scientist ever to receive the Nobel Peace Prize, and one of its least known recipients. I recall reflecting as he sat among many other laureates for the 100th Anniversary "class picture," that it was ironic that his name would be so little recognized, since compared to the other, more celebrated honorees, he had probably saved more lives than all of them put together.

In his laureate address, Borlaug stressed that the agricultural breakthrough achievements for which he was being honored were only providing a brief window of time during which the world must confront the specter of a burgeoning world population which would have to be fed. As a result, Dr. Borlaug's efforts did not cease or even slow after this recognition by the Nobel Foundation. While many individuals might consider retiring after receiving such recognition, Dr. Borlaug has worked even harder in the 30 plus years since his selection as the Peace Prize Laureate in the struggle against world hunger and malnutrition. In his ninth decade of life, Dr. Borlaug keeps a heavy travel schedule, pressing forward with projects in Africa, passionately advocating the primacy of science and technology in improving global food security, devoting significant time and energy to education, and promoting biotechnology as a way to preserve the environment.

Among Dr. Borlaug's greatest accomplishments since his selection as a Nobel Peace Prize Laureate are

- His leadership with former President (and fellow Nobel Peace Prize Laureate) Jimmy Carter and The Carter Center, of The Sasakawa

Global 2000 Program, which has promoted the production of Quality Protein Maize in sub-Saharan Africa, countering Marasmus and Kwashiorkor, and other forms of acute malnutrition for millions of at-risk children

- The founding of the World Food Prize, which exists to recognize and inspire Nobel-like achievements in food and agriculture
- His 30-year relationship with China as that country's agriculture was transformed
- His prodigious efforts to educate the next generation of students and leaders on the crucial importance of world hunger and food security

BRINGING THE GREEN REVOLUTION TO AFRICA

Since 1986, Dr. Borlaug has headed the Sasakawa Africa Association whose programs aim at defeating malnutrition and poverty in Africa. Its activities center on bringing science-based crop production methods to the small farms of sub-Saharan Africa. Proven agricultural technology is the key to overcoming widespread food shortages that condemn millions of people in Africa to lives of hardship and hunger.

Part of the Sasakawa Global 2000 endeavor, Sasakawa Africa projects are under way in a dozen African countries. In addition to its partnerships with ministries of agriculture, the Sasakawa Africa Association collaborates with NGOs, businesses, and international development agencies. Perhaps the most significant achievement of this effort is the successful development of highly nutritious corn—known as Quality Protein Maize—which offers great promise in preventing acute malnutrition among children in Ghana, Mozambique, and other African countries, as well as in Mexico. Perfected by a longtime Borlaug protégé at the Maize and Wheat Improvement Center in Mexico, Dr. Borlaug and the Sasakawa Africa Association have helped spread this lifesaving food into villages with immediate effect, enhancing and saving the lives of thousands and thousands of children.

THE WORLD FOOD PRIZE

One of Dr. Borlaug's most lasting contributions may be the creation of the World Food Prize. Norm has often said he believes he was nominated for the Nobel Peace Prize because there is no Nobel Prize for agriculture or efforts to counter poverty and hunger. Laureate Borlaug felt there should be, so shortly after receiving the Peace Prize, he approached the Nobel Committee urging the creation of a new Nobel Prize for Agriculture. But it was not possible. Not even Borlaugian grit and determination could change Alfred Nobel's will. Undeterred, Norm set out to create just such an honor. In 1986, with the assistance of Carleton Smith and the support of the General Foods Corporation, he established a new award to recognize exceptional achievement: the World Food Prize.

The Prize of $250,000 is now endowed by philanthropist and businessman John Ruan, himself with origins in a small Iowa town just like Borlaug. Ruan "rescued" The Prize when General Foods withdrew its sponsorship in 1989. He moved it to Des Moines, Iowa, Borlaug's native state, and established a foundation with a bipartisan Council of Advisors that includes former U.S. Presidents Jimmy Carter and George Bush; Pulitzer Prize winner Michael Gartner; former cabinet members Elizabeth Dole (now Senator from North Carolina) and Robert McNamara; and the Honorable Olusegun Obasanjo, now President of Nigeria.

Ruan, who serves as Chairman of The World Food Prize Foundation, stressed that he took this action because The Prize is now even more vital to inspiring a second green revolution that is necessary to prevent the possibility of future food crises. "Right now, close to one billion people still suffer from malnutrition, nearly one-sixth of the world population, primarily women and children, infants, and the unborn," Ruan points out.

In addition to the Laureate Award Ceremony, the World Food Prize holds an International Symposium and a Global Youth Institute each October to foster a dialogue on world hunger and related issues. Recent symposium topics focused on the safety of genetically modified crops, the relationship between food security and the potential for agroterrorism, and the coming global water crisis.

The World Food Prize has honored those who have made the most significant contributions to improving the quality, quantity and availability of food. Swaminathan, Barwale, Beachell, Khush, and Chandler all eventually became World Food Prize Laureates. Other recipients of The Prize include experts and scientists from China, Bangladesh, Switzerland, the United Kingdom, Denmark, Cuba, and the U.S.

OPENING TO CHINA

One of the World Food Prize Laureates was He Kang, the former Chinese Minister of Agriculture who was honored in 1993 for implementing the policies that moved China to self-sufficiency in grain in a remarkably short period of time. Minister He and many of the agricultural scientists in China look to Dr. Borlaug as an "old friend" who provided them significant assistance along this road to success. Borlaug's spring wheats went to China, via Pakistan, during the late 1960s. They were crossed with Chinese wheats, as well as directly selected for use. Borlaug himself led the way in establishing connections between China and the West. He has been going to China since 1974, one of the first scientists from the West to begin visiting there. He was there again in 1977 and at various times during the 1980s and 1990s. He has traveled extensively in the country, initially to wheat-growing areas, and over time to maize and other agricultural areas, talking with farmers and urging adoption of approaches he developed. During these visits, Borlaug has seen the tremendous improvements that have taken place since the collapse of the Cultural Revolution, particularly beginning in the watershed year of 1978. Borlaug often traces this success to the changes in Chinese nutrient management strategies, moving from a reliance on organic recycling of manure, through the building of small-scale plants and imports, to Chou en Lai's authorization of the purchase of multiple large-scale nitrogen plants.

Reflecting the friendship and spirit of cooperation he has always demonstrated in his relationship with Chinese scientists and policymakers, Norman Borlaug was made an honorary member of the Chinese Academy of Agricultural Sciences in 1994.

YOUTH EDUCATION

Dr. Borlaug has been committed to youth activities and education throughout his career. While pursuing breakthroughs in plant science in Mexico, he served as scoutmaster for his local Boy Scout Troop, and as a Little League baseball coach. Even today, he continues to devote himself to passing on to the next generation his passion for science and education as the means to uplift people mired in poverty. Thus, between attending conferences and giving lectures around the world, he continues to teach at Texas A&M University, where he holds a post as a Distinguished Professor in the Department of Soil and Crop Sciences.

To promote interest in global food security, in partnership with John Ruan, he created The World Food Prize Youth Institute, which is held in conjunction with The World Food Prize International Symposium each October in Des Moines. There, high school students interact with Dr.

Borlaug, World Food Prize Laureates, and other experts to discuss the potential solutions to world hunger and the roles they, the leaders of tomorrow, might play in making them a reality.

Under Dr. Borlaug's direction, the Youth Institute has developed an International Internship program, which sends exceptional high school students on eight-week internships to international agricultural research centers in Ethiopia, Kenya, India, Indonesia, the Philippines, Thailand, Mexico, Peru, Brazil, Malaysia, Costa Rica, Trinidad, and China. He wants them to have that same type of life-altering experience that he had when he heard Elvin Stakman speak on that cold Minnesota night in 1937.

When speaking to young people in the early years of the 21st Century, Borlaug often quotes Thomas Jefferson as rhetorically asking whether

> Ease and security—were these the drugs that abated the eternal challenge of the minds of men? . . . Did nations like men become lethargic when well fed and bodily comfortable?"

It is clear that Borlaug worries that this may be the case, particularly now that almost all young Americans are physically removed from farming, and the connection between our food supply and agriculture production is no longer so clearly understood. But no doubt, he takes heart when some of the students returning from their eight-week World Food Prize International Internships volunteer that coming face to face with third-world poverty was a life-changing experience, perhaps not unlike Borlaug's own epiphany as he listened to his mentor—Elvin Stakman—almost 70 years ago.

CONCLUSION—APPLYING THE LESSONS OF THE 20TH CENTURY

Exhibiting the virtues he learned growing up, Norm Borlaug is still going strong, traveling the world to promote greater attention to—and investment in—rural infrastructure (particularly roads and bridges), agricultural research, and education. Norm believes all these are essential if we are to have the next green revolution—the one that will lift the remaining one billion people out of the misery of malnutrition and end pandemic poverty.

In his speeches he advocates biotechnology and the crucial role he sees for it in feeding and enhancing the nutrition of those still in tenuous food-security situations, particularly in Africa. Genetically modified crops are controversial, but, never one to back away from a confrontation, Borlaug argues that we must rely on science and research to answer the questions about whether GMO foods pose any environmental risks. At the same time, he stresses that what is needed is not just miracle seeds and other agricultural inputs, but also the educational facilities to uplift the young and logistical infrastructure (such as roads and railroads) to make Africa prosper.

He laments the declining trend in support for public agricultural research, such as at CIMMYT, where the crucial discoveries that led to the first green revolution took place. In June 2002, he and all the living World Food Prize Laureates issued a statement at the World Food Summit in Rome calling for a reversal of this trend.

And when he concludes his remarks, something of the old forester comes to the fore. Borlaug points out that with the earth's population increasing exponentially, all these new people can be fed in only one of two ways. Either we significantly increase yields on the land now in production, or we plow under the remaining rainforests and other habitats for wild animals in order to have more land to farm. Biotechnology, he stresses, will help preserve the ecosystem while also reducing hunger and malnutrition, by providing these increased yields. In that way, he once told a group of Iowa high school students, he may be saving more trees as a plant pathologist than he would have as a forest ranger.

But, I believe Norm Borlaug's message may be just as relevant for those who seek to counter terrorism and bring a lasting peace in the Middle East and South Asia. Just as I saw the first green revolution evaporate political and military hostility more than 30 years ago in the Mekong Delta, it just may be that a new green revolution (and the roads to make it happen) represents one of the most potent forces available to this generation to dissipate the sources of terrorism, which breed and are sustained in the poorest parts of the world, such as Afghanistan and Somalia.

As the person who has probably saved more lives in the Islamic world than anyone who has ever lived, it would be only fitting if Norman Borlaug's 20th century message of using seeds and roads to reach across political chasms to uplift humanity would be the vehicle that brought peace and reconciliation to a deeply troubled and divided 21st century world.

NOTES

1. Kenneth M. Quinn is President of The World Food Prize Foundation, a nonpartisan organization dedicated to inspiring breakthrough achievements that can lessen hunger and malnutrition around the world. He served as U.S. Ambassador to Cambodia from 1996 to 1999. Readers may write him at The World Food Prize Foundation; 1700 Ruan Center; 666 Grand Avenue; Des Moines, IA 50309. More information about the World Food Prize and International Symposium is available on the web at http://www.worldfoodprize.org.

4

An Appeal by the 15 World Food Prize Laureates on the Occasion of the World Food Summit: Five Years Later
June 2002, Rome

Dr. M.S. Swaminathan
1987, India

Dr. Muhammad Yunus
1994, Bangladesh

Mr. B.R. Barwale
1998, India

Dr. Verghese Kurien
1989, India

Dr. Hans R. Herren
1995, Switzerland

Dr. Walter Plowright
1999, England

Dr. John S. Niederhauser
1990, United States

Dr. Gurdev Khush
1996, India

Dr. Evangelina Villegas
2000, Mexico

Dr. Nevin S. Scrimshaw
1991, United States

Dr. Henry Beachell
1996, United States

Dr. Surinder K. Vasal
2000, India

H. E. He Kang
1993, China

Dr. Perry L. Adkisson
1997, United States

Dr. Per Pinstrup-
Andersen
2001, Denmark

and

Dr. Norman E. Borlaug, 1970 Nobel Peace Prize Laureate
and Chair of The World Food Prize Council of Advisors

WE, THE WORLD FOOD PRIZE LAUREATES, wish to stress the importance of the conference, *The World Food Summit: Five Years Later* under the leadership of Jacques Diouf, Director-General of the Food and Agriculture Organization of the United Nations.

We recall that at the World Food Summit of 1996, Heads of State and Government from around the world pledged their political will and their joint and national commitment to achieving food security for all and to an ongoing effort to eradicate hunger in all countries. In the immediate view of those attending that 1996 Summit, the number of undernourished people should be reduced by half no later than 2015.

We note with dismay that, according to the latest information and analyses available, the number of undernourished people is falling only by about eight million a year, whereas it needs to fall by about 20 million a year if the World Food Summit target is to be achieved by 2015. Furthermore, while China achieved major reductions in the number of undernourished people during the 1990s, developing countries as a whole, excluding China, actually saw an increase in that number.

A continuation of the trend of the 1990s for the developing countries as a whole, excluding China, will result in a very significant increase rather than a decrease in the number of undernourished people. Nearly 90 million additional children are born each year, with more than 70 percent of them to poor and undernourished families.

This should be a major concern for everyone. Food, along with sanitary water and shelter, are the most vital of human needs; and the lack of food is a major barrier in achieving other human rights. Hunger is also often a cause and an effect of social instability and conflict. People debilitated by hunger are disposed to be less productive and more prone to infectious diseases, including HIV/AIDS.

The world as a whole cannot enjoy durable peace, social stability, and economic prosperity while hundreds of millions of people suffer from abject poverty and hunger.

The 15th Anniversary World Food Prize Foundation Symposium on "Risks to the World Food Supply," held in Des Moines, Iowa, U.S.A., October 18–19, 2001, addressed many of the crucial issues that must be dealt with if greater progress in reducing food insecurity is to be achieved. We also note the leadership provided by the 2001 World Food Prize Laureate Dr. Per Pinstrup-Andersen in formulating and implementing the 2020 Vision for Food, Agriculture, and the Environment, which, together with the work of FAO and others, offers a solid blue-print for actions to address these problems.

Since the large majority of the poor and the hungry in developing countries live in rural areas and rely for their livelihood on agriculture, including livestock and animal industries, and on activities dependent on agriculture, development policies of these countries as well as donor assistance should give priority to sustainable productivity increases in agriculture and rural development, including essential infrastructure such as rural roads, electrification, and markets.

In the coming decades a technological transformation of agriculture will occur that will be constrained by resource limitations and whose environmental implications will pose questions concerning the sustainability of food production adequate to feed the ever-increasing human population. Therefore, it is imperative that we work together to strengthen the research and policy framework underpinning the necessary productivity increases in agriculture, livestock, and aquatic resources in an environmentally sustainable manner.

We are greatly concerned that funding for international research centers and public agricultural research programs is being cut back. It was efforts at just such institutions—by Nobel Peace Prize Laureate Dr. Norman E. Borlaug and many World Food Prize Laureates—that produced the great gains in agricultural production during the 1970s, 1980s, and 1990s, averting famine in many areas. However, there is a danger of critically needed research capabilities being seriously eroded due to inadequate funding. The transfer and utilization of appropriate technologies and moving beyond traditional partnerships is also essential.

In regard to free trading policies, fairness is needed to provide markets to the poorer nations. We would like to emphasize the need for greater market access to food and agricultural products from developing nations. Today, it is often the case that heavy subsidies in industrialized countries and the imposition of non-tariff barriers by rich nations are closing the doors to the products of poor countries.

The causes of hunger are many and complex. We have outlined some of these causes, such as low food production, distribution, poverty, sustainability, and environmental degradation. We have emphasized the need to establish the international cooperation to confront these problems and work toward solutions on a worldwide basis. We believe that none of these efforts will provide a long-range solution to the problem of hunger unless we also dedicate our efforts to programs to promote population stabilization. Without population stabilization, our dedication to the production and distribution of food will only postpone the problem of even greater hunger in the world.

We therefore appeal to national governments, bilateral and multilateral development agencies, national and regional funding institutions, U.N. system organizations, and other organizations attending the FAO Conference in Rome to focus their efforts on these parallel goals, with the greatest possible transparency in defining the objectives, programs, and rate of progress of the organizations in reaching their ends.

THE WORLD FOOD PRIZE

Dr. Norman E. Borlaug conceived of the idea for a World Food Prize after receiving The Nobel Peace Prize.

The $250,000 World Food Prize is awarded each October in Des Moines, Iowa, U.S.A., to recognize the significant achievements of individuals who have reduced poverty, hunger, and malnutrition by improving the quality, quantity, or availability of food in the world.

The Prize was established in 1986 and, in 1990, Des Moines Businessman and Philanthropist John Ruan assumed its sponsorship, providing a generous endowment and moving The World Food Prize Foundation to Des Moines. Ambassador Kenneth M. Quinn currently serves as President of the Foundation.

Information about the World Food Prize Laureates and the annual programs conducted in conjunction with the Laureate Award Ceremony is available on the web at www.worldfoodprize.org, or by contacting Ambassador Quinn at

> The World Food Prize Foundation
> 1700 Ruan Center, 666 Grand Avenue
> Des Moines, Iowa USA 50309
> Phone: 515–245–3783; Fax: 515–245–3785
> E-mail: wfp@worldfoodprize.org

5

REDUCING HUNGER BY
IMPROVING SOIL FERTILITY:
AN AFRICAN SUCCESS STORY[1]
Pedro A. Sanchez[2]

Africa south of the Sahara is the only region where per capita food production has not risen over the last 40 years. The green revolution has resulted in sustained increases in per capita food production in Asia and South America. The adoption of dwarf, high-yielding rice (*Oryza sativa, L*) and wheat (*Triticum aestivum, L*) was the key factor in the success of the green revolution, particularly in Asia. Other key aspects were irrigation, fertilizer use, pest management, research and extension, and enabling government policies.

Between 1980 and 1995, per capita food production increased by 27 percent in Asia and 12 percent in South America, but decreased by 8 percent in sub-Saharan Africa. Moreover, the rates of adoption of improved crop varieties have been similar in Asia, Latin America, the Middle East, and sub-Saharan Africa. These varieties are responsible for 66–88 percent increases in crop yield in the first three regions, but only 28 percent in Africa (Special Panel on Impact Assessment 2001). In contrast to the sustained increases in per capita food production in much of the developing world, the situation of food insecurity in Africa is acute and a global concern.

The U.N. Food and Agriculture Organization estimates that for the years 1998–2000, there are 840 million undernourished/hungry people in the world with 799 million of those in developing countries. Approximately 180 million Africans, up about 100 percent since 1970, do not have access to sufficient food. Not only is per capita food consumption in sub-Saharan Africa declining, but also land is being depleted. Moreover, the starting point is extremely limited, with half the population being absolute poor (less than $1/day) and with the highest proportion of undernourished children.

LOSSES OF SOIL FERTILITY

The situation in sub-Saharan Africa is further accentuated by the depletion of essential plant nutrients from soils. It is estimated that the loss of nutrients from cultivated land in 37 African countries during 30 years is of the following magnitude:

660 kg nitrogen per hectare
75 kg phosphorus (P) per hectare
450 kg potassium (K) per hectare

Another way of describing the situation in sub-Saharan Africa (excluding South Africa) is the following annual aggregate figures. We have estimated that farmed fields in the region is losing the following annually:

4.4 million tons nitrogen
0.5 million tons phosphorus (P)
3.0 million tons potassium (K)

This can be compared to annual fertilizer use, which is as follows:

0.8 million tons nitrogen
0.26 million tons phosphorus
0.2 million tons potassium

Based on years of field research and observation throughout Africa, we have concluded that "Soil-fertility depletion in small-holder farms is the fundamental biophysical root cause of declining per capita food production in Africa, and soil fertility replenishment should be considered as an investment on natural resource capital" (Sanchez et al. 1996). This has become well recognized by the research community. Until very recently, the development community focused on other biophysical constraints (soil erosion and droughts) and the need for improved crop germplasm. We consider that no matter whether these other conditions are improved, "per capita food production in Africa will continue to decrease unless soil-fertility depletion is effectively addressed" (Sanchez et al. 1997).

The depletion of soil nutrients is due to grain and crop residue harvest removals, soil erosion, and to some extent leaching and gaseous losses. These processes are occurring without sufficient replenishment (by fertilizers, manures, or biological nitrogen fixation). Nutrient depletion has transformed fertile soils yielding ~3 tons maize/hectare to low-fertility soils with yields of less than 1 ton/hectare.

Fertilizer Use

Sub-Saharan African small-holders recognize the need for fertilizer, but usage is limited by cost, often greater than two to six times the international price (in Europe, North America, or Asia). For instance, spot checks show that a metric ton of urea costs the following (Sanchez 2002):

~$US90 FOB (free on board) in Europe
$120 delivered at the ports of Mombasa (Kenya) or Beira (Mozambique)
$400 in Western Kenya (700 km from Mombasa)
$500 across the border in Eastern Uganda
$770 in Malawi (transported from Beira)

Throughout sub-Saharan Africa, nitrogen and phosphorus are the limiting nutrients. While potassium depletion rates are higher, crop responses to potassium fertilizer are less common in Africa. Organic farming has been advocated as an alternative, but we disagree, because most forms of organic farming in Africa do not supply sufficient nutrients to food crops. Plants do not "know" the difference between organic plant nutrients and synthetic fertilizers. "In our view, African farmers do not need low-input, low-output systems that do not address poverty alleviation" (Sanchez et al. 1997). Therefore we need both mineral fertilizers and organic inputs—the mineral fertilizer–only approach is as insufficient as the organic fertilizer–only approach. Farmers in the First World routinely apply both types of nutrient inputs. African farmers should be given the opportunity to do the same.

Restoring soil fertility is a necessary but not sufficient strategy. It is also necessary to promote survival of children, cope with HIV-AIDS, improve governance, increase foreign investment, break trade barriers, and provide debt relief. Most of these development efforts do not, however, directly address agriculture, the economic sector engaging 70 percent of the population of sub-Saharan Africa. Africa's food insecurity is due to insufficient total food production. This is in contrast to the situation in South Asia and other regions where food insecurity is primarily due to landlessness and lack of purchasing power. African agriculture has performed dismally. Relatively little attention has been paid by African governments to rural areas. Some African states and developed countries are now considering restoring high priority to agricultural development. This is a great opportunity.

DEVELOPMENT OF SOLUTIONS

During the last 15 years, the International Center for Research in Agroforestry and its partners have developed a number of successful approaches using resources naturally available in Africa. Researchers, working with farmers in participatory research trials, have developed new approaches to replenishment of soil fertility. The farmers test and adapt these new practices to their particular production and resource situation. The approaches consist of three components that can be used in combination or separately: nitrogen-fixing leguminous tree fallows, biomass transfer of leaves of nutrient-accumulating shrubs, and indigenous rock phosphate applications.

With nitrogen-fixing leguminous trees/shrubs fallows, extensive research has demonstrated the value of inter-planting leguminous shrubs/trees into a young maize crop. The leguminous shrubs/trees are of the genera *Sesbania, Tephrosia, Crotalaria, Glyricidia,* and *Cajanus.* These are allowed to grow as fallows during dry seasons. In subhumid tropical regions of East and Southern Africa, this results in 100–200 kg nitrogen being accumulated per hectare over 6 months to 2 years. The quantities of nitrogen captured are similar to those applied as fertilizers by commercial farmers growing maize in developed countries. After harvesting the wood from the tree fallows, the nitrogen-rich residue (leaves, pods, and green branches) is incorporated into the soil by hoeing. This is done prior to planting maize at the start of the rainy season. Leaves and small twigs decompose along with tree roots, releasing nitrogen and other nutrients to the soil and overcoming the nitrogen deficiency. This approach has increased the yield of the staple food—maize—two- to fourfold.

For instance, in trials in Zambia, maize yields following two-year *sesbania* fallows were 5 Mg ha-1 compared to ~1.5 Mg ha-1 without fertilizer and 4.6 Mg ha-1 with fertilizer (Kwesiga et al. 1999). In East African areas where there is a bimodal rainfall pattern (two rainy seasons and two dry seasons per year), farmers use rotations of one year of trees followed by one crop of maize. In unimodal rainfall areas of Southern Africa, two years of trees are followed by two to three maize crops.

There are limitations to this soil fertility replenishment approach. Improved tree fallows have not yet proved attractive to farmers at the margins of humid tropical forests of the Congo Basin or shown applicability in the semi-arid tropics of Africa (where a longer dry season limits their growth and nitrogen-fixation potential). Fallows also do not perform well in shallow soils, poorly drained ones, or frost-prone areas.

With transfer of biomass (leaves of nutrient-accumulating shrubs) from roadsides, leaf biomass of the nutrient-accumulating shrub *Tithonia diversifolia* (the Mexican sunflower) is being transferred from roadsides and hedges into cropped fields to add essential nutrients. The green leaves of *Tithonia diversifolia* are high in nitrogen (3.5 percent), phosphorus (0.37 percent), and potassium (4.1 percent) (Jama et al. 2000). Biomass transfer has resulted in marked increases in the production of maize.

Maize yields are routinely doubled by *Tithonia* biomass applications, without fertilizer additions in Western Kenya. This source of nutrients has been found to be more effective than urea (at the same rate of nitrogen application) as other plant nutrients, particularly potassium and micronutrients, are also being added. With the high labor requirements for cutting and carrying the biomass to fields, the use of *Tithonia* as a nutrient source is profitable with high-value crops such as vegetables but not with relatively low-valued maize. *T. diversifolia* grows spontaneously on roadsides of subhumid tropical Africa.

These first two approaches also overcome another problem: the depletion of soluble carbon in many tropical soils that have been farmed. Soil microorganisms use this soluble carbon as their energy source. Organic inputs enhance nutrient cycling, mineralization rates, and the transformation of inorganic forms of phosphorus into more available organic ones. These agroforestry approaches sequester large quantities of carbon in the tree biomass and soil, 5–10 times as much as most other agricultural practices.

Replenishing phosphorus using indigenous rock phosphates is important because phosphorus deficiency is widespread in East Africa and the Sahel. For instance, in Western Kenya, 80 percent of the land held by small-scale owners is extremely deficient in phosphorus. Maize yield is greatly limited by this. The phosphorus sources available to sub-Saharan African farmers with phosphorus-deficient soils can apply expensive imported super-phosphates, but cannot rely on organic inputs to do the same. Organic sources, when applied at realistic rates can provide up to one-half of the phosphorus requirements of a 4-ton/hectare maize crop.

An alternative to traditional inorganic fertilizer is the use of indigenous rock phosphates. The mild acidity (pH 5 to 6) prevailing in most soils of subhumid Africa helps dissolve high-reactivity rock phosphates at a sufficient rate to supply phosphorus to several maize crops. These high-quality indigenous rock phosphates have proven to be ~90 percent as efficient as super-phosphate fertilizer over a five-year period.

After replenishing the soil's fertility, other practices now have a chance for success: high-yielding crop varieties, integrated pest management, conservation tillage, high-value trees, vegetable crops, and dairy cattle. These approaches also have positive environmental effects.

CONCLUSION

Tens of thousands of farm families in Kenya, Uganda, Tanzania, Malawi, Zambia, Zimbabwe, and Mozambique are today using combinations of fallows with leguminous nutrient-accumulating shrubs/trees, biomass transfers, and indigenous rock phosphates—with consistently dramatic results. Adoption of nutrient-enhancing options is taking place by farmer-to-farmer, village-to-village community-based organizations and knowledge transfer, together with national research and extension institutes, universities, nongovernmental organizations, and development projects. The on-farm production of firewood is reducing encroachment onto nearby forests, preserving biodiversity.

These "low-tech" but knowledge-intensive approaches have to *precede* the application of improved germplasm, either by traditional plant breeding or genetic engineering, because without adequate soil nitrogen and phosphorus, African farmers have no chance of succeeding. The challenge now is to accelerate the adoption rate to reach tens of millions of farm families. Large and sustained investment (requiring ~US$100 million a year for the next 10 years) is needed to scale up these approaches to tens of millions of African farm families. Attention is needed on such bottlenecks as the production of sufficient quality scrub/tree germplasm by community-based nurseries, access to rock phosphate, and increasing awareness of the three complementary approaches. Donor countries and organizations need to invest in community-based development projects that integrate the agriculture, education, and health sectors. These integrated approaches have a strong probability of success because the problems are interrelated.

African countries, with cooperating donors, need to identify constraints, to adopt policies, and to improve transport infrastructure that will greatly reduce the disparity between world market prices and the prices paid by African farmers for mineral fertilizers. Financial incentives need to be in place for tropical farmers to remove large quantities of carbon from the atmosphere through tree-based soil fertility replenishment technologies. This will also help alleviate poverty while reducing greenhouse gas emissions. It is most important for the leaders of African governments to show the political courage to adopt the necessary policies so that the rural development is a priority and both hunger and poverty are progressively eliminated.

REFERENCES

Jama, B., C.A. Palm, R.J. Buresh, A. Niang, C. Gachengo, G. Nziguheba, and B. Amadalo. 2000. *Tithonia diversifolia* as a green manure for soil fertility improvement in western Kenya: A review. *Agroforestry Systems* 49:201–221.

Kwesiga, F.R., S. Franzel, F. Place, D. Phiri, and C.P. Simwanza. 1999. *Sesbania sesban* improved fallows in eastern Zambia: Their inception, development and farmer enthusiasm. *Agroforestry Systems* 47:49–66.

Sanchez, P.A. 2002. Soil fertility and hunger in Africa. *Science* 295:2019–2020.

Sanchez, P.A., A-M.N. Izac, I. Valencia, and C. Pieri. 1996. Soil fertility replenishment in Africa: A concept note. p. 200–207. In *Achieving Greater Impact from Research Investments in Africa*. S.A. Breth, ed. Mexico City: Sasakawa Africa Assoc.

Sanchez, P.A., K.D. Shepherd, M.J. Soule, F.M. Place, R.J. Buresh, A.N. Izac, A.U. Mokwunye, F.R. Kwesiga, C.G. Ndiritu, and P.L. Woomer. 1997. Soil fertility replenishment in Africa: An investment in natural resource capital. In *Replenishing Soil Fertility in Africa*. Buresh, R.J., P.A. Sanchez, and F. Calhoun, eds. Soil Science Society of America, Special Publication no. 51, pp. 1–46. Madison, WI: American Society of Agronomy and Soil Science Society of America.

Special Panel on Impact Assessment. 2001. In *Contributions Made by the CGIAR and Its Partners to Agricultural Development in Sub-Saharan Africa*. Washington, D.C.: Consultative Group on International Research.

NOTES

1. This chapter is based in part on Sanchez, 2002.
2. Director of Tropical Agriculture, The Earth Institute at Columbia University; P.O. Box 1000; Palisades, NY 10964, U.S.A.; p.sanchez@cgiar.org.

PEDRO SANCHEZ—A LIFE OF COMMITMENT AND ACHIEVEMENT

Colin G. Scanes and John A. Miranowski

In 2002, Dr. Pedro Sanchez, a native Cuban, received the World Food Prize. He was selected for his groundbreaking contributions to reducing hunger and malnutrition throughout the developing world, transforming depleted tropical soils into productive agricultural lands. His achievements include the following:

- In Peru, he played a pivotal role to dramatically improve rice production within three years; Peru achieved among the highest rice yields in the world and food security by achieving self-sufficiency.
- In Brazil, his comprehensive approach to soil management enabled 30 million hectares (75 million acres) of marginal land (the Cerrado) to be brought into production. This was the single largest increase in arable agricultural land in the last half-century.
- In sub-Saharan Africa, as Director-General of the International Center for Research in Agroforestry (ICRAF) in Nairobi, he led the agricultural research program that is already providing 150,000 small farmers in Africa with the means to fertilize their soils inexpensively and replenish crucial nutrients in exhausted soils using agroforestry. Dr. Norman Borlaug has stated that "Dr. Sanchez' achievement gives hope that the green revolution can finally be extended to Africa."

Dr. Sanchez is currently a visiting professor of tropical resources at the University of California at Berkeley's College of Natural Resources and Chair of the United Nations Millennium Project Task Force on World Hunger.

Dr. Sanchez was born in Havana in 1940, the son of an agronomist. The inspiration for his career in agricultural research studies was his boyhood experiences traveling with his father around Cuba and keenly watching his father's efforts to convince farmers to use fertilizers more effectively. Dr. Sanchez was educated at Havana's Collegio de la Salle and Cornell University (obtaining B.A., M.A., and Ph.D. degrees).

BRINGING PERU THE HIGHEST RICE YIELDS IN THE WORLD IN THREE YEARS

Following his doctoral research at the International Rice Research Institute in the Philippines, Dr. Sanchez led North Carolina State University's Rice Research Team in Peru. He carried into Peru new "miracle rice" seeds, which he, working closely with Peruvian officials, helped spread around the country. The impact was immediate and dramatic, with Peru becoming self-sufficient in rice in just three years.

In the Peruvian Amazon, Dr. Sanchez encountered for the first time two issues that were to change his life:

- The widespread belief that tropical soils were useless for agricultural production
- The devastating environmental effects of large-scale land clearing and "slash-and-burn" agriculture

These convinced Dr. Sanchez to search for the means to transform tropical soils and obviate the need for bulldozer land clearing and slash-and-burn agriculture.

THE LARGEST EXPANSION OF AGRICULTURAL LAND IN THE LAST HALF-CENTURY

Dr. Sanchez led a program of soil management in the Cerrado, an enormous area of unproductive soil in Brazil, equivalent in size to Western Europe. Through painstaking research, Dr. Sanchez' team (including the newly established Brazilian Agricultural Research Program, EMBRAPA) discovered the approach that would permit the Cerrado to blossom. By calculating the precise depths and intervals at which to treat the soil with a carefully balanced package of minerals and fertilizers, the previously infertile tropical soils came to life. Thirty million hectares (75 million acres) were made productive. Average yields increased by 60 percent. Soybean production became on par with the United States, and the Brazilian grain harvest tripled.

BRINGING THE GREEN REVOLUTION TO AFRICA

In 1991, Dr. Sanchez became the head of ICRAF in Nairobi, Kenya. There, he quickly discovered that African agricultural production lagged due to the extremely depleted nature of the soil.

Dr. Sanchez' most enduring contribution to ending world hunger has been his development of the means to replenish crucial nutrients in exhausted soils, through the development and promotion of agroforestry (planting nitrogen-fixing trees on farms). This practice, when combined with adding locally available rock phosphate to the soil, has greatly increased yields (by 200 percent–400 percent above previous plantings) for African farmers. It has provided farmers in Africa with a way to fertilize their soils inexpensively and naturally, without relying on costly chemical fertilizers.

In response to this success, ICRAF plans to help African farmers plant 5.5 billion more trees over the next decade, the equivalent of another tropical rainforest. ICRAF's goal is to move 20 million people out of poverty and remove more that 100 million tons of CO_2 from the air with this project.

CONCLUSION

Reflecting his enormous contributions, Kofi Annan, Secretary-General of the United Nations has honored Dr. Sanchez by appointing him to Chair the U.N. Taskforce On World Hunger as part of the U.N. Global Millennium Development Project, and with the following statement on August 11 2002:

> I am very pleased that the 2002 World Food Prize has been awarded to Dr. Pedro A. Sanchez, a distinguished soil scientist and the Chair of the United Nations Millennium Project Task Force on World Hunger.
>
> Through his research in the Cerrado of Brazil, Dr. Sanchez developed ways to revitalize tropical soils, which had been considered extremely unproductive, thus greatly expanding Brazil's agricultural output. Subsequently, in East Africa, where Dr. Sanchez served as Director-General of the International Center for Research in Agroforestry, his innovative approach to restoring nutrients to severely depleted soils resulted in dramatic increases in crop yields, affecting hundreds of thousands of small farmers. It is clear that Dr. Pedro Sanchez's achievements offer great promise that the Green Revolution can be spread through sub-Saharan Africa. It is par-

ticularly fitting that the World Food Prize will be presented to Dr. Sanchez on 24 October, United Nations Day. Nothing could better reflect the direct connection between Dr. Sanchez's accomplishments and his new mission on behalf of the Millennium Project of the United Nations.

6

ACHIEVING THE 2020 VISION IN THE SHADOW OF INTERNATIONAL TERRORISM[1]
Per Pinstrup-Andersen[2]

The brutal terrorist attack on September 11, 2001, calls for two parallel campaigns. First, the ongoing campaign to identify and eradicate the organizations that sponsor international terrorism, and second, a campaign to eliminate poverty, hunger, malnutrition, and mismanagement of natural resources. Those are the issues that provide the perceived justification for fanatics such as bin Laden, and the many people who think like him, to do the kinds of things that they did to the free world in New York, Washington, and Pennsylvania on the 11th of September. No, the people who did that were not poor, but they got their justification from the extreme human misery and the large number of people who have nothing to lose—those who are willing to join in and provide the justification.

The 2020 Vision is a vision of a world that by year 2020 will be free of poverty, hunger, and malnutrition, and of a world where management of natural resources will be done sustainably. Can we reach this? Yes, we can. Will we? Depends on what we do between now and then. It is in our hands; it's in the hands of the people who can make the decisions to make it happen. It is a matter of priority, not a matter of whether it can be done or not.

Let me give you a few statistics, to illustrate the problem we are facing. The incomes of the richest one percent of the world's population are equivalent to the incomes of the poorest 57 percent. The richest one percent has an income that exceeds the income of half of the rest of the population. And the relative income distribution is getting worse. In 1960, the average per capita income in industrialized nations was about nine times the average per capita income in sub-Saharan Africa. Today it is 18 times greater. That is because incomes in industrialized countries have gone up, and incomes in sub-Saharan Africa have gone down.

Poverty, hunger, and malnutrition are widespread. Twenty percent of the world's population is trying to make a living on less than $1.00 per day. That corresponds to about 1.2 billion people. Eight hundred million people go to bed hungry; they do not know where the next meal is coming from. They are what we call "food-insecure." A hundred and sixty-six million children are malnourished. They do not grow to their full potential.

Now, just to put these numbers in perspective, the total population of the United States is about 280 million people. So we are talking about three times the total population of the United States being food-insecure, going to bed hungry. Five to seven million children die every year from nutrition-related illnesses—deaths that could be avoided. I would argue, based on history and based on common sense, that no society—national or international—can be stable with those kinds of inequalities and the widespread human suffering that exists.

So what does the future look like? Again, it depends on what we do. Projections are only as good as the action we assume will be taken. The most likely scenario, based on FAO's projections, is that we will be able to reduce the number of food-insecure people from 800 million to 600 million by year 2015. That is roughly half of what world leaders from more than 180 countries promised to do five years ago at the World Food Summit. The promise was—and the goal was set—that by year 2015 the number of malnourished people would be reduced by half, from 800 million to 400 million. In the best of cases, we are going to get halfway there. That is not nearly good enough. The most likely scenario for child malnutrition, based on IFPRI projections, is that the number of malnourished children will decrease from the current 166 million to about 135 million. That is a disgrace.

But there are viable alternatives to these most-likely scenarios. We have done a number of studies looking at what it would take to bring these numbers down faster. One scenario that we have looked at assumes a 40 percent increase in the annual investments in five key sectors: agriculture research, clean drinking water, primary education, rural roads, and irrigation. If developing countries would increase investments in those five sectors by 40 percent, the number of malnourished children would be reduced from the current 166 million to about 90 million over the next 20 years. That is a lot better than the most-likely scenario.

So the question is, how much are we willing to invest? Well, to put that in perspective, the annual investment that is needed to bring the number of malnourished children down to 90 million corresponds to less than five percent of the public expenditures in developing countries in 1997. It would be an even smaller percentage of public expenditures in the future. Seems to me that is not an excessive price to pay to have well-nourished children. Want to put it in a different perspective? What is needed in terms of annual investment corresponds to one week of global military expenditures—says something about where our priorities are.

Most of the action that is needed to attain sustainable food security —and implicitly, therefore, the 2020 Vision—is known. At least, we know enough to get going. What is different today is not so much the action itself, but that it has to be implemented in a new context. It has to be implemented within the context of a number of emerging forces or trends. At IFPRI, we have identified 10 such emerging trends. Let me cover six of these very briefly and talk about the action that is associated with each of them.

I refer you to IFPRI's book *The Unfinished Agenda*, which can be downloaded from our website at www.ifpri.org.

Let me move to the six emerging forces. Those I want to talk about are the following:

- The trends to globalization and the trade liberalization that is part of them
- Developments in science and technology
- The rapidly worsening health crisis
- Conflicts and instability
- The changing structure in farming and in food consumption
- The changing roles and responsibilities among the various stakeholders

Let's start with globalization trends. Globalization is like a knife—it can be very useful to slice your bread, or it can kill people. So the argument that globalization is good or bad for poor people is not a very constructive argument. The question ought to be, how can we make globalization beneficial for poor people? Based on a number of IFPRI studies, I suggest that there are two general areas where action has to take place.

First, developing countries need to put in place policies and investments that will make the agricultural and other sectors competitive in a free-market economy. That means investments in infrastructure, investment in primary education and primary health care, appropriate technology for small farmers, and, depending on the country, it may mean other things.

The second area to focus on in order for globalization to be beneficial for the poor is what I would call fairness in trade. But I do not want this to be misunderstood. It is really free trade with a human face. What it means is that we in the industrialized countries have to open up our markets for the products that developing countries can produce competitively on a level playing field—that is, without excessive support and subsidies or closure of our markets. We have a lot to do in this area. We are arguing that developing countries need to open up their markets for our products. In many cases they have done so; the Latin American countries have opened up their markets in a way that we have not seen in very many places for a long time. We, in Europe, North America, and Japan, are not opening up our markets for some key commodities that developing countries could export to us at a price that is considerably lower than the cost of producing it in our countries.

I grew up on a farm, and I know it can be extremely difficult to adjust to new price signals in the market. The question is, can we turn the agricultural subsidies in Europe and the United States around in such a way that they are not doing damage to developing countries, so that we can open our markets without excessive harm to our own agriculture?

I want to raise one other point related to globalization, and that is the food safety question. The Europeans, the Americans, and the Japanese are beginning to put excessive demands on imported food commodities, demands that developing countries cannot meet. Food safety regulations are now being used to keep commodities out of our markets, if it is not convenient for us to take those commodities at the particular time. We need to take a very close look at how we are using food safety regulations in the context of international trade. And I repeat, we cannot go to developing countries and tell them to open up their markets if we are unwilling to do that ourselves.

Let me give you another illustration. We have something called "escalating tariffs." That means, among other things, that the higher the degree of processing a developing country puts on its agricultural commodities, the higher is the rate of import duty. So on the one hand we are giving development assistance to developing countries so they can improve their processing industries in rural areas, generate employment, and add value to agricultural commodities. That is good. Then we turn around and say, "But don't you dare send it to our countries; and if you do, we are going to slap a very high input tariff on it." That is bad. So we have somehow got to come to grips with what exactly are the signals we are sending, and how serious we are about leveling the playing field for poor people in developing countries.

With reference to international terrorism and other instability, two concerns come up. One, are we going to use the prospect of terrorism in Europe, North America, and Japan to say, "It is too risky to import your food, because it may have been tampered with, so maybe we better become self-sufficient"? Many countries would pay dearly to become self-sufficient. So that is the one risk associated with the current situation.

The other risk is that other countries are going to say, "We do not want to import anything from the United States, particularly not if some food-related terrorism has been imposed on American food commodities." It hopefully will never get to that point. But if this current international climate is going to push people toward self-sufficiency in food, the economic losses will be tremendous. We have to keep trade open, and we have to make sure that it is fair and free.

Let me move to the second emerging force: developments in science and technology. Both the modern biological sciences and the modern information and communications technologies have a lot to offer poor people in developing countries. But the offers are not materializing at this point because most of the research is focused on solving problems that industrialized countries have. There is very little biotech available for developing countries that can be used by small farmers; the research has not been done.

One of the reasons it has not been done is that there is strong opposition in parts of Europe and parts of the United States and Japan that says, "Let us not use biotechnology—or at least let us not use genetic engineering for food and agriculture." We do not need it in Europe. We can do without it. We are eating quite well. But they need it in developing countries, and we must not tell developing countries that just because we do not need it, they cannot have it either. That would be like a group of Africans getting together in a meeting concluding that the Europeans could not use genetic engineering to treat lung cancer because there is a better way—they can just stop smoking. We would not like that in Europe. We do not think it is any of their business. And they should feel the same way when we tell them they cannot have access to modern science.

If you want to know more along that line, I refer you to a new book, *Modern Biotechnology for Food and Agriculture in Developing Countries*. The book's bottom line is that those who take the consequences of the decision should have the choice to *make* the decision. Let us consider the case of a the West African farmer who is trying to feed her six children and loses her corn crop every three or four years because of drought. If she wants a drought-tolerant corn variety and if that is best developed using genetic engineering, would any of us tell her that she cannot have it? Probably nobody in this room; but there are a lot of people outside this room who would say there are better ways. While we are figuring out what they are, the children are suffering.

Developing countries are currently under-investing in agricultural research, whether it is based on molecular biology or based on traditional approaches. Developing countries invest less than half of one percent of the value of their agricultural output in agricultural research. That compares to about two percent in North America and the European Union.

Now let's look at the third emerging force: rapidly worsening health crises. The interaction between HIV/AIDS and food security is important. There are many interactions, and the causality goes in both directions. This is something that we cannot ignore, even though we may be focusing on the food side of the equation. I need not tell this audience how important the HIV/AIDS crisis is in Africa; you know that. What may be new is that we now have ways of dealing with HIV/AIDS and food security if we understand how they interact.

There are many other health issues that are taking on importance, such as increased malaria and TB. And, again, they link very closely to food security in a number of ways.

The fourth emerging force is conflict and instability. There is a very close link between conflict and food security at the local and national levels. We talked about international instability already; I need not emphasize that. We have done research in a number of countries of Africa to show how governments can help improve food security and reduce conflict by looking at the interaction between the two. Some of these things are obvious—for example, farmers are not going to plant crops on land that is already planted with land mines. Of course not—some things are clear and obvious. But there are other things that could benefit from better understanding of how to move ahead.

I want to say just a couple of words about the changing structures in farming and in consumption. On the average, the South Asian farmer has less than a football field of land to grow his/her crops on. Are these small farmers viable in the long run—say, by year 2020? Probably not. Yet, we keep talking as though they are—at least some of us do. We are going to take a close look at that at IFPRI to see what is the most likely or the most desirable farming structure over the next 20 years. And what happens to these very small farms? Do they become part-time occupations, or do they merge with neighboring farms? I am bringing this up to draw your attention to the fact that although we can keep talking about small farms, I think that over the next 20 years the definition of a small farm is going to change quite dramatically.

The other structural change I want to mention has to do with consumption. We are in the middle of a livestock revolution. We have worked very closely with ILRI and FAO on documenting the increasing demand for livestock commodities and what we think will happen over the next twenty years. There is going to be a dramatic increase in the demand for livestock commodities, particularly in rapidly growing developing countries, including those of Asia. There will also be a dramatic increase in the demand for processed foods and for sugar and sugary-type foods. Again, I cannot go into a lot of detail here, but there is plenty of information on the IFPRI home page if you want to know more about these things.

The final emerging trend is changing roles and responsibilities of the various actors. It used to be that governments in many countries had a monopoly on agricultural input and output markets—if you wanted to buy fertilizers, you had to go to a parastatal to do so. The government decided how much fertilizer would come in, what the price would be, and when it would be delivered (sometimes too late). The parastatals did not work. They were not efficient and often not effective. They have been replaced by the private sector, which creates a completely new role for the government. There were people at one point in time who said, "Less government is better government. Let us get government out of all this." That turned out to be the wrong recipe. We need very strong government to do the kinds of things that only government can do. And that includes the creation of what we call "public goods," the goods that will not be produced by the private sector.

We need standards, measurements, and enforcement of contracts. A whole set of issues around the market reform question that; unless the government does it, it would not get done, and, worse, the private sector would not work. We have done about a dozen African studies that show that in most of those countries the market reforms have not worked to the extent that we had hoped. And part of the problem is that, in some cases, private monopolies have replaced state monopolies; in other cases, poor people in remote areas are not being served. So, governments have new roles. Only the government will invest in such clear public goods as primary education, primary health care, and infrastructure.

The private sector, of course, plays a much more important role now in developing countries than it did just 10 years ago. And, again, one of the issues here is what can the private sector do best, and what is best left for either the government or civil society, the nongovernmental organizations (NGOs)? One of the problems is that these groups are fighting among themselves. That is not helping poor people put food on their tables.

Related to the changing roles is one of the most serious reasons why we still have so many malnourished children and so many hungry people: The political will is not present in developing countries to give these things the highest priority. Now, it is easy to say "lack of political will." The problem is that governments in many developing countries really do not put high priority on poverty eradication and on hunger eradication and on eradication of child malnutrition. And as long as they do not, it is very difficult for development assistance to have much of an impact, even if development assistance were there. We need much better governance. There is still a lot of corruption in many of the developing-country governments, and we need better governance in a number of other way.

Let me say a couple more words on the private sector. We need a private sector that is socially responsible. Of course we need NGOs that are socially responsible as well, and that brings back the debate of genetic engineering in Europe where some NGOs are not socially responsible. But we cannot expect the private sector to produce the kind of goods and services they cannot make money on. At least I am not going to invest my pension plan in such companies. So clearly we have to define what the private sector can do and is willing to do. We need a new set of institutions to make sure that each of these groups of actors does what it does best.

The last point I want to make on changing roles and responsibilities has to do with the role of the United States, Europe, and Japan—in other words, the industrialized countries—not to forget, of course, Australia. We need to seek more engagement in the international community; we must not isolate ourselves. One of the potential risks associated with the terrorists' attack is that we build taller walls around ourselves, and we try to become self-sufficient not only in food, but also in other things. That would be a grave mistake. We need more rather than less engagement in the international community.

The United States is at the very bottom of the list of all the industrialized countries when it comes to development assistance measured in percent of the national income. The United States gives less than one-tenth of one percent of the national income in development assistance. Americans are willing to give a lot more development assistance, but that has not been communicated to the politicians.

Let me conclude simply by stating that we in the high-income part of world society have a choice. One option is to spend increasing amounts of money to protect ourselves, and we will not be as successful as we would like. We cannot really protect ourselves from all mad acts without taking away the kind of freedom that we want to keep—but that is one choice that we have. The other choice is that we spend the same amount of money on removing the root causes of international terrorism and other international and national instability, and that is what I have tried to outline today.

We will continue to live under a cloud of fear if we do not remove the foundation that the fanatics think they have for continuing to attack us —one way or the other.

NOTES

1. 2001 World Food Prize Laureate Address, Des Moines, IA, October, 2001.
2. Director-General of the International Food Policy Research Institute (IFPRI); 2033 K Street, N.W.; Washington, D.C. 20006–1002, U.S.A.; phone: 202–862–5600; fax: 202–467–4439; IFPRI@cgiar.org; web: www.ifpri.org.

PER PINSTRUP-ANDERSEN

Colin G. Scanes and John A. Miranowski

In 2002, Dr. Per Pinstrup-Andersen received the World Food Prize. He was selected for the critical research that enabled governments to reform their food subsidy programs and thereby dramatically increase food availability to the most poor. This laid the foundation for the Food For Education programs in which the families receive food subsidies when children stay in school.

Per Pinstrup-Andersen was raised on a farm in Denmark, leaving school after the seventh grade to become a farm worker. He was drafted into the Danish Defense Forces at age 18. After finishing military service, he returned to work as a farm laborer. Impelled by a desire to make a positive social contribution, he entered the Danish Agricultural University, receiving a bachelor's degree in Agricultural Economics. Motivated by the crucial issues he saw in global agricultural policy, he continued his education at Oklahoma State University (obtaining M.S. and Ph.D. degrees). Since then, Dr. Pinstrup-Andersen has devoted his life's work to bringing about the change that would permit those most afflicted by poverty and severe malnutrition to have access to the food they need.

Dr. Pinstrup-Andersen's professional career has been as an agricultural economist in Colombia (Centero Internacional de Agricultura Tropical); Alabama (International Fertilizer Development Center); Denmark (Royal Veterinary and Agricultural University); New York (Cornell University); and Washington, D.C., where he was the Director-General of the International Food policy Research Institute (IFPRI).

ACHIEVEMENTS

The following are among Dr. Pinstrup-Andersen's outstanding achievements:

- Initiating IFPRI's research and policy dialogue with the government of Pakistan, which was instrumental in changing that country's food and agricultural policy—particularly with regard to the rationing of wheat, greatly increasing the access to food by the most poor

- Carrying out research in Egypt which permitted that government to significantly reform its food subsidy system, again more effectively targeting the poorest and most disadvantaged in society
- Assisting the governments of Malawi and Uganda, which faced the specter of widespread famine, improving their capacities to implement relief distribution, thus avoiding mass starvation

CONCLUSION

Driven by a deep desire to alleviate the suffering of malnourished and starving children, Dr. Pinstrup-Andersen initiated a global effort to uplift those most at risk. Over the past decade, the 2020 Vision Initiative has alerted world leaders to potential crises in food security issues. This has helped reverse the trend of decreasing global developmental assistance. It also led to policy changes, bringing reduced world hunger and poverty levels. Dr. Pinstrup-Andersen has been categorized as a brilliant catalyst for policy change, which jarred the international community out of the complacency prevalent a decade ago—improving food security for millions in developing countries.

II

FRONTIERS IN FOOD

This section includes two seminal papers on new developments related to food. In the first, Catherine Woteki, former Undersecretary for Food Safety, and her colleagues discuss the Hazard Analysis and Critical Control Points (HACCP) system which is now used extensively in the United States of America and worldwide. In the second, Jean Kinsey considers scan marketing and other innovations in food marketing/grocery.

7

HACCP as a Model for Improving Food Safety[1]

Catherine E. Woteki,[2] Margaret O'K. Glavin,[3] and Brian D. Kineman[4]

Hazard Analysis and Critical Control Points, or HACCP as it is most commonly known, is an analytical, science-based tool that has revolutionized food safety in developed nations throughout the world. The HACCP approach can be modified and adapted by virtually any food industry to increase assurance that a product is safe from harmful contaminants, whether they are biological, chemical, or physical. HACCP enables manufacturers to focus on product safety as a top priority and plan for the prevention, control, and mitigation of problems accordingly. In this respect, it is a cost-effective approach, because it targets controlling hazards in a preventive way, and the benefits of preventing illness and death outweigh the costs of implementing and maintaining HACCP systems.

Although HACCP is often misunderstood as being a complex system, the basic philosophy behind its fundamental principles is very simple and straightforward and can be followed from production to the point of consumption. When HACCP is applied properly, the risk of manufacturing and releasing unsafe products to the public is significantly reduced. The adoption of HACCP by regulatory agencies, such as the Food and Drug Administration (FDA) of the U.S. Department of Health and Human Services and the United States Department of Agriculture's Food Safety Inspection Service (FSIS), has stimulated improvements in food safety practices in the food industry.

The current U.S. food safety system is based on legislation that dates back to the start of the last century. It is designed to address risks that were recognized at that time, such as filth and decomposition, and is not easily focused on more recently understood risks such as microbial pathogens. As microbial pathogens have emerged in recent years and been recognized as a significant food safety hazard, regulators in the FDA and FSIS have sought new approaches for addressing current food safety challenges within the limits of the existing legislation. Increasingly, these regulators have turned to HACCP as an important tool in a risk-based, prevention-focused food safety system.

This chapter reviews some of the concepts of the HACCP system, issues facing it today, the historical development of the program, how well HACCP has fulfilled its potential, and prospects for its future success. Discussion in this paper will be limited to food safety problems caused by pathogens. In addition to foodborne pathogens (bacteria, parasites, and viruses), chemicals, such as pesticides, environmental contaminants, and persistent organic pollutants (POPs), are also agents that can contaminate food and cause disease.

BACKGROUND

Foodborne disease is a significant public health problem in the United States, accounting for approximately 76 million illnesses, 325,000 hospitalizations, and 5,000 deaths in the U.S. each year (Mead et al. 1999). This is despite significant advances in food safety technology and public education as well as an expansion of government regulation over the past century. Important technological advances include pasteurization, the widespread availability of adequate refrigeration, improved methods of processing, and, more recently, the use of antimicrobials and microbial inhibitors in commercial food processing. Government has moved from regulating filth and the sanitary conditions under which food is produced and processed, to scientifically designed standards to control the level of chemical and microbiological contaminants in food and providing information to consumers through mandatory "safe-handling" labels and educational materials. Nevertheless, foodborne disease remains a health problem with large human and economic costs. And as the U.S. population ages, and the number of persons with compromised immune systems due to illness or medication increases, the problem is likely to grow.

THE ECONOMIC AND PHYSICAL
BURDENS OF FOODBORNE DISEASE

Symptoms of foodborne disease can range from mild flu-like symptoms to severe illness requiring hospitalization. In addition, chronic sequelae— such as Guillain-Barré syndrome, hemolytic uremic syndrome, Reiter's syndrome, and arthritis—are associated with certain foodborne illnesses. Infections such as listeriosis and toxoplasmosis can result in blindness, chronic neurological complications, mental retardation, and spontaneous abortion. Populations most at risk to foodborne disease include those that are typically dependent on the care of others: the very young, the elderly, and the immune-compromised.

To further assess the national problem of foodborne disease, the USDA's Economic Research Service (ERS), based on the most recent available data, estimated that illness associated with the top-reported bacterial pathogens (Table 7.1) costs the U.S. $6.9 billion, annually (Buzby and Roberts 1996). This estimate provides a good perspective on the scope of the economic problem of foodborne disease. However, until more accurate data are achievable, these figures can serve only as a conservative estimate of the total cost that the U.S. suffers each year as the result of foodborne illness. The bacteria featured in this study represent only a fraction of the over 40 documented foodborne pathogens. More importantly, the majority of cases are caused by yet-to-be identified pathogens (Mead et al. 1999).

Estimating the cost to society from foodborne disease is compounded by many complex issues. Many sectors of society, ranging from the industry to the consumer, are affected by the burden of illness. Costs considered in the ERS study, for instance, include medical expenses, productivity losses from missed work, complications from hemolytic uremic syndrome (associated with *E.coli* 0157:H7) and Guillain Barré

TABLE 7.1. Estimated annual costs and number of cases for 5 foodborne pathogens

Pathogen	# Annual Cases[1]	Annual Cost (Billion Dollars)[2]
Campylobacter spp.	1,963,141	1.2
Salmonella, nontyphoidal	1,341,873	2.4
E. coli O157:H7	62,458	.7
E. coli non-O157:H7 STEC	31,229	.3
Listeria monocytogenes	2,493	2.3[3]

[1]Based on data reported by Centers for Disease Control (www.cdc.gov/ncidod/eid/vol5no5/mead.htm).

[2]Based on ERS data.

[3]This value excludes less severe cases not requiring hospitalization.

syndrome (associated with *Campylobacter*), and an estimate of the value of premature death. Additional expenses that are accrued from foodborne disease can include travel costs; time lost from work in caring for the sick; business losses from company closings, product recalls, and mass slaughtering of carrier animals; and the expense of investigating outbreaks and cleanup costs.

Many people with foodborne diseases do not see a physician because the symptoms are mild and the long incubation periods of some pathogens makes the connection to a particular food source extremely difficult. Even when people seek treatment for a foodborne illness, physicians may treat the symptoms but not order confirmatory tests or report results to the state public health authority or the Centers for Disease Control (CDC) even when tests are ordered. Clearly, food safety, as much as it is a threat to public health, is a major economic concern to our society. The progress made in understanding the contributions of foodborne pathogens to the burden of illness and costs to society has led to broad support for the concept of preventing these costs through adequate food safety measures.

DEFINING AN OPTIMAL SYSTEM

The preventable nature of foodborne disease and its attendant human and economic costs highlights the need for improvements to the current food safety system. The system should be focused on reducing known risks and also be capable of responding to new risks as they emerge.

The current food safety system reflects the evolution of shared federal and state responsibilities for regulation and research (Table 7.2). At the federal level, six cabinet departments and more than a dozen agencies share responsibilities for regulating the food supply; performing research; and monitoring crops, livestock, and human health. The foodborne disease surveillance system, led by CDC and reliant on local and state health departments, provides useful data for decision-makers in designing improvements to the current U.S. food system. However, more data are needed on such issues as new or emerging threats, the points in the food production system where hazards may be introduced, the changes in hazard levels as food moves through the production system, and the level at which specific hazards become of public health concern.

There is some agreement on what an optimal system would look like. The consensus is that such a program should be risk-based, focused on

TABLE 7.2. The current U.S. food safety system

Agency	Responsibility
United States Department of Agriculture (USDA)	
Food Safety Inspection Service (FSIS)	Regulates all domestic and imported meat, poultry, and processed egg products
Animal and Plant Health Inspection Service (APHIS)	Controls food animal and plant disease
Agriculture Research Service (ARS)	Conducts food safety research
Cooperative State Research, Education, & Extension Services (CSREES)	Conducts research and educates the public about food safety issues
Economic Research Service (ERS)	Conducts research on cost benefit information on foodborne illness
Department of Health & Human Services (HHS)	
Food and Drug Administration (FDA)	Regulates all domestic and imported foods that are marketed in interstate commerce, except for meats, poultry, and processed egg products
Center for Veterinary Medicine (CVM)	Regulates all food additives and drugs given to animals
National Center for Toxicological Research (NCTR)	Conducts research on toxicity of products regulated by FDA
Centers for Disease Control (CDC)	Monitors and gathers information on foodborne illness in the U.S.
Department of Commerce	
National Marine Fisheries Service (NMFS)	Conducts a voluntary seafood inspection service for commercial seafood
Treasury Department	
Bureau of Alcohol, Tobacco, & Firearms (BATF)	Enforces regulation pertaining to alcoholic beverages
Other	
Environmental Protection Agency (EPA)	Provides tolerance levels for pesticide residues in foods and animal feed
Department of Defense (DOD)	Regulates all food eaten by military

prevention, and grounded in science. The National Academy of Sciences (NAS), in a 1998 report, described the goals of an effective food system, stating that "the mission of an effective food safety system is to protect and improve the public health by ensuring that foods meet science-based safety standards through the integrated activities of the public and private sectors" (NAS 1998). An optimal system would rank relative risks and opportunities for reducing those risks and use the rankings to allocate available resources. However, the tools for producing such rankings are not currently available. Research is needed to develop ranking tools and data on which to base risk decisions.

HACCP's Role in an Optimal System

HACCP requires a food processor to analyze the hazards reasonably likely to occur in its process and to design and implement a system to eliminate those hazards or reduce them to an acceptable level. Starting in 1995, regulators have required HACCP for seafood, meat and poultry, and juice, reasoning that a tool that eliminates or reduces foodborne hazards will lead to a reduction in foodborne disease. A recent report by the Institute of Food Technologists reflects this reasoning: "HACCP is a management tool used by the food industry to enhance food safety by implementing preventive measures at certain steps of a process. When HACCP principles are properly implemented, microbiological hazards that have the potential to cause foodborne illness are controlled, i.e., prevented, eliminated or reduced to an acceptable level." (Institute of Food Technologists 2002). In addition to the mandatory HACCP programs that have been put in place for seafood, meat and poultry, and juice, many food processors have voluntarily implemented HACCP systems for other food products.

So how did the HACCP system come to be? What factors led to its mandatory use in certain sectors of the food industry? To further understand the function of HACCP in the current food safety system, a brief review of the program's history is warranted.

HISTORY OF HACCP

The origin of the HACCP system of food safety can be traced back to the early days of the space program. In planning for manned space missions, some of the major questions confronting the National Aeronautics and Space Administration (NASA) program had to do with the food. The first U.S. endeavors to place humans in space—project Mercury (1961 and 1963)—were very short missions with a maximum duration of 34 hours, involving a crew of one. Food safety was not a primary concern on these missions. The major issue was one of ascertaining that the physiology of swallowing and peristalsis was not altered by microgravity. Emphasis was then placed on developing calorie-rich, nutritious, easily consumable, digestible foods (Lane and Shoeller 1999). With the Gemini missions, which had a maximum duration of almost 14 days, food storage and safety became important. Freeze dried, bite-sized foods were engineered to address concerns with storage and ease of eating in zero gravity. In response to the threat of microbial contamination in the foods, NASA and the Army Natick laboratories established quality standards for these missions, pinpointing certain specific pathogens. Foods used on the Gemini missions were screened for *Salmonella* sp. and *Staphylococcus* sp., for which a zero tolerance standard was set. Additionally, the foods had to meet a low total coliform count (<10,000/g) (LaChance 2002).

Hazard control was not unique to food. NASA required a similar approach for all components of the Apollo program (e.g., rocket, spacecraft, and suits). NASA imposed a Reliability Program Provision on its contractors (NASA NPC 250–1), which dictated an engineering requirement to identify critical control points. To meet this requirement, Natick labs developed an engineering system—Failure, Mode, and Effect Analysis (FMEA)—to establish mandatory Critical Control Points (CCPs) for all activities pertaining to Apollo systems. FMEA was an effective system to use because it successfully predicted possible failures in a process and in the planning for adequate response systems (Mortimore and Wallace 1998). For the production and delivery of the entire food system for Apollo, NASA contracted with the Pillsbury Company, after an initial subcontract with the Melpa Corporation (LaChance 2002). The Pillsbury team eventually combined the concept of hazard analysis, imposed upon them by NASA as a producer of some of the food items used in the Gemini missions, with the CCP engineering system already in place and required for the Apollo project (LaChance 1994, 1998). With this fusion of principles, HACCP was born.

In 1971, a representative of the Pillsbury Company presented the concepts of HACCP to the public for the first time at the National Conference of Food Protection where it was recommended for widespread use (FDA 1972) (Bauman 1974). An outbreak of botulism in New York

City that year traced to canned soup prompted the FDA to contract with Pillsbury to train government personnel on the HACCP system (Colmore et al. 1971) (Kauffman 1974). The FDA employed HACCP ideas in developing the Good Manufacturing Practice (GMP) regulations for low-acid canned foods in 1973 (Kauffman 1974).

In the years following HACCP's development, Pillsbury utilized the system in the manufacturing of its products. Gradually, throughout the 1970s and the early 1980s, more food companies became aware of the HACCP system and voluntarily applied its principles to production. Approaches for food safety using HACCP were further modified and recommended for canned foods (Ito 1974), frozen foods (Peterson and Gunnerson 1974), and for entrée food service systems (Bobeng and David 1977, 1978). Eventually, the need for HACCP systems for all food service operations was advocated (Bryan 1981a, 1981b).

It was not until 1985, however, that HACCP truly caught on with the food industry. In that year, the National Academy of Sciences published "An Evaluation of the Role of Microbiological Criteria for Foods and Food Ingredients." In this landmark report, the Academy endorsed HACCP, stating that the system "provides a more specific and critical approach to the control of microbiological hazards in foods than that provided by traditional and quality control approaches" (NAS 1985). Two events, which also occurred in 1985, focused the food industry's and regulators' attention on this report. First, a salmonellosis outbreak involving 16,000 confirmed cases in six states occurred—the largest foodborne salmonellosis outbreak in U.S. history (Sum 1985). Second, Mexican-style cheese, contaminated with *Listeria monocytogenes,* was responsible for numerous stillborn deaths in Southern California (Linnan et al. 1988).

In 1992, the National Advisory Committee on Microbiological Criteria for Foods (NACMCF) endorsed the HACCP system and established seven principles (see sidebar) for the development and implementation of effective HACCP plans (NACMCF 1992). These principles form the backbone of HACCP today. The HACCP approach also began to gain international acceptance as details of the system were published by the Codex Alimentarus Commission shortly thereafter (CCFH 1993). Despite the global recognition of HACCP, the only regulations utilizing its theories remained the low-acid canned food regulation of 1973. This was soon to change, however, due to a series of dramatic events.

SEVEN PRINCIPLES OF HACCP

- **Principle 1:** Conduct a hazard analysis. Prepare a list of steps in the process and identify where significant problems can occur. In addition, devise control measures that can be used to meet these problems.
- **Principle 2:** Determine the Critical Control Points (CCPs). These are places in the process identified in step one, where control is critical to assuring the safety of the product.
- **Principle 3:** Establish critical limits. This describes the difference between safe and unsafe products and requires that measurable parameters be established to determine this.
- **Principle 4:** Establish a system to monitor control of the CCP. Procedures have to be developed that allow for the effective monitoring and accountability of the process.
- **Principle 5:** Establish the corrective actions to be taken when a particular CCP is not under control.
- **Principle 6:** Establish procedures for verification to confirm that the HACCP System is working correctly.
- **Principle 7:** Establish documentation concerning all processes in the procedure.

The public was becoming more concerned about foodborne illness traceable to meat, seafood, and poultry products. Reports of foodborne outbreaks and documentation from various consumer groups supported the consumer's demand for more effective regulation of the food industry. Public pleas for intensified regulation of the meat industry galvanized in 1993, when an outbreak of foodborne illness occurred in the Pacific northwest of the U.S. The culprit of the outbreak was undercooked ground beef contaminated with *E. coli* O157:H7; the consequence—hundreds became ill and four died (Bell et al. 1994). In response, the FSIS in 1996 promulgated the Pathogen Reduction/ HACCP rule for all meat and poultry slaughter and processing facilities (FSIS 1996). The regulation required all processors to implement a series of programs beginning in 1997. All meat and poultry plants, regardless of size, were required to have standard sanitary operating procedures (SSOP) in place by 1997. Beginning in January 1998, the rule phased in implementation of HACCP plans over three years. The largest plants (those employing 500 or more people) began operating under HACCP plans in January 1998, midsize plants (10–499 employees) in January 1999, and small plants (610 employees) in January 2000. As part of the regulation, the FSIS set pathogen reduction performance standards for *Salmonella* and required testing for generic *E. coli*. The FDA also responded to the threats posed by lack of adequate regulation in the seafood industry. With support from the Congress, the FDA in 1995 required that all seafood processors establish and implement HACCP programs by December 1997 (FDA 1995).

Even though juice, by and large, was considered safe for the general population, an incident in 1996 exposed potential hazards in this industry as well. Unpasteurized, commercially sold apple juice, contaminated with *E. coli* O157:H7, resulted in a nationwide outbreak (CDC 1996). One 16-month-old child died as the result of consuming the tainted product. The following year the FDA's Center for Food Safety and Applied Nutrition conducted a study, which concluded that although the majority of contaminants are probably introduced in juice during harvesting, it is very possible that contamination can occur during manufacturing. As a follow-up, the FDA required all large juice processors to have an HACCP system in place by January 22, 2002, with a January 21, 2003, deadline for small businesses (500 employees or fewer) and January 20, 2004, for very small companies (<100 people) (Lewis 1998).

It is clear that HACCP can be further expanded to encompass more of the food industry in the future.

MOVING TOWARD AN IMPROVED SYSTEM

While the current U.S. food safety system has many strengths, important questions concerning the goals of the system, allocation of responsibilities within the system, accountability, and optimal resource allocation need to be considered if the system is to succeed in reducing the burden of foodborne disease.

There is agreement on the need for a risk-based, prevention-focused system founded on science, and agreement that HACCP is a valuable tool in such a system; but important issues remain. A fundamental question is, what should we expect from our food safety system? What is the level of protection that Americans expect? No food safety system is likely to achieve the eradication of foodborne disease, but an understanding of the level of protection that the system is expected to achieve is essential for policymakers and for the food industry. Without such an understanding, it is not possible to set performance goals, to make rational resource allocation decisions, or to measure success. Policymakers have no road map, and the food industry is left to guess what the next regulatory demand will be. Today there is no settled understanding of the level of protection we expect from the U.S. food safety system. A process for arriving at this understanding is urgently needed. Policymakers, the food industry, and consumers must come together to address the question in order to provide a common framework for food safety improvements. With such a framework, HACCP can be fine-tuned to optimize its effectiveness, and evaluated against the goals and expectations of the food safety system.

ROLES AND RESPONSIBILITIES

Translating public health goals into requirements for individual processors is crucial to *meeting* the public health goals. Achieving this translation requires defining the appropriate roles and responsibilities of government, industry, and the consumer in assuring a safe food supply. FSIS addressed the issue in its 1995 Pathogen Reduction/HACCP rule for meat and poultry when it set performance criteria and performance standards that plants producing raw meat and poultry must meet. FSIS attempted in the rulemaking to define an acceptable level of pathogens on raw meat and poultry, a translation of the public health goal of reducing foodborne disease into a regulatory requirement for individual processors. The agency concluded that an infectious dose approach to defining an acceptable pathogen level was not viable. This was because of the lack of data on what constitutes an infectious dose [questions on what population should be used as the metric (healthy adults, children, immune-compromised individuals)]. Moreover, the level of a pathogen on a food measured at one point in the farm-to-table continuum is not necessarily the level at a subsequent point. In lieu of an infectious dose standard, FSIS chose to set microbiological performance criteria and performance standards at the national baseline prevalence of contamination, a level that reflected what was currently achievable using available technology. For example, the *Salmonella* performance standard for ground beef was set at a prevalence of 7.5 percent, meaning that no more than 7.5 percent of the samples of ground beef from a particular plant could test positive for *Salmonella* (Table 7.3).

FSIS reasoned that its approach furthers the public health goal of reducing foodborne disease by requiring all processors of raw product to meet the current national baseline prevalence level of a particular pathogen (*Salmonella*). The policy is based not on a quantitative assessment of the risk posed by a particular level of contamination nor on the determination of a "safe" prevalence level, but on the belief that it is feasible for all processors to meet or exceed the current baseline prevalence level, thereby reducing overall pathogen prevalence. Over time, the baseline prevalence would be re-assessed and moved downward, leading to further reductions in prevalence of pathogens in products.

A different approach to translating public health goals into requirements for individual food processors is proposed in an Institute of Food Technologists Expert Report "Emerging Microbiological Food Safety Issues" (IFT 2002). The authors propose using the concept of Food Safety Objectives as an approach to achieve public health goals, and state that Food Safety Objectives offer "a practical means to convert public health goals into values or targets that can be used by regulatory agencies and the industry." The report describes a Food Safety Objective as "a statement of the maximum frequency and/or concentration of a microbiological hazard in a food at the time of consumption that provides the appropriate level of protection." This approach provides a way to arrive

TABLE 7.3. FSIS *Salmonella* performance standards for raw products

Product	Baseline Prevalence (%)
Broilers	20.0
Market hogs	8.7
Cows/bulls	2.7
Steers/heifers	1.0
Ground beef	7.5
Ground chicken	44.6
Ground turkey	49.9

at specific targets that can be used by regulators and applied to individual food processors.

ACCOUNTABILITY

A successful food safety system must have means to hold individual processors accountable both for producing safe product and for reducing the level of hazards in the food supply. The government mandated HACCP programs place responsibility on the individual food processor to identify and control hazards, with government ensuring that processors are addressing the hazards that are of public health concern. The FSIS Pathogen Reduction/HACCP rule for meat and poultry also sets specific pathogen reduction standards (i.e., the *Salmonella* performance standards described above) for plants producing raw meat and poultry. The performance standards are a means of holding individual processors accountable for designing and implementing an effective HACCP system and for achieving a reduction in the level of hazards in the food supply. Processors have questioned whether they should be held accountable for controlling and reducing hazards that arrive on incoming product e.g., *Salmonella* on raw meat) or that are ubiquitous in the environment (e.g., *Listeria)* and for which available interventions are limited, or whether their responsibility should be limited to controlling those hazards which they introduce and those that would render the product adulterated.

There have been both political and legal challenges to the authority of the regulatory agencies to hold the industry accountable for hazard control and risk reduction. In 1999, FSIS attempted to withdraw inspection from a Texas processor of raw ground beef products when the processor failed to meet the *Salmonella* performance standard (withdrawal of inspection has the effect of closing the plant since it is illegal to ship meat products that have not been inspected and passed). The processor went to court, arguing that the government did not have the authority to withdraw inspection for failure to meet the standard since *Salmonella* contamination is not considered an adulterant on raw meat, and since the *Salmonella* was on the raw beef the processor purchased, not the result of contamination during processing. In December 2001, the Fifth Circuit Court of Appeals found that the government had exceeded its statutory authority on the grounds that the performance standard improperly regulated the *Salmonella* levels of incoming meat. The

Court of Appeals stated that "the performance standard is invalid because it regulates the procurement of raw materials." (Supreme Beef Processors, Inc., versus United States Department of Agriculture, Appeal from the United States District Court for the Northern District of Texas, December 6, 2001).

The questions of measurable standards, accountability mechanisms, and processor responsibility for hazards that they do not introduce require further debate and development.

RESOURCE ALLOCATION

In addition to the need for a common understanding of the appropriate roles and responsibilities in a food safety system for government, industry, and consumers and the need for practical accountability mechanisms, a risk-based food safety system requires the risk-based allocation of resources. Given finite resources, who decides which risks are to be targeted? What factors ought to be considered in making this determination? Should the hazard that causes the most illnesses or the hazard that results in the most deaths receive the most resources? Is the severity of illness a factor that should be considered? How should those deciding resource allocation consider the disparities in how a hazard affects different populations? Should the general population or the most vulnerable be the focus?

Comparative risk rankings are a first step, but risk-based resource allocation also requires systematic approaches to comparing the effectiveness of available risk reduction interventions. This is an area where little data exists and few tools are available. Until risk-based allocation tools are available to decision makers, HACCP cannot fully meet its promise to reduce foodborne disease.

MEASURING HACCP'S SUCCESS

Policymakers who have come increasingly to rely on HACCP to achieve improvements in food safety have a number of tools for judging its success. HACCP, coupled with performance standards, provides an opportunity to assess both the performance of individual food processors and the performance of the food safety system as a whole. A significant benefit of HACCP is the availability of real-time performance data. Periodic verification is one of the foundations of HACCP. Individual processor performance can be measured by regulators through verification activities, including observations, record reviews, and product testing. In addition to providing a measure of a processor's performance, verification is a mechanism for holding the processor accountable for controlling hazards and producing safe food. When HACCP is coupled with mandatory performance standards, it becomes possible to assess and compare the performance of all processors.

Measuring the success of HACCP in reducing foodborne disease is more difficult but some data exist. FSIS reports that *Salmonella* prevalence in all product categories with performance standards was lower in 2001 than in agency baseline studies and surveys conducted before PR/HACCP implementation (FSIS 2002). And CDC reports that the estimated incidence of infections from *Salmonella* decreased 15 percent from 1996 to 2001. The CDC report notes that "the decline in the incidence of these foodborne infections occurred in the context of several control measures, including implementation by the U.S. Department of Agriculture's Food Safety Inspection Service (FSIS) of the Pathogen Reduction/Hazard Analysis Critical Control Point (HACCP) systems regulations in meat and poultry slaughter and processing plants. The decline in the rate of *Salmonella* infections in humans coincided with a decline in the prevalence of *Salmonella* isolated from FSIS-regulated products to levels well below baseline levels before HACCP was implemented." (CDC 2002)

Well-designed product sampling programs can identify trends in the prevalence and level of hazards and point to areas needing further effort. Disease surveillance can provide insight into whether food safety efforts are having an effect on the level of foodborne disease and can identify new or emerging hazards. Regulators and policymakers can make use of verification data, product sampling schemes, and disease surveillance to hold processors accountable, to measure progress on reducing foodborne disease, and to continuously improve the design and execution of the food safety system.

CONCLUSION

HACCP is an important tool in the existing food safety system and has the potential to bring that system to a new level. A comprehensive dialogue on expectations of the system and the appropriate roles and responsibilities of government, industry, and consumers in assuring a safe food supply would help policymakers design enhancements to realize the full potential of HACCP. Accountability within a HACCP-based system also must be addressed. Performance standards, such as the *Salmonella* performance standard in the FSIS Pathogen Reduction/HACCP rule, are one approach to accountability, but the legal authority for this approach has been challenged. HACCP without accountability will remain just a process control system and not an engine for significant improvements in food safety.

REFERENCES

Bauman, H. 1974. The HACCP concept and microbiological hazard categories. *Food Technology* 28:30–34 and 74.

Bell, B., M. Goldoft, P. Griffin, M. Dacis, D. Gordon, P. Tarr, C. Bartleson J. Lewis, T. Barrett, and J. Wells. 1994. A multistate outbreak of *Escherichia coli* 0157:H7-associated bloody diarrhea and hemolytic uremic syndrome from hamburgers. The Washington experience. *Journal of the American Medical Association* 272:1349–1353.

Bobeng, B. and B. David. 1977. HACCP models for quality control of entrée production in foodserve systems. *Journal of Food Protection* 40:632–638.

Bobeng, B. and B. David. 1978. HACCP models for quality control of entrée production in hospital foodservice systems. *Journal of the American Dietetics Association* 73:524–529.

Bryan, F. 1981a. Hazard analysis of food service operations. *Food Technology* February 78–87.

Bryan, F. 1981b. Hazard analysis control point approach: epidemiologic rationale and application to foodservice operations. *Journal of Environmental Health* 44:7–14.

Buzby, J.C., and T. Roberts. 1996. ERS updates U.S. foodborne disease costs for seven pathogens. *FoodReview.* "http://www.ers.usda.gov/publications/foodreview/sep1996/sept96e.pdf".

Centers for Disease Control. 1996. Outbreak of *Escherichia coli* O157:H7 infections associated with drinking unpasturized commercial apple juice— British Columbia, California, Colorado, & Washington, October 1996. *MMWR* 45(44):975.

Centers for Disease Control. 2002. Preliminary FoodNet data on the incidence of foodborne illness—Selected states, U.S., 2001. *MMWR* 51(15):325.

Codex Committee on Food Hygiene. 1993. *Training considerations for the application of the HACCP System to food processing and manufacturing.* WHO/FNU/FOS/93.3 II. Geneva: World Health Organization.

Colmore H., H.C. Neu, J.J. Goldman, D. D' Archangelis, J. Goafor, A.R. Hinmin, and H.S. Ingraham. 1971. Botulism associated with commercially canned vichyssoise. *MMWR* 20(26):1242.

Food and Drug Administration (FDA). 1972. *Proceedings of the 1971 National Conference on Food Protection.* Washington, D.C.: U.S. Department of Health, Education and Welfare.

Food and Drug Administration. 1995. Procedures for the safe and sanitary processing and importation of fish and fish products, final rule. *Federal Registrar* 60:#242.

Food Safety Inspection Service. 1996. The final rule on the pathogen reduction and hazard analysis and critical control point (HACCP) system. *Federal Registrar* 61:#144.

Food Safety Inspection Service. 2002. Progress report on *Salmonella* testing of raw meat and poultry products, 1998–2001. Washington, D.C.: U.S. Department of Agriculture.

Institute of Food Technologists. 2002. Emerging microbiological food safety issues: Implications for control in the 21st century. February.

Ito, K. 1974. Microbiological critical control points in canned foods. *Food Technology* 28:16.

Kauffman, F. 1974. How FDA uses HACCP. *Food Technology* 28:51 and 84.

LaChance, P. 1994. *Nutrition in Space from Modern Nutrition in Health and Disease, 8th ed.* Philadelphia: Lea and Febiger. pp. 686–703.

LaChance, P. 1998. How HACCP started—Letter to the editor. *Food Technology* 51:35.

LaChance, P. 2002. Interview. June 25, 2002.

Lane, H. and D. Shoeller. 1999. Nutrition in space flight and weightlessness models. Boca Raton: CRC Press.

Lewis C. 1998. Critical controls for juice safety. *FDA Consumer Sep–Oct.* FDA 99–2324. "http://www.fda.gov/fdac/features/1998/598_juic.html".

Linnan, M., L. Mascola, X. Lou, V. Goulet, S. May, C. Salminen, D. Hird, L. Yonekura, P. Hayes, R. Weaver, A. Audurier, B. Plikaytis, S. Fannin, A. Kleks, and B. Broome. 1988. Epidemic listeriosis associated with Mexican-style cheese. *New England Journal of Medicine* 119:823–828.

Mead, P.S., L. Slutsker, V. Dietz, L.F. McCaig, J.S. Bresee, L. Shapiro, P.M. Griffen, and R.V. Tauxe. 1999. Food-related illness and death in the United States. *Emerging Infectious Diseases* 5:607–625.

Mortimore, S. and C. Wallace. 1998. *HACCP: A Practical Approach.* Gaithersburg, MD: Apsen Publishers.

National Academy of Sciences. 1985. *An Evaluation of the Role of Microbiological Criteria for Foods and Food Ingredients.* Washington, D.C.: National Academy Press.

National Academy of Sciences. 1998. *Ensuring Safe Food from Product to Consumer.* Washington, D.C.: National Academy Press.

National Advisory Committee on Microbiological Criteria for Foods. 1992. Hazard analysis and critical control point system. *International Journal of Food Microbiology* 16:1–23.

Peterson, A., and R. Gunnerson. 1974. Microbiological critical control points in frozen foods. *Food Technology* 28:37–38.

Sum, M. 1985. Illinois traces cause of *Salmonella* outbreak. *Science* 228:977–978.

Supreme Beef Processors, Inc., vs. United States Department of Agriculture. No. 00–11008 (5th circuit 12/06/2001)."http://www.biotech.law.lsu.edu".

NOTES

1. We wish to thank Dr. Paul LaChance for graciously providing us with detailed information on the early development of the HACCP program.
2. College of Agriculture, Iowa State University; Ames, IA 50011; agdean @iowastate.edu.
3. Resources for the Future; 1616 P St. N.W.; Washington, D.C. 20036; Glavin @rff.org.
4. College of Agriculture, Iowa State University; Ames, IA 50011; bkin @iowastate.edu.

8

LEADING CHANGES IN FOOD RETAILING: SEVEN STEPS TO A DEMAND-DRIVEN FOOD SYSTEM
Jean D. Kinsey[1]

The food and agricultural industry is a giant in our national economy. It makes up 9 percent of the gross domestic product; 60 percent comes from wholesale/retail activity. The industry employs over 14 percent of all workers and 54 percent of all retail workers. Retail food stores, restaurants, and bars sell over $900 billion of food and drink each year, which is more than one-quarter of all retail sales.

In most economic and industrial sectors, supply chains are managed to fill the orders of customers and the perceived demand of consumers. Food and agriculture supply chains have traditionally been managed to find markets for agricultural commodities. Final demand was assumed. After all, people always needed food. Supply was always demanded even if the government was the purchaser and distributor of last resort. During the 20th century, new production technologies increased yields and decreased costs. Markets expanded around the world. Firms in the supply chain became experts at handling large volumes of bulk commodities. Efficient distribution systems helped drive down the overall price of food in the United States. Meanwhile, wholesale and retail firms operated largely with local or regional monopolies. These firms had little incentive to innovate; their customers were glad to have a variety of food at reasonable prices in their local retail food stores.

By the 21st century, the supply chain between farm and fork was splitting up into fragmented channels. New retail firms, more responsive to consumers' changing tastes and preferences, were finding ways to operate more efficiently. Retailers sent signals back up the supply chain about what and how much they would purchase, influencing what should be produced. What made this different was that orders and forecasts were based on real-time, comprehensive data about sales and product movement. The supply-push food and agricultural system turned 180 degrees into a consumer-driven, demand-pull system that was more flexible, more in tune with consumers' demands, and more diversified in delivery channels.

No single event, invention, disaster, or person can claim credit for this turnaround. A confluence of at least seven separate but related forces has realigned the food sector of the economy. They occurred in science and technology, in business management, and in social and cultural movements. Some of them were largely unnoticed at the time, but they have cumulatively resulted in a fundamental change in the way this industry conducts business. These seven forces have converged to change the landscape of rural and urban America forever:

- More diverse consumer characteristics and tastes
- The Universal Product Code (bar code) and all the information technology that followed
- Wal-Mart, the early adopter of information technology and the mother of efficient supply chain management
- Efficient Consumer Response, a defensive response to Wal-Mart's expansion
- Concentration of retail store ownership
- Global concentration of food processing and manufacturing
- New business models

These phenomena do not deal with the scientific and technological developments related to genetic mapping, DNA tracking, or bioengineering. Though enormously important to the productivity of agriculture and the quality and safety of food, they represent a separate and parallel force in the industry. They interact most directly with the forces identified above through consumer fears, changes in tastes and preferences, and promises of new, health-enhancing foods.

CONSUMERS

Many changes in consumer behavior and preferences have been attributed to changing demographics. Households are less homogeneous than ever. They are smaller, richer, and spend less time in their homes than older generations. Almost 60 percent of women (compared to 75 percent of men) are in the labor force, and many achieve positions of responsibility and wealth. Sixty-eight percent of wives are in the labor force full-time year-round (U.S. Census Bureau, 2001, Table 674). Dual-earner households have higher incomes than other households, contributing to an income distribution skewed toward the upper end. Almost half (47.2 percent) of all income was earned by the top 20 percent of families in 1999, compared to 41 percent in 1980. Even more telling is that 20.3 percent of all income is earned by the top 5 percent of households, compared to 14.6 percent in 1980 (U.S. Census Bureau, 2001, Table 670). Diverse and upwardly mobile consumers demand new types of places to shop and eat. Increased exposure to ethnic foods and an aging population play their part in creating new, diverse demands. Busy lives accompanied by discretionary income leads to a continuing quest for convenience, new experiences, and ways to substitute capital and leisure for household time in food preparation and eating.

The time we spend cooking/assembling evening meals at home fell from 2 hours in 1967 to less than 45 minutes today (Nielsen 1998); other reports show that the average time we expect to spend preparing a meal is 15 minutes (Mark Clemens 1995). Although 73 percent of eating occasions were at home in 1995, only 21 percent involved food prepared totally at home. Twenty-eight percent of these at-home eating occasions used food fully prepared and packaged away from home. Twenty-seven percent of the time people ate away from home (Farello et al. 1996).

Greater diversity in types of households, reflecting a variety of lifestyles, leads to a variety of preferences for food products and services. Table 8.1 shows that the percent of U.S. households made up of married couples with children present increased between 1987 and 2000 from 28 to 32.8 percent. After children reach age six, mothers are more likely to be in the labor force (77–80 percent). Married couples with wives in the labor force have a median annual income of $66,529; if husbands and wives are employed full-time year-round, and have no children at home, median income reaches $76,459, making them the highest-income earners of any household type on Table 8.1. One-fourth of our households have only one person, and 27 percent of them are women over age 65. The median income in single households is only two-thirds that of the average married couple.

Consumers are increasingly segmented into niche markets. Ethnic mixing of foods and people have created new tastes and preferences for out-of-the-ordinary food. Spicy sauces and Asian, Italian, and Mexican

TABLE 8.1. Household types, women in the labor force and income: 1999–2000

Household Type	Percent of All Household[1]	Percent of Women in Labor Force	Median Income (2000)
Single person	25.5[2] (24)[3]	68.6	$19,917
Single mothers	7.0 (7)	75.5	$26,164
Married, no children at home	20.0 (30)	54.7	
Married, children at home	32.8 (28)		
Under age 6	14.3 (NA)	62.8	$56,827 (all married HH)
Ages 6–13	18.5 (NA)	75.8	$66,529 (married, wife in LF)
Ages 14–17		80.6	
Married, husband and wife both in labor force full time/full year			
No children	7.4 (NA)	100	$76,459
Children under age 18	8.5 (NA)	100	$70,515
Unmarried Couples	4.2 (NA)	NA	NA

Source: Statistical Abstract of the United States, 2001: Tables 52,54, 577,578, 666,674.

[1]Numbers in this column do not add up to 100% due to overlapping categories.

[2]Thirty-six percent of single households are over age 64. Single female households over age 64 make up 27% of single households over age 64 and 7% of all single households.

[3]Numbers in parentheses refer to 1987.

foods are common. Yet a popular niche is old-fashioned comfort food favorites like mashed potatoes and ice cream. The resurgence of retro–comfort food was heightened in response to terrorism in 2001 and a general patriotic nostalgia.

Consumers' taste and preferences for food are often shaped by their social and cultural beliefs. This leads to "socially responsible" demand and agricultural production. Strong and growing food niches dictated by the socially conscious include food produced in an environmentally friendly way. This led to the sale of organic food to increase more than 20 percent annually in the late 1900s and early 21st century. Though some of that growth is due to concern about pesticide residues and their affect on human health, surveys reveal that strong preferences for organic food are tied to preserving the natural environment and a rural lifestyle.

Concern over animal welfare grows along with increased concentration of animal farming. So strong are the voices of concerned citizens that by 2002, global fast-food companies (McDonald's, KFC, and Burger King) told their meat suppliers that the animals who provide the meat for their restaurants must be grown in more space, given better treatment, and suffer less pain. Cattle, hog, and poultry producers agree that it pays to meet the product specifications of these large retailers. It is a premier example of demand-driven supply chains.

A variety of influences—aging populations, new scientific discoveries, affluence—converge to build a notable market niche for health-enhancing foods. Foods on the market that seek to fill this demand are organic and "natural" foods. Foods laced with herbal and energy-enhancing substances known as functional foods or neutraceuticals are marketed to health-conscious consumers. New food fortification with established vitamins and minerals like calcium in orange juice and balanced nutrition in "power bars" also serve this niche market. The promise of blending food and medicine further segments the demand chain and leads to segmentation of raw agricultural commodities at the source, which leads to various new technologies for traceability and identity preservation. A segmented supply chain adds costs to food production and distribution, but many consumers are willing to pay for information about the source and characteristics of their food. In addition, traceability makes the distribution channel more transparent, efficient, and safer. By responding to consumer demand, benefits are realized by producers, manufacturers, and distributors.

We often tout the ever-declining portion of disposable income that American consumers pay for food, down from 25 percent in 1960 to around 11 percent in 2002 (Kinsey and Senauer 1996). However, in real terms (inflation-adjusted dollars), between 1990 and 1999 Americans' food spending rose more than their incomes. Real, after tax, income increased 7.6 percent over that time, but real expenditures for food to take home increased 12 percent and expenditures for food in food service places increased 26 percent (U.S. Census Bureau, 2001, Tables 648,664). This trend supports the often-observed convergence of the proportion of U.S. food dollar spent on food from a grocery store and food purchased ready-to-eat in a food service place. Oddly enough, these two trend lines have never converged, and seem to have leveled off at about 53 percent of the food dollar being spent for food to eat at home (purchased at retail food stores). However, with one-third of full service and fast food restaurant sales of about $356 billion going for take-out food, the portion of the total food dollar spent on food to eat at home (even if it is not prepared at home) is much higher (The Food Institute Report 2001). Using the one-third estimate and applying it to food-away-from-home expenditures, and adding that to the dollars reportedly spent for food eaten at home, results in an estimated two–thirds (versus 53 percent) of consumers' food dollars spent on food eaten at home.[2] Competition among grocery stores, other food sellers, and various types of restaurants for a share of the American stomach seems to enforce a mixture of places to procure food and places to eat.

INFORMATION TECHNOLOGY

Consumers' food tastes evolve slowly, and so does the adoption of information technology. The development of the bar code in 1972 was a giant leap forward. Developed by the Uniform Product Code Council, it allowed retailers to stop putting price stickers on every item and to scan the bar code at the checkout stand. To realize its benefits, new computerized equipment had to be installed at checkout counters in 170,000 food stores and in the offices and headquarters of retail stores. Some stores have still not met this challenge.

New, computerized cash registers revealed the prices and told the store manager when inventory was getting low. But this was only the beginning. The flood of electronically generated information led to new management practices, such as inventory control, automatic ordering and replacement of products, and examining—in great detail—what types of consumers prefer what types of food at what time of day, week, or year. Bar codes now appear on case lots, truck loads, and shipping containers. They are used to pick foods in warehouses and drop them on the right pallets for delivery to the right stores, at the right time. In some cases they are even used to track food through the grocery store so a vendor can bill the store only after product has been sold (scanned at the checkout stand). The power of the information held in the computer systems of retailers is just now being realized by a large number of firms in the food chain. The possession of unprecedented knowledge about their customers and sales shifted negotiating power to large retailers. By 2000, half of food retailers implemented some type of frequent shopper or customer loyalty program. Many developed Internet pages and on-line shopping opportunities. On-line, home-delivered sales are less than two percent of grocery sales, but they are growing steadily as both consumers and businesses learn how to order and deliver efficiently.

Information technology launched e-commerce, which is a way to conduct business at arm's length with anonymous parties in the privacy of one's office, home, or hotel room. It requires superb trust and induces fierce competition into the market. There are thee forms of e-commerce relevant to the food industry:

    Business to consumer (B to C)
    Business to business (B to B) for shopping
    Business to business (B to B) for coordination and planning among
        retailers and vendors

Business to consumer e-commerce sprinted ahead in the late 1990s only to see many companies fail by 2001. Stand-alone, Internet grocery companies simply could not make a profit competing with traditional grocery stores. Only those who delivered products and service that met customers' highest expectations survived. If they cost consumers more time or money than conventional shopping, and did not deliver supreme product on time, consumers did not order a second time. However, since 2001, a slow and steady increase in on-line shopping from established supermarkets has developed. Called "clicks and bricks," these supermarkets are experimenting with new customer services in selected locations. They have new partners and more experienced on-line technicians, marketers, and management.

Business to business e-commerce takes two forms, one for businesses to buy and sell products and services, and one for sharing point-of-sale and inventory data and forecasts. The first form sprinted onto the scene in the late 1990s; only a few of these companies survived. Primarily, these companies promised much more than they could deliver. Also, many large manufacturers bought supplies on contract to ensure quality and delivery and would not use an auction market except in an emergency. What many of the on-line marketers failed to understand is that many buyers are not necessarily looking for the lowest price on a generic commodity. They need more information about and control over product quality.

Business to business e-commerce involving the sharing of point-of-sale and inventory data and forecasts led to fully coordinated ordering and replenishment between vendors and retailers. Collaborative Planning, Forecasting, and Replenishment (CPFR) is flourishing and its cost-saving benefits are growing, but only a few of the largest retailers (including Wal-Mart and Target) and their suppliers (such as Johnson & Johnson, Proctor and Gamble, and Kimberly Clark) are actively using the CPFR process. Trends driving the adoption of CPFR include retailer consolidation, the use of store-specific forecasting, and sales-driven replenishment systems such as category management (Garry 2002). These benefits of CPFR will become obvious in the following sections.

WAL-MART

A behemoth retail company, called Wal-Mart, was an early adopter of information technology. It built its business on knowing exactly what consumers were purchasing in all their stores, all the time. Wal-Mart avoided tying up cash in inventory by working with vendors to drive down the cost-of-goods-sold (COGS) and the cost of moving goods from manufacturers to consumers. They were the first to track inventories with computers (1969), adopt bar codes (1980), use electronic data interchange to coordinate with suppliers (1985), and use wireless scanning guns (late 1980s). They created the "big-box" format, every-day-low-prices (EDLP), electronic data interchange with vendors (early form of CPFR), and a hub-and-spoke system of building stores around a central distribution center, which they owned. By 1987, with just 9 percent of the retail market they were 40 percent more productive than their competitors, as measured by real sales per employee. This productivity gap grew to 48 percent by 1995, boosted in part by selling slightly higher-priced merchandise to slightly more affluent consumers (Johnson 2002). All this innovation enabled Wal-Mart to offer lower prices and capture an ever larger share of the market. By 2002, they were ranked as the largest company in the world in the Fortune 500. They had 4,485 stores, sales of $219.8 billion, and employed more than 1.38 million people, most in non-union positions (Fortune 2002). They captured 11.1 percent of all U.S. grocery market total on sales of $57.2 billion in 2001. The top five supermarket chains—Kroger, Albertson's, Safeway, Wal-Mart, and Ahold captured almost 40 percent of U.S. grocery sales in 2001 (Food Marketing Institute 2002).

Wal-Mart is considered the company to study, to emulate, to sell to, and to buy from—but not to compete with on prices. They continue to grow and innovate and compete around the world. They are the only American food retailer to open a significant numbers of stores in other countries, including England, Germany, and China.

It would be difficult to exaggerate the influence of Wal-Mart on the structure of the rest of the retail industry, to say nothing of supply chain management in the entire food and agricultural industry. They forced the rest of the industry to become more efficient and more organized. They brought lower prices to consumers in small towns where local food stores often had a monopoly. Consumers realized real benefits from Wal-Mart's presence in the market, directly through lower prices and indirectly through the emulation by other companies trying to compete with Wal-Mart. In the end, all companies are getting smarter, using more information technology and/or finding a market segment where consumers prefer something other than big stores and low prices.

EFFICIENT CONSUMER RESPONSE (ECR)

By 1992, the rest of the retail food industry realized the need to emulate the Wal-Mart model. Organizing through trade associations they proposed a system of sharing information between retailers and vendors that would allow slimmer inventory holdings and lower delivery costs. The system, known as Efficient Consumer Response (ECR), was designed to introduce the rest of the retail food industry to management practices that employ the power of information technology and exploit the data held by retailers. Since there are positive spillover effects of having more participants involved, a massive educational program was undertaken. A flurry of conferences, lessons, software developments, and a whole new set of relationships between retailers and vendors was spawned. Implementing electronic data interchange (EDI) for ordering and reordering merchandise, slimming down the offerings in each category in order to streamline delivery, and reducing costs requires, however, that computer systems of retailers and suppliers be compatible. This requires considerable investment capital and operating skill.

Sharing private sales data with vendors also requires a new form and level of trust. This asked a lot of an industry made up of 130,000 disparate stores, all operating on thin margins and accustomed to treating their suppliers as adversaries. In 1998, only 6.8 percent of retail stores reported to *Supermarket News* that they used EDI for some purposes (Blair 1999). In 2001, 86 percent of stores in the Supermarket Panel of The Food Industry Center at the University of Minnesota reported using EDI of some sort to transmit data to headquarters or to major suppliers. Eighty-two percent used EDI to transmit orders to vendors or suppliers (King et al. 2001).

In the late 1990s, while other retail food stores were struggling to enter the information age, the next generation of ECR was being tested by Wal-Mart. In 1996 they joined with Warner Lambert to test a new system of information exchange and studied the effects on the sale of Listerine. This system led to the CPFR processes; a manufacturer and a retailer each forecast sales over some future period of time, shared their forecasts, and then committed to a schedule of orders and deliveries. The supplier is responsible for keeping the shelves stocked in this vendor-managed inventory (VMI) system. Scanner data is transmitted in real time to suppliers via an Internet interface, avoiding many of the problems with incompatible software. "The whole intent of CPFR is to establish trust between retailers and manufacturers" (Robinson 1999). Each party faces lower risk of excess inventory or stock-out in this system and sales tend to increase (Margulis 1999).

Meanwhile a group of retail food industry leaders were working with the Uniform Code Council to develop a lower cost, Internet-based platform to facilitate real-time transmission of electronic data between trading partners (Amato-McCoy 1999). As more companies use UCCNet (or other Internet-based systems), business to business e-commerce should increase in the food industry. The continuous flow of information from retailers to manufacturers and eventually to farm input suppliers pulls from that supply chain a continuous flow of product designed to meet consumers' preferences at the retail end. Figure 8.1 illustrates this system. In 1998, only 9 percent of retailers and 26 percent of wholesalers were trying CPFR programs. In 1999, 26 percent of retailers and 44 percent of wholesalers were planning to try CPFR (Blair 1999). By 1999,

Information Technology Demands
COORDINATION/COMPATABILITY

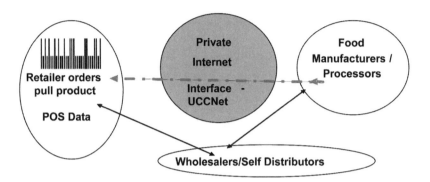

FIGURE 8.1 Information flows in the food supply chain.

Wal-Mart was using CPFR with over 7,000 of their suppliers (IGD). In the 2001 Supermarket Panel, only 23 percent of stores reported using scanner data for automatic inventory refill and half said they had no plans to do so. Most of the stores who shared scanner data with suppliers were in chains with more than 60 stores (King et al. 2001).

Scan based trading (SBT) is a part of CPFR and is moving into the industry rapidly. Under this system, the store is not billed for its inventory until after it is scanned and sold to a consumer. In effect, the manufacturers lend stores an interest-free loan. Manufacturers replenish shelves as they learn what sells in real time. The store saves on labor costs and does not take a risk on buying inventory that will not sell. It improves retailers' cash flow, reduces the need for capital, and improves the return on assets. Most of the products being sold with SBT are delivered to retailers by manufacturers; they tend to be concentrated in the snack and beverage items. By 1999, 59 percent of retailers and 16 percent of wholesalers reported plans to test an SBT system (Blair 1999). In the 2001 Supermarket Panel, 22 percent of stores reported actually using SBT (King et al. 2001).

MERGERS AND ACQUISITIONS

Improved information technology led to rampant mergers and acquisitions in all segments of the food chain. The adoption of information technology facilitated vertical coordination and invited horizontal mergers. For example, at the retail level, the percent of national grocery store sales held by the top four grocery chains grew from 16 in 1990 to more than 31 in 2001. Mergers were motivated largely by the need to drive down the costs of delivery and lower the cost of goods sold. There are economies of scale in these mergers. The economies come in the form of bargaining power with manufacturers, more efficient use of transportation and ordering systems, and ability to utilize information technology to manage inventory throughout the chain. Estimates by Nielsen show that retail mergers can reduce the cost of goods by 0.5 percent and save another 2.5 percent throughout the distribution chain (Nielsen 1998).

Wholesalers basically assemble food from a variety of manufacturers, reorganize it according to orders from retailers, load it on trucks, and deliver it. There are three distinct methods of operating at this link in the chain:

- The self-distributing retailer owns its own distribution centers, buying directly from manufacturers or producers. These types of operations make up about 35 percent of the companies that perform retail food assembly and transportation services (A.T. Kearney 1998). They include all the largest retail food chains and several smaller ones. Their labor expenses as a percent of sales at inventory costs are 0.9 percentage points lower than third-party wholesalers' costs, and their non-labor expenses are 1.32 percentage points lower. Their operating costs per case are $0.18 less. They select 39 cases per hour more and their costs as a percent of sales are 1 percentage point lower than traditional wholesalers (Krochersperger 1998). Retailer-owned distribution centers lower the cost of handling food on its way to the consumer.
- Direct store delivery (DSD, 27 percent of the market) is when manufacturers deliver their own products directly to individual stores and often arrange it on the shelves. They are in a good position to service the needs of a retailer and are the main parties using scan-based trading.
- Third-party wholesalers are the traditional food assemblers. They buy food from manufacturers and resell it to retailers, making profits on the price spread and payment for services to stores. This type of food assembly and delivery business is shrinking in the industry as self-distributing chains and direct store delivery increase in size. To

survive, larger wholesalers of this type have increased their ownership of retail operations and formed what some call "virtual chains."
This type of wholesale business conducts about 37 percent of the retail food distribution business to food stores (not food service or
restaurants). Mergers among wholesalers have resulted in the two
dominant firms serving retail food stores (Supervalu and Flemming)
and two dominant firms serving restaurants and institutional food
service (Sysco and Ahlod). Multiple specialty wholesalers continue to
operate on local and regional bases and are likely to play an important role in niche markets.

GLOBAL CONCENTRATION
OF FOOD MANUFACTURERS

As retailers became larger and more concentrated, food manufacturers
continued to merge partly to concentrate their negotiating power with
retailers, but more importantly to achieve larger and larger shares of the
world market. The source of raw commodities and the customers of
food manufacturers reside around the world. Manufacturers will purchase their raw materials from the lowest-cost producer, subject to quality of the input. Given that food manufacturers want to produce a
consistent quality and quantity of product year around, they seek inputs
from worldwide sources. The global market changes consumption
habits, transportation modes, seasonal prices and supply, and the location of primary production.

An example of how global markets change consumption and expectations is in the area of fresh fruits and vegetables. We are now accustomed to having almost all varieties of fresh produce available all year,
meaning that seasonality of consumption is gone and fresh produce
comes from various parts of the world. This tends to smooth out the total seasonal supplies and dampens the variation in prices (adjusted for
transportation costs). It also means that there are many substitutes for
domestic supplies, and short-run agricultural "disasters" have a smaller
effect on the price of finished food. There simply are several other places
to obtain raw commodities and manufacturers do not always need to locate close to agricultural production regions.

On the other hand, food manufacturers are likely to locate processing plants closer to their retailers. This has the advantage of saving on
transportation costs; it is less expensive to ship bulk commodities than

to ship cans and bottles of final food products. Also, it is easier to accommodate the preferences of local consumers as well as meet various countries' laws regarding safety, labeling, and packaging. They can deliver fresher product, give retailers better service, and gather better marketing intelligence if they are close to their customers.

The ownership of the top five world food manufacturers is split between the U.S. and European companies. As of 2001, two of the top six food processors were European; Nestle of Switzerland is number one and Unilever of the Netherlands is number 6. The rest, in order of size, are U.S. companies: Phillip Morris (Kraft), Con Agra, General Mills, and Cargill (Olson 2001). In the case of global retailers, the proportion is reversed. Two of the top six food retailers are U.S. owned; Wal-mart is number one and Kroger is number six. Those in between are, in order of size, Carrefour of France, METRO AG Gruppe and Tengelmann of Germany, and Tesco of the U.K. Of these six retailers Kroger is the only one that operates in only one country, the U.S. (http://www.Stores.org).

The alternating consolidation moves by retailers and manufacturers are seen by many as a game of poker, each playing for a larger share of the market, upping the ante with each new merger. Larger stores tend to decrease the number of their suppliers, develop their own private label products, and have their own distribution centers. All these factors threaten the market share of international branded products.

NEW BUSINESS MODELS

The previous six developments in the food industry outline culminate in a new way to conduct business in this economic sector. Consumer demands and preferences, fears, and politics drive the business decisions by virtually every company and farmer who contributes to the final food product.

Perhaps one of the easiest ways to illustrate the demand chain is with two stories. One is a story about Wheaties, long advertised as the Breakfast of Champions. Formerly, if you poured a box of Wheaties on a table and sorted out the flakes, some were curly and some were flat. Consumers prefer whole curly flakes to flat flakes that crumble, because the former stay whole, they are crunchier and tastier, and the box stays full in storage. Consumer preferences told the manufacturer to produce curly flakes. It turns out curliness is genetically determined. So the food manufacturer contracted with farmers to produce the variety of wheat that produces curly Wheaties, ensuring an adequate supply (Olson 1999). The manufacturer is willing to pay more, but the farmers and grain handlers must produce and store this identity-preserved wheat in separate containers all through the supply chain, lest an errant kernel of flat-flake wheat gets mixed with those that make nice crisp curly flakes.

The second story is about "bagged lettuce." This phenomenally successful new product has changed lettuce specifications from size and shape of lettuce heads to specifications controlled by the manufacturing process. More varieties are demanded, and safety and sanitation concerns increase. The carefully controlled process reduces waste and levels out margins for the packer; farmers' prices are determined by contracts rather than a fluctuating market of supply and demand. Needing to have bags of lettuce in the retail stores every day of the year drives these manufacturers to obtain lettuce from many growing regions of the world and to encourage greenhouse production.

Retailers and manufacturers of the food products, striving to meet consumer's demands, pull the right type of product off the farm. Thus, the term "food demand chain" seems appropriate for the food system of today and tomorrow.

REFERENCES

A.T. Kearney. 1998. *Strategies 2005: Visions for the Wholesale Supplied System.* Washington, D.C.: Food Distributors International p. 1.

Amato-McCoy, Deena. 1999. Six leaders, UCC working on low-cost e-commerce tool. *Supermarket News.* February 23.

Blair, Adam. 1999. The big picture. *Supermarket News.* February 22, 3A.

Farello, Michael, Robert Mitchell, and Kari Alldredge. 1996. Satisfying America's changing appetite. *The McKinsey Quarterly* 4:193–200.

The Food Institute Report. 2001. *Restaurants Find Success with Takeout.* April 19.

Food Market Institute, *Facts and Figures.* http://www.fmi.org.

Fortune Magazine. 2002. The Fortune 500. April 15:F-52.

Garry, Michael. 2002. Focus on the consumer. *Supermarket News.* June 24, pp. 17–18,20,34.

IGD (International Grocery Distribution). 1999. *Wal-Mart in the U.K.* England: Letchmore, Heath Watford.

Johnson, Bradford. C. 2002. Retail: The Wal-Mart effect. *The McKinsey Quarterly* 1:41–43.

King, Robert P., Elaine Jacobson, and Jonathan Seltzer. 2001. *The 2001 Supermarket Panel Annual Report.* St. Paul, MN: The Food Industry Center, University of Minnesota.

Kinsey, Jean, and Benjamin Senauer. 1996. *Food Marketing in an Electronic Age: Implications for Agricultural Producers.* TRFIC Working Paper 96–02. St.

Paul, MN: The (Retail) Food Industry Center. http://foodindustrycenter.umn
.edu.

Krochersperger, Richard H. 1998. *1997 Wholesale/Retail Distribution Center
Benchmark Report*. Washington, D.C.: Food Marketing Institute and Food
Distributors International. p. 10.

Margulis, Ronald. 1999. One more acronym: CPFR takes a quick response to
next level. *ID* August:33.

Mark Clemens Research, Inc. 1995. What America eats: Shopping, preparation,
eating, nutrition. *Parade*, 9.

Nielsen, A.C. 1998. *The Second Annual A.C. Nielsen Consumer and Market
Trends Report*. August:7,10.

Olson, Ronald. 1999. Speech to Twin Cities Agricultural Round Table Lun-
cheon, by vice president for the Gain Division, General Mills. Minneapolis,
MN. October 27.

Olson, Ronald. 2001. *Prepared Foods*. July, pp. 13–16.

Robinson, Alan. 1999. The circle is broken. *Food Logistics* June 15:43.

U.S. Census Bureau. *2001 Statistical Abstract of the United States* http://www
.census.gov/prod/2002pubs/01statab/stat-ab01.html.

NOTES

1. Professor, Applied Economics and Co-Director, The Food Industry Center,
   University of Minnesota.
2. For example, in the Minneapolis/St.Paul area in 2001, the average household food expenditure for food at home is recorded at $3050 or 53 percent
   of total food expenditures. If .33 of food-away-from-home expenditures is
   for food eaten at home, then the proportion of the total food dollar spent
   for food eaten at home is 68 percent. ($2728 × .33 = 900; 900 + 3050 =
   3950 divided by 5778 = .68) (Source: The Food Institute Report, 5/7/2001
   using BLS data).

III

FRONTIERS IN ANIMAL AGRICULTURE

This section includes a series of chapters related to livestock in both western and global contexts. The first paper, authored by Hank Fitzhugh [formerly Director-General of International Livestock Research Institute (ILRI), Nairobi, Kenya] and his colleague, provides an overview of the critical importance of livestock to so many people in the developing world. The second paper is by two prominent animal scientists with extensive records addressing environmental impacts of animal agriculture. The third paper considers the societal impact of horses. This is authored by Karyn Malinowski, Director of the Rutgers University's Equine Research Center and of Cooperative Extension, together with Norman Luba who is very prominent in the horse industry.

9

THE IMPORTANCE OF LIVESTOCK
FOR THE WORLD'S POOR
R. R. von Kaufmann and Hank Fitzhugh[1]

Animal agriculture has great potential to alleviate poverty, but this is not generally realized. Today the relatively affluent populations of developed countries are predominately urban and well fed, plus they are two generations or more away from any firsthand experience with livestock production. If asked their views about livestock, they are likely to repeat hearsay allegations that livestock compete with humans for grain and degrade the environment, and that consumption of meat, milk, and eggs is bad for human health.

These views about livestock are very different from those held by their grandparents, when most people in what are now developed countries were rural dwellers and relatively poor. At that time, livestock played important roles in the success of most family farms: generating income, building assets, relieving drudgery, fertilizing fields, and—through meat, milk, and eggs—providing the essential amino acids and micronutrients to the family diet that ensured the good health and the physical and cognitive development of the children.

Today, livestock play a similar role in the lives of the rural poor in most developing countries, where the majority of the populations live in rural small-holder farming communities that have limited opportunities to improve their livelihoods. Just as they did a century ago in what are now developed countries, livestock provide one of the few means for resource-poor families to generate income, build assets and move out of subsistence agriculture.

Population growth, industrialization, urbanization and income growth are promoting significant changes in the livestock sectors of developing countries, where demand for meat and milk has more than doubled since the mid-1970s and is expected to double again by 2020. This "livestock revolution" offers one of the rare significant opportunities for sustainable poverty reduction in developing countries. However with these opportunities, the livestock revolution will also bring challenges: to ensuring availability of staple food grains at affordable prices, to providing fair and equitable market access for small-holders, to protecting the environment, and to safeguarding human health. Failure to meet these challenges would have serious consequences for future generations, in developed as well as developing countries.

In the following sections, we address the following:

- The contributions of livestock to the livelihoods of poor people
- Concerns that livestock are competing with the poor for basic food requirements
- The opportunities and challenges posed by the livestock revolution and its consequences for livestock production and marketing systems
- The priorities for livestock-related research and policy analysis

Our focus is on livestock production in developing countries, but there are also ramifications for developed countries, including opportunities for export of feed grains and meat.

LIVESTOCK AND LIVELIHOODS

Livestock production systems and the contributions of livestock to livelihoods are more diverse in developing countries today, and arguably much more important, than in developed countries. These differences stem largely from the fact that the majority of livestock keepers in developing countries are rural and relatively poor.

Four-fifths of the world's population of six billion live in developing countries. Of these more than a billion people are desperately poor: They, especially the children, are almost always malnourished and vulnerable to disease, and have little hope for a better life in the future. The largest numbers of these very poor people are in Asia but, as a percentage of total population, poverty is most severe in sub-Saharan Africa (see Table 9.1).

Livestock provide one of the few means for the poor to generate income, acquire assets, and escape from the poverty trap. Sales of livestock, animal-source food, hides and fibers—through both formal and informal markets—make major contributions to household income. Evidence from in-depth field studies in Asia and Africa indicates that livestock contribute as much as 60 percent of household incomes, and they generally contribute a higher percentage to the incomes of poorer house-

TABLE 9.1. Regional poverty distribution in 2000

Region	Population (Millions)	Proportion Below Poverty Line[1](%)	Proportion Poor Livestock Keepers (%)
East Asia	1304	9	2
South Asia	1337	43	15
Southeast Asia	505	31	12
Central and South America	506	40	11
West Asia–North Africa[2]	467	27	10
Sub-Saharan Africa	627	54	26
Total	4746	32	12

Source: Thornton et al., 2002.

[1]Based on national poverty lines used by Randolph et al., 2001.

[2]Includes newly independent states in Central Asia and Caucasus region.

holds than they do to the incomes of wealthier households (Delgado et al. 1999).

Far from competing with the poor for food grain, most livestock in developing countries, especially ruminants, convert human-nonedible feeds, including crop residues, household wastes and forage, into high-value meat, milk, and eggs. These animal-source foods provide high-quality, readily absorbed protein and micronutrients. Because of their amino acid composition, eggs are the nutritional standard for valuing biological availability of food proteins, which range from 90–100 percent for meat, milk, and eggs, compared to 50–70 percent for proteins from plant sources (CAST 1999).

Micronutrients from animal-source foods are especially important in the health of women of reproductive age, and to the cognitive development and school performance of children (Neumann and Harris 1999). Micronutrients with high bioavailability that are provided from animal-source foods include minerals, calcium, iron, phosphorus, zinc, magnesium, and manganese and vitamins thiamine B_1, riboflavin B_2, niacin, pyridoxine B_6, and B_{12} (CAST 1999).

In 1993, the amount of calories from animal-source food in developing country diets ranged from 7 percent in sub-Saharan Africa to 18 percent in Latin America, with an overall average of 11 percent. Animal-source foods also provided significant proportions of the protein in

Table 9.2. Percentage of food calories and protein from
animal products, 1983–93

Region	Percentage of Calories from Animal Products[1]		Percentage of Protein from Animal Products	
	1983[2]	1993	1983	1993
China	8	15	14	28
India	6	7	14	15
Other South Asia	7	9	19	22
Southeast Asia	6	8	23	25
Latin America	17	18	42	46
West Asia and North Africa	11	9	25	22
Sub-Saharan Africa	7	7	23	20
Developing world	9	11	21	26
Developed world	28	27	57	56
World	15	16	34	36

Source: Delgado et al., 1999, from FAO, 1997.

[1]"Animal products" includes meat, dairy, and egg products, and freshwater and marine animal products.

[2]Each cell is a three-year moving average centered on the year listed.

human diets, ranging from a low of 15 percent in India to a high of 46 percent in Latin America (see Table 9.2), with an average of 26 percent for the developing world. But this is less than half the average found in developed country diets.

Livestock are valued for their capacity to raise and diversify smallholder incomes. They convert crop residues and by-products, household scraps, and forage from road and hillside vegetation into high-value food, fiber, skins, and hides. These products can be sold directly or be processed into butter, cheese, leather, wool, and other value-added goods. These are important opportunities for women, for whom processing and marketing livestock products earn "butter and egg money," often their sole source of independent income.

Manure and urine from about two-thirds of livestock in developing countries contribute to small-holder crop production by providing soil nutrients, maintaining soil structure, and improving soil moisture retention. By providing power for cultivation and transportation, livestock also reduce the hard labor and drudgery of farm work, which in developing countries is disproportionately women's work.

Livestock keeping requires less manual labor than crop cultivation, and draught power reduces the drudgery of cropping. This is especially important to households headed by women and those that, due to HIV/AIDS, have limited access to adult labor.

In market economies, one or two crop failures can impoverish farm families who have no savings. Investing in farm animals is one of the few options available for small-holders to accumulate assets that they can realize to tide them through the inevitable adverse situations. The importance small-holders place on livestock as assets is indicated by their readiness to assume debts to purchase livestock.

These multiple and diverse benefits are appreciated by the resource-poor farmers and pastoralists who keep livestock in all regions of the developing world (see Figure 9.1).

LIVESTOCK PRODUCTION SYSTEMS

There are three principal livestock production systems: pastoral, mixed crop-livestock, and industrial. Sometimes livestock progress through elements of two or all three systems in what are termed stratified production systems (CAST 1999). The contributions from these systems to the

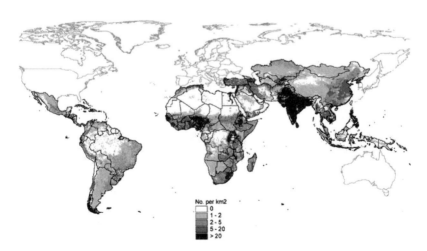

FIGURE 9.1. Density of resource-poor livestock keepers in developing regions (Thornton et al. 2002).

Table 9.3. Percent of global production from livestock production systems

Product	Million MT	Pastoral	Mixed Crop-Livestock	Industrial
Meat				
Cattle	53	23	65	12
Buffalo	3	0	100	0
Sheep/goat	10	30	69	1
Swine	71	1	60	39
Poultry	43	2	24	74
Milk	473	8	91	1
Eggs	40	1	31	68

Source: Adapted from CAST, 1999, based on Seré and Steinfeld 1996.

global production of animal-source foods are shown in Table 9.3. Production from stratified systems is included as part of that from industrial systems, which are the end point for most stratified systems.

Pastoral Systems

Pastoral systems utilize the native range and other grasslands that cover about 30 percent of the global land surface. These grasslands are generally not well suited for cropping, but do play an important role in carbon sequestration, which has potentially important ramifications for global climate change. Ruminants (primarily cattle, sheep, and goats), camelids and equines are the principal species kept by pastoralists to produce meat, milk, and fiber for sale and local consumption.

Mixed Crop-Livestock Systems

Mixed crop-livestock systems (generally family farms) take advantage of the complementarities between crop and livestock production. Livestock add value by converting low-cost crop residues and by-products into high-value milk and meat, by recycling nutrients to improve soil fertility and crop productivity, and in many developing countries by providing the draught power for crop cultivation. On a global basis more than 50 percent of all meat, and more than 90 percent of the milk, is produced from mixed crop-livestock systems (CAST 1999).

Industrial Systems

Until the mid-20th century most animal-source foods were produced from either pastoral or mixed crop-livestock systems. However over the last fifty years there has been phenomenal growth in production from grain-based industrial systems. By the mid-1990s, 79 percent of poultry meat and 39 percent of pig meat came from industrial systems (de Haan et al. 1997). This trend has accelerated with expansion of industrial poultry and swine systems in developing regions, especially in China and Southeast Asia, responding to the huge increase in urban demand for low-cost meat.

These systems generally confine stock in small areas, feed grains and other concentrates, and impose tight controls over livestock management to improve productivity and reduce wastage (CAST 1999). The U.S. and other developed countries apply industrial systems to large-scale dairy cattle operations and beef feedlots, but globally, industrial systems are primarily used for grain-based poultry and pork production (see Table 9.3).

The robust market forces will continue to promote intensive, specialized systems with high-input livestock management and disease control practices, and dependence on feed grains and protein supplements. This shift to industrial pig and poultry production is underpinned by vertical integration of input, production, processing, and marketing facilities, in which the role of small-holders as animal breeders or feed producers is uncertain.

Industrial production systems have brought immediate benefits to consumers by reducing prices. However, there are hidden costs and subsidies that unduly favor the larger, capital-intensive industrial units to the disadvantage of small producers.

For example, the technologies for industrialized systems are readily transferred and, encouraged by weak regulatory controls and policies, industrial systems are expected to expand rapidly in developing countries, thereby creating major environmental concerns, especially where they are overly concentrated (Steinfeld 1998). Another disadvantage is the fact that manure and wastewater from large-scale industrial operations create risks of soil and ground water pollution, for which they have not in the past been required to pay. Moreover, concentrations of large numbers of livestock under confined, often unsanitary conditions, are incubators for diseases, including zoonoses such as influenza (Perry et al. 2002).

Stratified Systems

In order to take advantage of different land, labor, and feed resource and market opportunities, animals in stratified systems are moved from one system to another during their life cycle, often with a change in their ownership. One example involves the movement of young cattle born and raised for 1–2 years in pastoral herds to industrial beef feedlots for a few months of grain feeding to improve carcass quality before slaughter. Stratified systems offer significant potential for improving productivity of livestock populations in developing countries, including the production of feeder pigs and chickens from small-holder family farms, as well as lambs and calves from pastoral herds (ILRI 2000).

Resource-Poor Producers in Developing Countries

The different systems of production are used by almost 700 million resource-poor people in all regions of the developing world in keeping livestock (see Table 9.4). There are growing numbers of landless livestock keepers whose livestock graze roadsides and other public lands and scavenge for household and commercial food wastes in urban areas. There

TABLE 9.4. Number and types of resource-poor livestock keepers in developing countries

| | Type of System Used by Resource-Poor Livestock Keepers | | |
Agro-Ecology	Pastoral	Mixed Crop-livestock Farmers	Landless
Arid and semi-arid	63	213	. . .
Temperate, including tropical highlands	72	85	. . .
Subtropical, humid, and subhumid	. . .	89	. . .
Total	135	387	156

Source: ILRI, 2000, updated from Livestock in Development, 1999.

are also a large number of pastoralists, nomadic and transhumant, in the dryer rangelands, including those in temperate ecologies, where rainfall is so limited and unpredictable that cropping is not an option. However most resource-poor people are mixed crop-livestock farmers, and they are producing increasing quantities of meat, milk, and eggs.

THE LIVESTOCK REVOLUTION

Demand for meat, milk, and eggs in the developing world is expected to double by 2020, while increasing only slightly in the developed world (Delgado et al. 1999; Rosegrant et al. 2001). This huge increase in demand for animal-source foods is the livestock revolution, which will have profound effects on the world's livestock industry, bringing changes in production, processing, and marketing of livestock products.

In fact, the livestock revolution has been under way for some time. Since the late 1960s, growth in meat demand in developed regions has slowed, even becoming negative in the 1990s. By contrast in developing regions, growth in meat demand has accelerated over the last three decades (see Table 9.5), with the result that meat consumption more than doubled from 11 to 25 kilograms per person on average in developing regions.

Projections for 2020 were calculated using the International Food Policy Research Institute's (IFPRI) International Model for Policy Analysis of Agricultural Commodities and Trade (IMPACT). This covers 36 countries and regions, accounting for virtually all global production and consumption of cereals, soybeans, roots and tubers, fruits, vegetables, oils, oil cakes, meals, meat, milk, and eggs (Rosegrant et al. 2001).

Delgado et al. (1999) noted extraordinary changes in the diets of the emerging middle class in developing countries and concluded that these trends would continue as incomes rise in the future. This is because the demand for nutrient-rich, easy-to-cook meat and dairy foods is driven by increased purchasing power and increased urbanization in developing regions, which will reach 50 percent in 2020 as compared with 30 percent in 1980. Incomes are projected to increase 3 percent per annum in sub-Saharan Africa and 6 percent in Asia and Latin America (Cranfield et al. 1998). Moreover Rosegrant et al. (2001), citing Bhalla et al. (1999), anticipated that in developing countries with low but increasing per capita incomes, meat demand will grow even faster than per capita income.

TABLE 9.5. Growth in per capita demand for meat

Region	Growth Rate, % per yr				Demand, kg/capita		
	1967–82	1982–90	1990–97	1997–2020	1967	1997	2020
Developed	1.5	1.2	1.1	0.6	60	76	84
Developing	2.0	3.4	5.2	2.8	11	25	35
World	1.1	1.5	1.4	2.0	26	36	44

Source: Rosegrant et al., 2001, and Delgado et al., 2002, using 1967–97 data from FAO, 2000 and 2020, estimated using IMPACT model.

Taking population growth into account, total meat consumption in developing countries has increased from only 36 percent of world total in 1983 to 53 percent in 1997 (see Table 9.6). Since 1967, poultry's share of total meat consumption has doubled from 15 to 29 percent in developed regions, and from 12 to 26 percent in developing regions (Rosegrant et al. 2001). In North America and Europe, this increased share for poultry was primarily at the expense of beef. However in developing countries, especially in East and Southeast Asia, growth in poultry consumption is primarily at the expense of pork consumption (Rosegrant et al. 2001). Almost all the increase in poultry meat production, in both developed and developing regions, was from chickens produced from grain-based industrial systems.

Demand for milk products has also grown rapidly, at over 3 percent per year in the developing regions, while remaining almost unchanged in developed regions. By 2020, demand for dairy products in developing countries will be almost 50 percent greater than the total demand in developed countries (see Table 9.7).

Over the next two decades, demand for meat is expected to grow only 0.6 percent per year in developed countries, compared with 2.8 percent in developing countries. Meat consumption in developing countries will increase from 112 million metric tons in 1997 to 213 million in 2020, accounting for almost two-thirds of global consumption.

Two-thirds of the increased demand will be for pork and poultry meat, but with significant differences among regions, largely because of cultural and religious restrictions (see Table 9.7). Most of the increased demand for pork will be in East and Southeast Asia, with little increase in South and West Asia, and North Africa. (Beef consumed in Asia includes meat from water buffalo as well as cattle.)

COMPETITION FOR GRAIN AND OTHER HUMAN-EDIBLE FOODS

Livestock consume about one-third of the global cereal production (CAST 1999). This has provoked concerns that livestock are competing with the poor in developing countries for basic food requirements. These concerns have been reinforced by the steady increase in grain feeding in developing regions over the past two decades, primarily for

TABLE 9.6. Meat and milk consumption trends for developed and developing regions[1]

Region	Per Capita Consumption, kg			Total Consumption, MMT			%World Total Consupmtion		
	1983	1997	2020	1983	1997	2020	1983	1997	2020
Developed Countries									
Beef	27	23	25	32	30	34	67	52	40
Pork	29	28	29	34	36	39	63	44	33
Poultry	16	22	28	19	28	38	66	49	36
Meat	74	76	84	88	99	114	64	47	35
Milk	195	195	203	233	254	276	66	56	43
Developing Countries									
Beef	5	6	9	16	27	52	33	48	60
Pork	6	10	13	20	46	80	37	56	67
Poultry	3	7	11	10	29	67	34	51	64
Meat	14	25	35	50	112	213	36	53	65
Milk	35	44	61	122	198	372	34	44	57

Source: Delgado (personal communication)

[1]Meat is uncooked bone-in weight, including mutton and goat as well as beef, pork, and poultry. Milk is liquid milk equivalent weight, including milk products but not milk fed to calves. Million metric tons (MMT) and kg are three-year moving averages centered on year shown.

TABLE 9.7. Within region totals for production and demand for different types of meat and milk in 1997 and 2020, million MT[1]

Region	Beef 97 P	Beef 97 D	Beef 20 P	Beef 20 D	Pork 97 P	Pork 97 D	Pork 20 P	Pork 20 D	Poultry 97 P	Poultry 97 D	Poultry 20 P	Poultry 20 D	Sheep/Goat 97 P	Sheep/Goat 97 D	Sheep/Goat 20 P	Sheep/Goat 20 D	Meat 97 P	Meat 97 D	Meat 20 P	Meat 20 D	Milk 97 P	Milk 97 D	Milk 20 P	Milk 20 D
DVED	31	30	35	34	36	36	41	39	29	28	41	38	3	3	4	4	100	98	120	114	329	291	411	344
USA	12	12	14	14	8	8	10	9	15	12	22	17	<1	<1	<1	<1	35	32	47	40	239	183	297	227
EU15	8	7	8	8	17	16	18	16	8	8	10	9	1	1	1	1	34	32	37	35	37	38	39	40
FSU	5	6	5	6	17	16	18	16	1	2	1	3	1	1	1	1	10	12	11	13	7	7	10	8
DVING	27	27	51	52	47	47	79	80	29	29	64	67	8	8	13	13	110	111	206	213	266	295	441	508
LA	13	12	21	19	4	4	6	6	9	9	19	18	<1	<1	1	1	27	26	47	45	74	76	123	118
SSA	2	2	5	5	1	1	2	2	1	1	2	2	1	1	2	2	5	5	11	11	26	29	46	52
WANA	1	2	2	3	<1	<1	<1	<1	3	3	6	7	2	2	3	3	6	7	11	13	9	18	13	28
S. Asia	4	4	8	8	1	1	1	1	1	1	3	3	2	2	3	3	7	7	15	16	13	14	19	19
ESE Asia	6	6	14	16	42	41	69	71	14	14	35	37	2	2	4	4	64	64	121	127	143	159	240	291
WORLD	58	57	85	85	83	83	119	119	59	57	105	105	11	11	17	17	210	208	327	327	596	586	852	852

Source: Rosegrant et al., 2001; 1997 data from FAO, 2000, 2020 data estimated using IMPACT model.

[1] Amount traded is difference (P–D) between production and demand; negative values indicate amounts imported to meet demand. Three-year averages centered on year shown.

Table 9.8. Cereals fed to livestock in million MT and as % of total cereal consumed in the region[1]

Region	Cereals Fed, Million MT				% Fed of Total Consumed	
	1983	1993	1997	2020	1997	2020
Developed	465	442	425	493	59	60
Developing	128	194	235	432	21	26
East Asia	52	81	119	233	29	39
Southeast Asia	6	12	15	27	13	16
South Asia	3	4	3	6	2	3
Latin America	40	55	58	98	42	46
WANA[2]	24	29	36	59	28	30
SS Africa[3]	2	3	4	8	5	5
World	592	636	660	925	36	37

Sources: Delgado (personal communication), and Rosegrant et al., 2001.

[1]Cereals include wheat, rice, maize, barley, sorghum, millet, rye, and oats. Maize and other coarse grains are the principal grains fed to livestock. Three-year averages centered on year shown.

[2]West Asia and North Africa (WANA).

[3]Sub-Saharan Africa.

poultry and swine in Latin America, and East and Southeast Asia, and for poultry in West Asia and North Africa (see Table 9.8). According to Rosegrant et al. (2001), these trends in cereal feeding will accelerate through 2020, inducing significant imports of yellow maize and other coarse grains to feed livestock in West Asia and North Africa (30 million MT), and East and Southeast Asia (57 million MT). However, despite this growth, the percentage of cereals fed out of the total cereals consumed in developing regions will be much less than in developed regions. For South Asia and sub-Saharan Africa, the proportion of cereals fed to livestock will remain below five percent of total cereal consumption (Rosegrant et al. 2001).

Farmers are motivated by profitable prices for grain to produce cereals, soybeans, and other crops to meet human requirements, However, prices fall quickly whenever there are surpluses. Governments often respond with subsidies and price supports, and by storing surpluses. These are typically expensive, ineffective, and prone to creating perverse distortions in production and marketing. Grain-based feeding systems on the other hand can play a role in stabilizing grain prices by encouraging surplus production without financial loss in good years. Planting with the aim, in normal and good years, of producing grain surplus to human requirements results in smaller food deficits in poor crop years because, as yields fall, the livestock feeders switch to human-nonedible feeds (Fresco and Steinfeld 1998).

Some people believe that feeding grain to livestock is inevitably an inefficient use of the crop and object in principle to the quantities of grain consumed by them. However, the two major classes of food-producing livestock, ruminants and monogastrics, are distinguished by their digestive anatomy and the consequent differences in the feeds they primarily consume. The digestive capability of monogastrics (swine, chickens, turkeys, ducks, and others) is similar to that of humans, and therefore monogastrics tend to compete for human-edible nutrients from cereal grains, soybeans, cassava, sweet potatoes, and other crops. The multicompartment stomach of ruminants (cattle, buffalo, sheep, goats, camelids, and others) and the microbes carried primarily in the largest compartment, the rumen, enable ruminants to utilize high-fiber diets and non-protein nitrogen, and to produce essential micronutrients.

There is concern about grain fed to cattle and other ruminants because they are not as efficient in converting grain and other human-edible foods as are monogastrics. This is because the ruminant digestive system is adapted to high-fiber diets, and fewer slaughter stock are produced each year per breeding female, whose nutrient requirements must also be met. However, objections that ruminants are inefficient grain converters miss the point that ruminants primarily consume human-nonedible nutrients, forages, crop residues, agro-industrial by-products, etc. Even in stratified systems where beef cattle are finished on grain in feedlots, the nutrients used to maintain the beasts and to produce most of their growth to slaughter weight are from human-nonedible feeds (Fitzhugh 1998).

There are other concerns based on the important differences that affect the life cycle efficiency of the two classes of food producing livestock. Monogastrics, especially poultry, have a shorter generation interval and substantially higher reproductive rates per breeding female than do ruminants; these characteristics are major advantages for increasing meat offtake per year. On the other hand, ruminants produce

TABLE 9.9. Use of human-edible grain for meat production by different livestock species on a life cycle basis in developed and developing countries

Species	Grain Consumed per Meat Produced, kg/kg	
	Developed	Developing
Cattle	2.6	0.3
Sheep/goats	0.8	0.3
Swine	3.7	1.8
Chickens	2.2	1.6

Source: CAST, 1999.

almost all the milk consumed by humans, and this dual production of meat and milk improves their efficiency for food production.

As shown in Table 9.9, the conversion efficiencies of human-edible grain to meat, especially for ruminants in developing countries, are remarkably high because, on a life-cycle basis, most feed consumed by ruminants is from forages and crop residues, not from grain.

TRADE AND MARKETING OF MEAT, MILK, AND FEED GRAINS

Meat production and demand figures are presented by region in Table 9.7, with the volumes traded indicated by differences between production and consumption. Most consumption, especially of milk, will continue to come from local production because of the costs of refrigerated transport, which are increased by poor infrastructure and the costs of meeting health and safety standards for cross-border trade. However, the United States will have opportunities for exporting pork and especially poultry products, which are convenient ways of converting grain produced in the U.S. East and Southeast Asia will continue to import livestock products. In West Asia and North African countries, the demand for poultry meat is in excess of local production, but the patterns of grain imports to the region suggest that they will prefer to increase local production as a means of ensuring that slaughter procedures meet Islamic requirements.

Noteworthy are the expected imports by East, Southeast and West Asia, and North Africa from North America and Australia. The 15 European Union states plus New Zealand (not shown) will be major exporters, with all developing regions being importers, even South Asia. East and Southeast Asia will be the largest importer, but West Asian and North African imports will be larger relative to local production.

The feed grain data in Table 9.8 indicate that the U.S. will be the major exporter of maize, with East and Southeast Asia being major importers. In relation to the quantity produced within the region, the West Asia and North Africa region is relatively the largest importer of maize and especially of other coarse grains. Almost all of it is used for chicken feed, but some is used for intensive feeding of ruminants. Maize imports by sub-Saharan Africa will probably continue to be for human consumption.

CONTROL OF LIVESTOCK DISEASES THAT AFFECT THE POOR

Livestock diseases affect resource-poor small-holders and large-scale farmers in different ways (Perry et al. 2002). The diseases that most concern large-scale farmers are those that restrict trade, such as foot-and-mouth disease. Poor farmers on the other hand are much more concerned about diseases that cause death because, having only a few animals, the loss of a single animal is important to them. Small-holders are concerned about diseases that cause infertility, neonatal mortality, and reduced milk and meat yields, and those that prevent them from adopting more productive breeds, which are usually more susceptible to endemic diseases than the indigenous breeds.

Developing controls for livestock diseases that affect the poor requires public investment, because the low capacity to pay of the poor restricts the potential for commercial companies to get adequate returns on investment in research. But public research institutions have little capacity for production and marketing, so innovative public-private partnerships must be developed.

Livestock diseases have greater impacts on the poor than on other farmers because of the combination of a number of factors. In the tropical regions, where most poor farmers live, there are more diseases associated with freer movement of livestock, and less effective formal disease surveillance and control. Ironically, these factors indicate that disease control in developing countries, which have weaker technological capacities, has to depend more on technical solutions, such as vaccines, than on policies and regulations, such as quarantines.

Achieving effective control of livestock diseases is not only desirable for reducing poverty. Developed countries also have strong vested interests due to the linkages between human and animal diseases: Of the 1709 organisms that cause disease in people, 832 are zoonotic, naturally transmitted from animals to people; and of 156 emerging diseases, 114 are zoonotic. Examples include influenza, brucellosis, BSE, TB, and rabies.

Fortunately, advances in science are producing new approaches to controlling livestock diseases that promise to be safe and cost-effective. Advances in genetics and genomics are being applied to finding new vaccines. An example of this is the ongoing quest for a better vaccine for East Coast fever (ECF), which is estimated to kill one cow a minute in small-holder herds in Eastern and Southern Africa. Through collaboration with The Institute for Genomic Research (TIGR), the International Livestock Research Institute (ILRI) will soon obtain the complete genomic sequence of the causative organism, *Theilaeria parva*. This has greatly enhanced the chances of finding an effective, safe, and easy-to-administer vaccine.

Another approach that takes advantage of new science is the utilization of the genetic capacities for controlling parasites inherent in the "unimproved" breeds indigenous to developing countries. The Red Maasai sheep have, for example, genetic capacity to control internal parasites that in other breeds require treatment with anthelmintics (Baker et al. 2002). This is particularly significant, given the declining efficacy of the drugs and more stringent resistance to wide-scale application of drugs in livestock production. Genetic ability to tolerate trypanosomes, which cause the equivalent of sleeping sickness in livestock, has been identified in the N'Dama and other breeds of cattle in West and Central Africa (d'Ieteren et al. 2000).

New epidemiological tools, based on geographic information systems, are being developed that will enable governments to monitor the spread of diseases and institute the most cost-effective responses. However, as with the development of drugs and vaccines, this will require new public-private partnerships that will provide the means to conduct the necessary research with public funds, and will rely on the private sector for the things it does best, including but not limited to production and marketing.

THE NEED FOR PRO-POOR LIVESTOCK RESEARCH

The potential for agricultural research to improve rural income and food production was most dramatically demonstrated by the green revolution, which has staved off famines in India and Pakistan that were thought to be an inevitable consequence of South Asia's high population pressure, peasant agriculture, and harsh environments. It is less well known that, even though the green revolution focused on increasing grain production, there was at the same time a marked increase in the consumption of meat, milk, and fish. At world prices, the additional meat consumed had a market value three times the value of the extra cereals consumed (Delgado et al. 1999).

The growing demand for livestock products is robust, and production will increase to respond to this demand. Unlike the green revolution, demand, not technology, will drive the livestock revolution, which Delgado et al. (1999) refer to as the next food revolution. This situation raises a number of concerns about the interests of the poor. On past trends, it is concerning that the financial incentives for expanding industrial units may overwhelm equity and environmental considerations to the detriment of poor producers.

International public goods research in developing technologies and enabling policies will be required to respond to these concerns and build the competitive capacity of small-holder and pastoral livestock owners to take advantage of the growing demand for livestock products. If the required innovations are not forthcoming, the industry will be increasingly polarized: Large-scale industry will capture the benefits and the small-holders will suffer the consequences of inappropriate intensification. That would exclude millions of small-holders from an exceptional market-driven opportunity to pull themselves out of poverty.

Small-holders should also benefit from the important new applications for other technological advances, including geographic information systems, satellite imagery, climate forecasting, monitoring of climate change, and spatial analysis—for example, in mapping the incidence of poverty—determining recommendation domains, assessing impact, setting research priorities, and planning for development. Ensuring that these applications are relevant and validated will require strong links between laboratory and field-based research. Advances in informatics and multimedia applications will make it possible for research to have quicker and wider impact, because information and its products will be processed and distributed more efficiently. However, without publicly funded research, poor livestock producers in developing countries will not be beneficiaries of these advances, and the potential for animal agriculture to alleviate poverty will not be realized.

CONCLUSIONS

The perceptions in developed countries that livestock inevitably degrade the environment and compete with humans for grain, and that consumption of meat, milk, and eggs is bad for human health are wrong. In developing countries, livestock complement crop production or, as in the vast rangeland areas, may even be the only means of sustainable small-holder food production, income generation, and asset building. And increased consumption of animal-source foods would be beneficial to the health of millions of malnourished people and especially of their children.

In developed countries where per capita consumption of animal-source foods is already high, per capita consumption of poultry meat continues to increase, but at the expense of beef and, to a lesser extent, pork consumption. These trends in per capita consumption, combined with low population growth, indicate little growth in total consumption of meat in developed countries over the next two decades.

By contrast, developing countries are a growth market for livestock producers. Per capita demand for meat and milk will increase by 50 percent, with the largest growth in demand being for poultry meat. As incomes and urban populations increase, meat consumption generally increases as well; and both growth in income and urbanization are expected for most developing countries in the future. These per capita increases will be multiplied by the continuing growth in human populations so that total demand for meat and milk will more than double in developing countries.

The robust increased demand will stimulate increased production of meat and dairy products. Most of these must be produced in the same countries where they will be consumed because of the limited capacity in most developing countries for maintaining cold chains for processing, storing, and marketing meat and dairy products. Other factors favoring local production include cooking and taste preferences, lower quality and sanitary standards, and non-tariff barriers to imports. The net result is that, if the doubled demand for meat and milk is to be met, total production in developing countries must double as well.

Since only a little of the expected increase in meat and milk production in developing countries can come from expanding livestock numbers, the bulk of the increase must be derived from higher per head yields of meat and milk. This will be possible by reducing waste from reproductive inefficiency, by lowering losses from disease and parasites, and by providing better nutrition and health care through the use of better-adapted genotypes and more efficient processing and marketing.

Feeding of concentrates, including feed grains and oilseed meals, is expected to double in developing countries by 2020. Crops adapted to the tropics—such as sorghum, sweet potatoes, cassava, other coarse grains, roots, and tubers—that are now primarily subsistence food crops, will be increasingly used for livestock feed, creating market opportunities for farmers in developing countries. There will also be major opportunities for exports of yellow maize and other feed crops from North America, Australia, Brazil, and Argentina.

The combination of growing market demand and significant opportunities for improving productivity makes animal agriculture one of the most promising sectors for alleviating poverty in developing countries. If poor livestock producers are enabled to participate, by appropriate research and policy support, there will be significant opportunities for sustainably improving livelihoods through increased incomes secured by accumulated assets. The diversity of livestock products will also provide opportunities for improving equity because of the role of women in processing and selling livestock products. Children, and, therefore, the human potential for future economic development, will benefit from the increased availability of affordable animal-source foods.

REFERENCES

Baker, R.L., S.L. Rodriguez-Zas, B.R. Southey, J.O. Audho, E.O. Aduda, and W. Thorpe. 2002. Resistance and resilience to gastro-intestinal nematode parasites and relationships with productivity of Red Maasai, Dorper and Red Maasai × Dorper crossbred lambs in the sub-humid tropics. *Animal Science* (in press).

Bhalla, G.S., P. Hazell, and J. Kerr. 1999. *Prospects for India's Cereal Supply and Demand to 2020. Vision for Food, Agriculture, and the Environment*

Discussion Paper 29. Washington D.C.: International Food Policy Research Institute.

CAST. 1999. Animal Agriculture and Global Food Supply. Ames, IA: Council for Agricultural Science and Technology.

Cranfield, J.A.L., T.W. Hertel, J.S. Ealles, and P.V. Preckel. 1998. Changes in the structure of global food demand. American Journal of Agricultural Economics 80(5):1042–1050.

de Haan, C., H. Steinfeld, and H. Blackburn. 1997. Livestock and the Environment: Finding a Balance. Suffolk, U.K.: WREN Media.

Delgado, C., M. Rosegrant, H. Steinfeld, S. Ehui, and C. Courbois. 1999. Livestock to 2020: The Next Food Revolution. Food, Agriculture and Environment Discussion Paper 28. Washington, D.C.: IFPRI/FAO/ILRI.

d'Ieteren, G.D.M., E. Authie, N. Wissocq, and M. Murray. 2000. Exploitation of resistance to trypanosomes and trypanosomosis. In Breeding for Disease Resistance in Farm Animals. R.F.E. Axford, S.C. Bishop, F.W. Nicholas, and J.B. Owen, eds. Wallingford: CAB International, pp. 195–215

FAO (Food and Agriculture Organization of the United Nations). 1997. FAO Statistics Database. http:faostat.fao.org.default.htm. Accessed Summer 1998.

Fitzhugh, H.A. 1998. Competition between livestock and mankind for nutrients: Let ruminants eat grass. In Feeding a World Population of More than Eight Billion People—A Challenge to Science. J.C. Waterlow, D.G. Armstrong, L. Fowden, and R. Riley, eds. Oxford University Press in association with The Rank Prize Funds. pp. 223–231.

Fresco, L.O., and H. Steinfeld. 1998. A food security perspective to livestock and the environment. In Proceedings of the International Conference on Livestock and the Environment, 16–20 June 1997. A.J. Nell, ed. Wageningen, the Netherlands: International Agricultural Centre. pp 5–12.

ILRI. 2000. ILRI Strategy to 2010: Making the Livestock Revolution Work for the Poor. Nairobi, Kenya: International Livestock Research Institute.

LID (Livestock In Development). 1999. Livestock in Poverty-Focused Development. Crewkerne, Somerset, England: LID..

Neumann, C.G., and D.M. Harris. 1999. Contribution of Animal Source Foods in Improving Diet Quality for Children in the Developing World. Commissioned paper for World Bank, internal document and Web site article, February 18.

Perry, B.D., T.F. Randolph, J.J. McDermott, K.R. Sones, and P.K. Thornton. 2002. Investing in Animal Health Research to Alleviate Poverty. Nairobi, Kenya: ILRI.

Rosegrant, M., M.S. Paisner, S. Meijer, and J. Witcover. 2001. Global Food Projections to 2020: Emerging Trends and Alternative Futures. (Occasional Papers and Books). Washington, D.C.: International Food Policy Research Institute.

Seré, C., and H. Steinfeld. 1996. World Livestock Production Systems: Current Status, Issues and Trends. (FAO Animal Production and Health Paper No. 127) Rome: Food and Agriculture Organization of the United Nations.

Steinfeld, H. 1998. Livestock production in the Asia and Pacific region: Current status, issues and trends. World Animal Review 90(1):14–21.

Thornton, P.K., R.L. Kruska, N. Henniger, N. P.M. Kristjanson, R.S. Reid, F. Atieno, A.N. Odero, and T. Ndegwa. 2002. *Mapping Poverty and Livestock in the Developing World*. Nairobi, Kenya: ILRI.

NOTES

1. International Livestock Research Institute (ILRI) Nairobi, Kenya.

10

AGRICULTURAL AND ENVIRONMENTAL ISSUES IN THE MANAGEMENT OF ANIMAL MANURES

H. H. Van Horn[1] and W. J. Powers[2]

One of the major challenges facing society is how to improve environmental quality. Some think animal agriculture especially is challenged, but in reality all facets of society are interrelated. If one facet can utilize another's waste, that waste becomes a resource. Agriculture is uniquely able to utilize wastes that contain residual nutrients needed by animals, microbes, and plants. Agricultural soils contain natural biological systems for processing organic wastes and intestinal microbes, i.e., soil microbes, insects, and worms. And harvested plants or plant crops offer the primary mechanism to recycle elemental nutrients economically.

Regions of the world with intensive domestic livestock production have begun monitoring farms to ensure that losses of nutrients to the environment are minimized. Of special concern are losses of nitrogen (N) to groundwater and the atmosphere and phosphorus (P) to surface waters. Emissions of odorous compounds are regulated in all U.S. states through nuisance legislation and, in several states, through odor measurements taken at the property line. Additional regulations sometimes include standards for volatile emissions of ammonia (e.g., in The Netherlands), and methane emissions possibly could be regulated in the future.

Production of animal foods permits utilization of some of the world's fibrous feed resources that are indigestible by humans and use of many by-products of the human plant food chain. Animal and human manures are unavoidable. The objective of this paper is to review many of the manure-related issues and some of the environmentally accountable manure management alternatives. The outline chosen is the following:

1. Nutrient management to ensure water quality
2. Air quality issues

3. Partitioning and processing alternatives
4. Discussion: debits and credits associated with animal manures

The future likely will include other manure issues that may be emerging, e.g., pathogens, antibiotic residue, and endocrine disruptors. Food safety is critical and, while we have a safe food supply in the U.S., we must protect our food from any source of threat. Potential development of antibiotic resistance as a result of consumption of animal products warrants further scrutiny by the industry (Manie et al. 1999; Arvanitidou et al. 1998; D'Aoust et al. 1992). Limited and unduplicated data suggest that antibiotics are emitted from animal facilities (Zahn et al. 2001). The effects of endocrine disruptors on fish and bird populations have been widely studied and land application of animal excretions might be a potential source of estrogens (Finlay-Moore et al. 2000).

NUTRIENT MANAGEMENT
TO ENSURE WATER QUALITY

Predicting Manure Amount Composition

Manure is what is excreted in the form of urine and feces after the animal has utilized as much as possible of the absorbed (digested) nutrients in the diet. Apparent digestibility is the difference between amounts fed and amounts recovered in feces. Previous nutrition research has given us good estimates of apparent digestibility of ingredients that can be combined to estimate total ration digestibility and, inversely, fecal amounts. Amounts in urine are residuals of metabolic activity to maintain life; produce offspring; gain body weight; and produce milk, eggs, wool, etc. Manure amounts of elemental nutrients (feces plus urine) are functions of dietary nutrient inputs and nutrient outputs in products produced. Farmers, often with help from consulting nutritionists, formulate rations that give known nutrient intakes and produce known nutrient outputs in animal food products. Input-output data were utilized by Powers and Van Horn (2001) to estimate fresh-manure nutrient excretions of confined food animals consuming typical rations for the U.S. inventory of food animals in 1997. Table 10.1 presents the estimated amounts of manure dry matter (DM), nitrogen (N), phosphorus (P), potassium (K), and fertilizer value of manures from confined food animals in the U.S. in 1997. Amounts in daily or life-cycle manure, predicted from nutritionally based input-output models less expected losses of N, were ex-

Table 10.1. Estimating manure quantities, nutrient concentrations, and annual fertilizer resources available from confinement animal feeding

	Units	Numbers Below Expand from Daily Averages to Years			Numbers Below Based on Life Cycle Grow-Out		
		Dairy Cows	Beef Steer	Hens	Broilers	Turkeys	Hogs
Animals/day or animals/grow-out	No.	1	1	1000	1	1	1
Dry matter (DM):							
Input: kg DMI	kg	21.8	9.5	94.8	3.9	24.3	298.0
Output (feces) = kg DMI—(digestibility × DMI)	kg	7.6	1.9	16.1	0.6	4.4	53.6
Output (urine) = 5% of DMI	kg	1.1	0.5	4.7	0.2	1.2	14.9
Total DM output = feces + urine DM =	kg	8.7	2.4	20.9	0.8	5.6	68.5
Manure DM output, % of input =	%	40.0	25.0	22.0	21.0	23.0	23.0
Manure DM/yr or grow-out period =	kg	3183	870	7612	0.82	5.59	68.54
Estimated DM% of fresh manure	%	14	16	20	20	20	16
Yearly manure (wet)	kg	22734	5435	38,062	4.1	28	428
% of manure collected (% of time in collectible area)	%	100	100	100	100	100	100
Cubic feet of wet manure stored/day or grow-out	cu ft	2.2	0.5	3.7	0.15	1.0	15.1
Cubic ft = (lb wet manure/day or grow-out)/(8.346 lb/gallon × 7.48 gallons/cu ft)							
N kg excreted yearly or per animal grow-out	kg	167	59	612	0.068	0.551	5.195
P kg excreted yearly or per animal grow-out	kg	30	10	187	0.018	0.130	0.869
K kg excreted yearly or per animal grow-out	kg	81	27	186	0.019	0.124	1.736
Excreted N recovered (40% of excreted)	kg	67	23	245	0.027	0.220	2.078

continued

Table 10.1. Continued

	Units	Numbers Below Expand from Daily Averages to Years				Numbers Below Based on Life Cycle Grow-Out	
		Dairy Cows	Beef Steer	Hens	Broilers	Turkeys	Hogs
Manure N% of DM (excreted)	%	5.24	6.73	8.04	8.34	9.86	7.58
Manure P% of DM (excreted)	%	0.93	1.19	2.45	2.22	2.32	1.27
Manure K% of DM (excreted)	%	2.53	3.08	2.45	2.33	2.22	2.53
N% of DM if 40% of N recovered, 20% DM reduction	%	2.62	3.37	4.02	4.17	4.93	3.79
P% of DM if 20% DM reduction	%	1.16	1.48	3.06	2.77	2.90	1.59
K% of DM if 20% DM reduction	%	3.16	3.85	3.06	2.91	2.78	3.17
N:P ratio predicted in recovered manure	ratio	2.25	2.27	1.31	1.50	1.70	2.39
U.S inventory (yearly) or numbers slaughtered/yr:							
Cows, steers, or hens/d or animals slaughtered/yr	millions	9.35	13.21	297.48	7,598.00	301.38	92.39
U.S yearly manure tonnage (extrapolated from above):							
DM recovered if 20% DM reduction	1000s of tons	23807	9190	1812	4991	1348	5066
N recovered (40% of excretion)	1000s of tons	623.7	309.4	72.8	208.0	66.4	192.0
P recovered	1000s of tons	277.0	136.2	55.5	138.3	39.1	80.3
K recovered	1000s of tons	752.8	354.0	55.4	145.1	37.4	160.4
$ value (billions)	1031	504	141	370.5	109	287	Yearly
DMI of USDA number of animals	1000s of tons	74,398	45,950	10,293	29,708	7,324	27,532
Estimated corn consumption (DM) by USDA number	1000s of tons	22,319	27,570	7,720	22,281	5,493	22,026

Reproduced with permission from Powers and Van Horn (2001).

trapolated to yearly amounts for national inventory numbers of animals. Manure value was calculated from amounts of N ($.66/kg), actual P ($1.34/kg), and actual K ($.33/kg). Some error is certain, but Table 10.1 data gives an approximation of manure nutrient excretions by food animals in confinement and the fertilizer resource in manures that potentially can spare use of commercial inorganic fertilizers.

Nutrition-based input-output models predict the amounts of nutrients in fresh manure excretions more accurately than collections from animal pens because of the dynamic state of manure after excretion. For example, usually 40–50 percent of the excreted N will occur as urea or uric acid in the urine component for ruminants (Tomlinson et al. 1996) and up to 75 percent for swine (ASAE 1994; Carter et al. 1996). Urease enzyme, which is of bacterial origin and is nearly ubiquitous in the environment, converts urea and uric acid N to ammonia that can be lost to the atmosphere. Also, anaerobic digestion that begins in the large intestine of animals before feces are voided continues after excretion if environmental conditions permit. Or a shift to oxidative fermentation may take place, e.g. composting and degradation on soil surfaces. Either way, volume reduction takes place as carbon compounds are emitted to the atmosphere, primarily carbon dioxide from aerobic degradation and methane, carbon dioxide, and odorous volatiles from anaerobic degradation. Additionally, variation in composition of manure collected occurs because physical separations may take place in animal pens and within the manure management system. For example, urine or urine plus added water may drain away from fecal residues, thus making solids collected from animal pens different from original excretion and some systems deliberately separate solids. Measures of excreted and collected amounts are both important because differences give estimates of losses that occurred after excretion.

The major advantage of showing that manure nutrient production is a function of ration and performance is that it is easy to visualize the importance of ration management to minimize excretions. Eliminating dietary excesses, where they exist, is the easiest and first step to take to reduce on-farm nutrient surpluses. For example, supplementation of limiting amino acids permits reduction of total dietary protein and, hence, reduces excretion of N (e.g., Carter et al. 1996). For every percentage unit that dietary protein can be reduced, predicted N excretions (Table 10.1) would be reduced by 8–10 percent (average of 8.5 percent), which would reduce manure N to manage nationally (assuming 40 percent recovery of excreted N) by 124,000 metric tons actual N. Because manures become more and more P-rich as more N volatilizes, ration management to minimize dietary P concentrations will become especially important. Utilization of phytase enzymes in poultry and swine rations makes organic P available to those animals and permits reduction of dietary P (Yi et al. 1996; Kornegay et al. 1996; Carter et al. 1996). Phytase enzyme is inherent in ruminant rations because ruminal microorganisms provide it, so dietary addition is not necessary (Morse et al. 1992). However, dairy and beef producers often feed more dietary P than animals require (e.g., NRC 2001, for dairy cattle) and, thus, excretions can be reduced by dietary reduction. For example, if ration P as percent of dry matter were reduced .1 percent in all rations used to derive Table 10.1, P excretions for different species would be reduced by 19–35 percent (average of 29.5 percent) and the amounts of P in manures from confined livestock operations nationally could be reduced by 193,000 metric tons actual P. Care must be taken to not overinterpret the implications of reducing dietary nutrient content. Sufficient nutrients must be maintained in the diet to meet animal biological functions and performance goals. Therefore there are lower limits for each nutrient, below which detrimental effects to the animal would be observed. Obviously, complete elimination of nutrient excretion is not possible.

LOSSES IN STORAGE AND FROM SOIL

Nutrient losses vary tremendously, especially for N, from farm to farm depending on type of storage and handling systems used. Losses of N due to volatilization of ammonia from manure are likely and variable. Use of anaerobic lagoons and any form of long-term storage where aerobic or anaerobic degradation occurs contribute further to atmospheric N losses. In most animal manure management systems, more than 50 percent of the excreted N is lost to the atmosphere before it can be recycled for fertilizer use by plants (Van Horn et al. 1996, 1997; Patterson and Lorenz 1996). In addition to ammonia volatilization, airborne losses include denitrification, an anaerobic process in which soil bacteria convert nitrate to nitrous oxide and then to N gas. Denitrification is a major process of N removal from soils. However, enhancing N loss by denitrification on farms is not feasible, and nitrous oxide, an intermediary gas that may be emitted in the process, has been implicated as a contributor to global warming. Moore and Gamroth (1995) found denitrification losses from 7–28 percent of applied N in wet or poorly drained soils, and Newton et al. (1995) observed denitrification losses of 20–43 percent. Losses of other nutrients should be small, comparatively.

Whole-Farm Nutrient Budgeting

Manure N and P have been of most concern environmentally with respect to water quality. Most often, N has been chosen as the most critical risk, because excess application of N to fields leads to N conversion to nitrate in soil and, thus, leaching of nitrate to drinking water sources is possible. Budgets consider manure nutrient excretions, crop production requirements, expected losses of nutrients, and, if necessary the amounts of manure nutrients that must be exported off-farm for use as fertilizer elsewhere (Van Horn et al. 1997, 1998; Janzen et al. 1994).

The nutrient budgeting process is most used for farms that produce sufficient crops to utilize much of the manure nutrients on-farm. Most dairies, for example, have potential to utilize all of the manure on-farm if manure applications are permitted based on N budgets, thus accepting overapplication and soil accumulation of P. If, however, manure applications were limited to agronomic P recommendations, most dairies would not have sufficient crop production and, thus, would be forced to export manure nutrients. If they did have sufficient crop production acreage for P budgeting, production would need to be directed partially to feed grains or to forage for sale off-farm, because manure P would fertilize more forage production than their livestock could consume. The first step to cope with P budgets is to reduce dietary P to the fullest extent possible. The benefits will be reduced ration cost and reduced manure P excretion.

Many intensive livestock producing farms—perhaps most—export most or all of the manure nutrients and, thus, the manure budgeting process applies primarily to the user of the fertilizer they export. For example, poultry producers, beef cattle feedlots, many swine producers, and some dairy farmers specialize in animal production only, and therefore must plan to export most of the manure produced to be used in crop production elsewhere.

Grazing Budgets

One reason grazing budgets are so important is that manures exported off farms often are used on pastures for beef production, especially poultry manures in intensive poultry production areas in the South and

Southeast. Total nutrient budgets for pasture conditions seldom show excessive nutrient applications from manure distributed by the animals grazing. Reviewing pasture budgets for dairy and beef cattle developed by Van Horn et al. (1994, 1996) shows that supplemental N fertilization from commercial fertilizer, or from N fixation via legumes, will be necessary to prevent N depletion and to maintain forage production even with dairy cows fed more than half their DMI from imported concentrates. The case for P, however, is one showing that accumulation is likely, even when no commercial fertilizer P is applied to the pastures. Therefore, use of P fertilizers on pastures should be avoided after soil storage levels reach desired fertility levels, and dietary P concentrations should be held to the minimum needed for optimum performance. When considering application of imported manures on pastures, a nutrient budget must first be developed that considers the grazing animals and nutrient retention in those animals. Imported manures can be used in the pasture setting only as a source of supplemental nutrients to meet pasture growth needs beyond that provided by the grazing animals such that a balance is maintained.

Air Quality Issues

Odor emanating from animal manures and their perceived presence are the source of much friction between non-farm rural residents and food-animal producers. Additionally, ammonia and methane emissions are being scrutinized carefully to determine if regulatory oversight of these units will become necessary.

Odor Control

Volatile odorous compounds emitted from manure during transport, storage, treatment, and disposal have become an acute public relations problem for animal agriculture. Odorous compounds usually are present at such low levels (some in the parts-per-trillion range) that they are not considered by most to be toxic at the concentrations found downwind of livestock production facilities. Thus, the problem depends largely on subjective factors, how much the smell bothers people or the "nuisance value" of the odor. Often, flies add to an odor nuisance, and the two problems may be difficult to separate in the minds of complainants. Odor complaints range from casual comments, indicating displeasure, to major lawsuits and court orders that have the potential to terminate the affected food-animal enterprises (Sweeten and Miner 1993).

The U.S. Environmental Protection Agency does not regulate odors. However, odor is regulated as a nuisance in every state in the U.S. States regulate odors as a public nuisance through air pollution control and public health protection statutes. Definitive measures of odor are needed to evaluate the extent of an odor nuisance. One problem is in defining what to measure. Manure odors are caused principally by intermediate metabolites of anaerobic decomposition. Over 75 odorous compounds, in varying proportions, were long ago identified around manure storage areas (e.g., Sweeten 1988), and the number continues to expand as methodology permits identification of more metabolic intermediates. Phenols, volatile fatty acids, and sulfides are thought to be the major odor-causing compounds, with total phenols the largest factor associated with odor intensity, as identified by human panelists, of dairy manure products removed from an anaerobic environment (Powers et al. 1999). However, typical chemical analyses measure concentrations of only a small number of constituents in the complex mixture that contribute to the odor people identify by smell. Correlation of concentrations of individual or multiple analytes with human panelist evaluations have been too low to accept as an accurate assessment of odor or degree of nuisance. Consequently, a key facet of odor measurement technology utilizes sensory methods (i.e., using the human nose) with on-site scentometers or employs sampling methods to bring odorous air or fabric samples, placed on-site to absorb odors, back to a laboratory for panel evaluation (human assessment). From a regulatory perspective, the challenge is determining what intensity and duration of an odor constitutes a nuisance. No research is available at this time to help assess these factors. Typically, policy has used as a guideline that some odor is acceptable in an agricultural area. However, public pressure continues to challenge the animal industry to adopt odor control measures.

Odor-control methods fall into three broad categories:

- Control of odor dispersion
- Odor capture and treatment
- Treatment of manure

Control of odor dispersion is primarily a function of site selection, system design and construction, and manure handling methods (e.g., sprayfield application may provoke more odor drift than soil incorporation). Odor capture and treatment methods include containment, wet scrubbing, packed-bed adsorption, and soil filter fields and vegetative filter strips (tree borders at the property line) (Sweeten 1988). Manure treatment methods include anaerobic digestion, aeration, and biochemical treatment. Commercial chemical and biological manure and feed additives available to date have not reliably reduced odors.

Ammonia Emissions

The major source of ammonia is urea from urine, or uric acid in the case of birds, which can be easily converted to gaseous ammonia (NH_3) by bacterial urease, found naturally throughout the environment. In animals and birds, urea and uric acid are the product of protein breakdown, respectively. In aqueous solution, NH_3 reacts with acid (H^+) to form an ion (NH_4^+), which is not gaseous. Thus, the chemical equilibrium in an acid environment promotes rapid conversion of NH_3 to NH_4^+ with little loss of NH_3 to the atmosphere. However, most animal manures, lagoons, and feedlot surfaces have a pH >7.0, making H^+ scarce and, thus, permitting rapid loss of ammonia to the atmosphere. As a consequence, N losses from animal manures can easily reach 50–75 percent, most as NH_3 before NH_3 is converted to NO_3^- through nitrification.

In Europe, atmospheric ammonia concentrations have become a public concern due to their perceived contribution to acid rain resulting in the degradation of forests (e.g., Apsimon and Kruse-Plass 1991). Consequently, European livestock and poultry operations are being required to utilize practices to minimize ammonia losses to the atmosphere as an alternative to reducing animal numbers.

Ammonia can be toxic to cells, and the potential exists for plant damage if excessive ammonia is released after manure application. Also, excessive ammonia concentrations in closed buildings used to house large numbers of animals may reduce animal performance and may be a potential health hazard for workers. Atmospheric ammonia has been known to cause blindness in chicks and turkey poults. Thus, it is important to avoid ammonia buildup where animals are confined and people work. Beyond occupational exposures to ammonia is a more recent concern with community exposure to emissions from animal agriculture. In 2002, the state of Iowa also adopted standards for hydrogen sulfide and ammonia at a residence neighboring a CAFO. Odor standards were considered but not adopted because of a tenuous association in the literature between exposure to livestock odors and negative health impacts. Under Senate File 2293, the state Department of Natural Resources will conduct a comprehensive field study to monitor the level of airborne pollutants—particularly hydrogen sulfide, ammonia, and odor —emitted from animal feeding operations in the state. Following, the department may develop comprehensive plans and programs for the abatement, control, and prevention of airborne pollutants if the baseline data from the field study demonstrates to a reasonable degree that airborne pollutants emitted by an animal feeding operation are present, downwind, and at levels commonly known to cause a material and verifiable adverse health effect. Enforcement of an air quality standard will not occur prior to December 1, 2004. Unlike other states, any air quality standard established should rely on measurements taken at the residence because this regulation is based on human health impacts of emissions from animal feeding operations and will reflect exposure levels consistent with negative impacts under chronic exposure (>365 d). Therefore, measurements must be taken where chronic exposure conditions will occur. A report released by the University of Iowa and Iowa State University in February 2002 adopts ATSDR recommendations for hydrogen sulfide and ammonia chronic (>365 d) exposures. These numbers correspond to 15 ppb hydrogen sulfide and 150 ppb ammonia when both compounds are present. The committee did not reach consensus on an odor recommendation. The Minnesota Department of Health has begun to consider adoption of inhalation standards for hydrogen sulfide that are health-based rather than nuisance-based standards.

Ammonia emissions are indirectly related to particulate matter. Particulate matter poses perhaps the greatest challenge for animal agriculture. Direct emission sources of PM_{10}, the coarse particulates, arise from, primarily, combustion processes (EPA 1998). Direct emissions of $PM_{2.5}$, respirable particulates, are also primarily the result of combustion processes (EPA 1998). In addition to direct emissions, secondary processes, whereby SO_x, NO_x, and NH_3 react in the atmosphere to form ammonium sulfate and ammonium nitrate fine particles, contribute to as much as half of the $PM_{2.5}$ measured in the U.S. The EPA estimates that 86 percent of the national ammonia emissions are from miscellaneous sources that include livestock and fertilizer (EPA 1998). Livestock agriculture accounted for 83 percent of all emissions in the miscellaneous category with fertilizer application comprising the remainder. In 1997, the Clean Air Act was amended and a new criteria pollutant was proposed, $PM_{2.5}$. However, a 1999 federal court ruling blocked the implementation of this addition citing an unconstitutional delegation of legislative power. Despite this setback, regulation of $PM_{2.5}$ is in the foreseeable future for the U.S.

Most volatilized ammonia is dissolved in water vapor in the lower atmosphere and washed back to earth by rainfall. During this process, ammonia neutralizes the acidity of the rainwater. In industrial regions with somewhat acidic rainfall, e.g., Pennsylvania, neutralization is one potential benefit of ammonia release (Elliott et al. 1990). Most of the N that is redeposited from the atmosphere is to nonagricultural areas such as forests that may benefit from increased soil fertility.

METHANE EMISSIONS

Methane emissions from animal production systems do not present an odor-related problem because methane is odorless. The concern with methane relates to its role as a greenhouse gas and as a potential contributor to global warming (EPA 1989; Johnson et al. 1992; Waggoner et al. 1992). Carbon dioxide is the most abundant greenhouse gas and is being added in the greatest quantity; carbon dioxide is expected to cause about 50 percent of the global warming occurring the next half century. Methane is generally held to be the second most important greenhouse gas and is expected to contribute 18 percent of future warming (Johnson et al. 1992). Indeed, molecule for molecule, methane traps 25 times as much of the sun's heat in the atmosphere as carbon dioxide. Thus, methane is estimated to contribute 18 percent of future warming from <1 percent of the total greenhouse gas emissions. In addition to warming effects, increased atmospheric methane will likely be detrimental by increasing ozone in the stratosphere, which shields the earth from harmful solar ultraviolet radiation.

The origin of methane produced by animals is microbial action in the gastrointestinal tract, which occurs to varying degrees in all animals. Major fermentative digestion, allowing utilization of fibrous dietary components, occurs in ruminants. This, coupled with large body sizes, high dry matter intakes, and animal numbers, results in 95 percent of animal methane emissions arising from ruminants.

Energy losses resulting from methane production in the rumen are usually 6–8 percent of gross energy intake in cattle consuming high forage diets; greater losses occur when forage is of low digestibility (EPA 1989). Dairy cows fed moderately high concentrate diets convert about

5 percent of their gross energy intake into methane and belch this methane into the atmosphere. The methane produced by animals and animal manures constitutes about 16.4 percent of estimated annual methane emissions (from Johnson et al. 1992), which translates roughly to 2.9 percent of the estimated contribution of all greenhouse gases to global warming (i.e. 16.4 percent of 18 percent, the projected contribution of all methane sources). Although an extremely small part of the total, the feasibility of reducing animal-related methane emissions is being investigated.

PARTITIONING AND PROCESSING ALTERNATIVES

Large food-animal producing units vary greatly in land resources that are available on the same farm to produce crops that will consume the manure nutrients. Many farms must export nutrients. When possible to do so, exporting fresh manure or solid manure that is scraped from lots usually is the most efficient method of export. Frequently, however, there are nuisance problems associated with hauling manure off-farm to be spread on other cropland. Odors and flies associated with manure are the most common nuisance complaints. Therefore, some form of pretreatment may be necessary to facilitate handling of manure on-farm to prepare for export of needed amounts of nutrients off-farm. The methods most often utilized include composting, solids removal, and composting of selected solids.

SOLIDS SEPARATION BY SCREENING OR SEDIMENTATION

The primary reason for solid-liquid separation is to produce two products that are more manageable than the original slurry. When cropland is available nearby, often the liquid can be more readily applied to croplands through an irrigation system, and the solids can also be spread on croplands or more easily exported off-farm than the wet product that precedes solid-liquid separation. Other benefits from solid-liquid separation include the following:

• Removal of fiber, which may serve another productive use such as bedding material, use in compost, or, after undergoing further processing, use in building materials or feed
• The possibility of odor reduction in the holding pond or lagoon, as well as increasing the surface life of the storage area

Removal of solids allows a slower sludge accumulation rate, thereby reducing the loading rate of a lagoon or holding pond. Reduced loading improves organic matter digestion, maintains useful volume and designed retention times much longer before cleanout is necessary, and reduces odors in effluent.

Composting

A significant amount of dried manure, composted manure, composted solids separated from manure, or some combination of these is bagged and sold as organic fertilizer or a soil amendment. Composting is a logical way to process wet manure solids (but not slurries unless the slurry can be added to drier materials) when animal producers must create a product that easily moves off-farm and is stable enough so that suburban users or agricultural users near urban centers will want to use it. Composting is relatively costly and labor-intensive, and much of the most valuable constituent, N, is driven off to the atmosphere during processing. Therefore, dairies and feedlots usually consider the process only if a marketable product that will help them remove excess nutrients —especially P—from the farm can be generated even if income does not equal processing and handling costs. Several advantages include the following: aerobic composting reduces volume and converts biodegradable materials into stable, low-odor end products; thermophilic temperatures of 54°C (130°F) to 71°C (160°F), achieved in this process, kill most weed seeds and pathogens. If moisture content is too high, anaerobic conditions develop and odorous compounds can be produced. Obviously, high-quality compost has much greater value in horticultural and urban markets than simply assessing N, P, and K value.

Manure as an Energy Source

The gross energy in most feedstuffs consumed by animals varies little, but there may be large variation in the amount of energy that can be digested and utilized by animals. For example, cellulose is undigested by nonruminants; that and undigested carbon sources in any ration pass out of the digestive tract as a potential energy resource. Lignin in feeds will burn but is unavailable to animals, even ruminants, and is relatively unavailable to anaerobic microbes. Figure 10.1 (from Van Horn et al. 1994) shows how a dairy cow producing 22.7 kg milk/day partitions DM, volatile solids (VS = organic matter), C, and Mcal[3] of gross energy (GE) during digestion and metabolism. In this example, approximately

Flow of DM and OM (volatile solids)

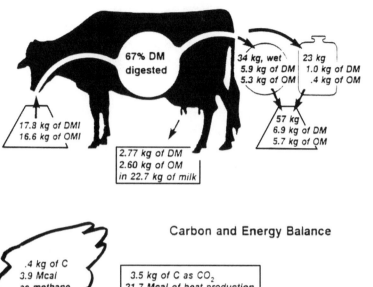

Carbon and Energy Balance

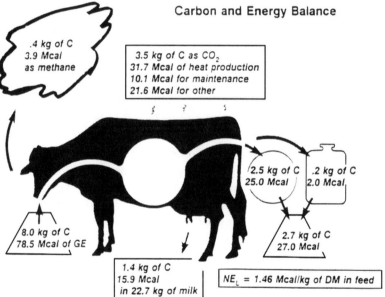

FIGURE 10.1 Estimated daily flow of DM, organic matter, energy, and carbon through typical Holstein cow (typical of year-round amounts when extrapolated to 365 d). Reproduced with permission from Van Horn et al. (1994) (DM = dry matter, DMI = DM intake, OMI = organic matter intake).

5 percent of the gross energy consumed is eructated (belched) from the rumen as methane, 20 percent secreted into milk, 40 percent lost as heat (maintenance energy plus heat of fermentation), and 35 percent excreted in manure, of which approximately 93 percent is in feces. These percentages differ somewhat between species and between rations within species.

BURNING

Manure in a relatively dry form, may be burned directly as fuel. The use of manure as fuel is an ancient practice that is still utilized in many developing countries. In the 1800s, westward pioneers crossing the prairies of the U.S. used buffalo chips for fuel. The gross energy of manure, for burning, varies little between species and most of that variability can be explained by differences in ash content. Manure energy content is similar to the dairy manure represented in Figure 10.1 that was suggested to contain 27.0 Mcal in 6.9 kg DM. Thus, 1000 kg dairy manure DM would contain 27.0 Mcal/6.9 kg DM × 1000 kg DM × 4.184 GJ/1000 Mcal = 16.4 GJ/metric ton DM or, at 90 percent DM (air-dry), 14.7 GJ/metric ton (about 55 percent of the energy value of coal). Manure from less digestible rations will contain similar energy to this to contribute fuel to a fire but, if low digestibility was due to lignin content, much of that energy may not be readily converted to combustible gas (methane) by fermentation. Energy in manure from non-ruminants, however, contains little lignin and the undigested cellulose and pectins are digestible by microorganisms and, hence, may be utilizable for digestion by rumen microbes or for digestion and methane production by microbes in anaerobic digesters.

The first large-scale resource recovery project in the world to burn cattle manure as a fuel to generate electricity was in the Imperial Valley of southern California (Van Horn 1990, personal communication with Western Power Group and National Energy Associates, El Centro, CA). It was designed to utilize manure from the many beef cattle feedlots in the valley. Poultry litter is being utilized for fuel extensively in the United Kingdom and up to 80 percent of the litter produced is expected to be burned in the near future (Badger et al. 1995). While caloric value of poultry litter depends on moisture, Dagnall (1993) indicates for air-dried samples it is typically 13.5 GJ/metric ton. This is slightly less than the estimate above of 14.7 GJ/metric ton for 90 percent DM dairy manure (represented in Figure 10.1), probably because of greater ash content.

Small-scale recovery units can be used on-farm as a source of heat and power, e.g., Dagnall (1993) reported use of a 1.6 MW (megawatt) biomass precombustor, retrofitted to an existing boiler, supplying steam to a central heating system servicing the poultry houses; electricity was generated by means of a steam engine coupled to a generator. The project demonstrated an ability to dispose of poultry litter reliably, safely, and efficiently without the aid of supplementary fuels. The U.K. also is utilizing a large-scale electric power plant that receives poultry litter from a 40 km radius around the site (Dagnall 1993). Finding users for large amounts of steam that might have been generated in a combined steam and electric power plant was difficult in the large-scale plant, so electric power was accepted as the most practicable option. When manure is burned, the ash nutrients still need to be managed accountably.

FEED

Animal manures sometimes serve as a viable feed source, usually manure from nonruminants serving as a ruminant feed (Fontenot 1991). Manure from ruminants is lower in digestible energy value than that from nonruminants due both to the inclusion of forages in the diet and because it already has passed through an anaerobic microbial fermentation in the rumen. Based on in vitro cell wall digestibility, Smith and Wheeler (1979) rank the feeding value, from high to low, of wastes from various species as follows: young poultry excreta, deep litter of young poultry, laying hen excreta, hog feces, hog and layer manure solids, and cattle excrement. Cell wall digestibility of wastes from all-concentrate-fed cattle is similar to cell wall digestibility of nonruminant wastes.

Animal manures must be processed prior to feeding in order to destroy pathogens, improve storage and handling characteristics, and maintain or enhance palatability to obtain adequate intakes (Fontenot 1991). Processing normally involves heating as a means of destroying any pathogens present. The most commonly used methods of processing include heated drying, composting or deep stacking, and ensiling. Each method has its own advantages and disadvantages (Arndt et al 1979). While drying requires input costs of energy to heat and results in nitrogen loss, the equipment, space, and technology investment is minimal relative to other methods, such as composting. Ensiling manures in combination with forages has been done successfully and offers an advantage in that additional feedstuffs, such as molasses, can be added that will increase palatability. Chemical treatments—such as formalin to destroy pathogens and reduce odor and sodium hydroxide as a solvent to hydrolyze fiber and extract protein—also have been used as a means of processing animal manures.

Feeding poultry wastes to ruminants is the most researched refeeding practice and is most successful with broiler litter fed to beef cows and growing cattle (Fontenot 1991). Layer waste has higher ash content and lower energy.

METHANE GENERATION

Anaerobic digestion (or degradation) is initiated in the lower digestive tract of animals and continues in feces droppings, manure piles, and storage facilities. The digestion consists of a series of reactions that are catalyzed by a mixed group of bacteria that hydrolyze polymers—such as cellulose, hemicellulose, pectin, and starch—to oligomers or monomers that are then metabolized by fermentative bacteria with the production of hydrogen (H_2), carbon dioxide (CO_2), and volatile organic acids such as acetate, propionate, and butyrate. The volatile organic acids other than acetate are converted to methanogenic precursors (H_2, CO_2, and acetate) by the syntrophic acetogens. Finally, the methanogenic bacteria produce methane from acetate or H_2 and CO_2. Generally, the final products of microbial degradation of carbonaceous material in an anaerobic ecosystem are methane and carbon dioxide, which are odorless. Odorous compounds are intermediates in anaerobic digestion; thus, when anaerobic digestion is disrupted and not allowed to degrade the odorous intermediates on to methane and CO_2, odor is emitted.

As a manure treatment, anaerobic digestion stabilizes manure against further decomposition, reduces odorous intermediates, reduces volume of manure as CO_2 and methane are released, and maintains most of the nutrient value of the manure (much now in bacterial cells) for land application. Ammoniacal nitrogen is increased and, if not lost to the atmosphere, is readily available for nitrification and crop uptake. A further benefit of anaerobic digestion is the potential to capture and utilize the methane gas that was produced as an alternate energy source.

In general, increasing the operating temperature of an anaerobic reactor produces an increase in the rate of digestion (Varel et al. 1977). Upper limits to increased rates of digestion due to higher operating temperature appear to occur at about 60°C (Varel et al. 1977). Thermophilic conditions also have been found to result in a higher concentration of methane in the biogas produced, e.g., 18 percent increase in methane production (Mackie and Bryant 1981) and a 30 percent increase in gross energy production (Pain et al. 1988).

How much of the potential energy value in manure organic matter (volatile solids, VS) that can be recovered by means of methane capture from anaerobic digestion is of primary economic importance. Fabian (1989) estimated biogas production of .35 L of biogas/g of VS input when hydraulic retention times (HRT) are >20 d. Thus, with VS production as in Figure 10.1 (Van Horn et al. 1994), biogas potential for the typical cow was estimated at

$$5.7 \text{ kg VS/d} \times 1000 \text{ g/kg} \times .35 \text{ L/g VS} = \sim2000 \text{ L biogas}$$

Considering the biogas to contain 60 percent methane and methane to contain 8.9 kcal/L (Fulhage 1980), daily production of megacalories would be 8.9 kcal/L $\times$ 1 Mcal/1000 kcal $\times$ .60 $\times$ 2000 L = 10.7 Mcal or 40 percent of the potential 27.0 Mcal in the manure. Manure from poultry and swine is more fermentable than dairy or fattening beef cattle (Hobson 1984).

Hydraulic retention time of influent material determines total volume and, hence, it has the largest influence on fixed costs of a digester system. Manure collected from swine facilities is often stabilized with a 10–12-day retention in conventional methane generation systems (Hobson 1990; Summers and Bousfield 1980). Hobson (1990) recommended a HRT of 20–25 days at mesophilic temperatures for cattle waste because the waste already has been subjected to bacterial degradation in the rumen, and that which remains is recalcitrant material.

For ruminant wastes, some form of pretreatment of the fiber present can be beneficial in reducing HRT. Powers et al. (1997) and Rorick et al. (1980) observed an increase in methane production (L/g VS fed) when some cattle manure solids were removed by a commercial separator, supporting the suggestion that a significant portion of cattle manure is lignified, indigestible material. Methane production remained higher when digested at 60°C versus 40°C for both the whole manure and the separated manure, but, at both temperatures, separation resulted in better utilization of the solids present in the manure (Rorick et al. 1980).

In conventional digester systems, biomass is removed from the digester with effluent removal, thereby reducing the numbers of bacteria present to digest manure constituents. However, addition of a media for bacterial attachment aids in the retention of bacterial populations through biofilm development on the media. These systems, often called fixed-film digesters, have a faster rate of digestion of a given feedstock, which reduces digester storage requirements (Lo et al. 1984; Sanchez et al. 1992, Powers et al. 1997). Usually, screening out of poorly digested fibrous solids is done to avoid clogging of the media by fibrous solids. Lo et al. (1984) estimated that a fixed-film digester required 86 percent less volume than a conventional digester with solids screening pretreatment.

With dilute screened dairy waste, Powers et al. (1997) concluded that, at an operating temperature of 22° C, a laboratory-scale fixed-film digester at a 2.3 d HRT produced methane as efficiently as a conventional digester operated at 10 d HRT when fed the same feedstock. At higher HRT (>6 d RT), the advantages of a fixed-film system were not evident from either a gas production or COD/TS removal standpoint (Lo et al. 1984). Sanchez et al. (1992) found that the fixed-film digestion of screened cattle manure was most efficient, on a L CH_4/g VS digested basis, at 6 d RT compared to 0.9, 1.5, 2.1, and 8 d.

While anaerobic digestion serves as a valuable treatment method for stabilization of animal manures and recovery of energy in methane, the nutrient value of the manure is maintained. Thus, producers who are land-limited must seek additional means of removing manure nutrients.

DEBITS AND CREDITS ASSOCIATED WITH ANIMAL MANURES

A key to the public continuing to enjoy relatively cheap, nutritious animal-derived foods and other products is for the extra costs of environmental safeguards imposed on animal producers to be recoverable in value added to their operations by the more efficient use of marginal resources. For example, better use of manure nutrients will save the purchase of equivalent commercial fertilizer nutrients, and the installation of an anaerobic digester for improved odor control in a suburban setting will be partially offset by the energy value of the methane utilized from it.

The public sector should not conclude that agriculture will be the major problem relative to water quality. In agriculture, sufficient land and crop production potential that is suitable for recycling waste nutrients exists. Waste nutrient flows from human activities are similar to waste nutrient flows from farm animals, but processing differs and nutrient recycling is more difficult. Additionally, urban use of commercial fertilizer is appreciable even if small relative to commercial agriculture. Frink (1971) also underscored that municipal wastes and urban runoff were major contributors to N and P loads in surface waters.

The urban population may benefit from an assessment of the ability of agriculture to process urban wastes. That avenue has potential to reduce costs of processing urban wastes and, at the same time, provide better environmental accountability. This already is happening, with some municipalities managing agricultural land or contracting with farmers to utilize treated wastewater (reclaimed water) and sewage sludge (residuals) monitored to contain safe levels of heavy metals. If most consumers really understood and evaluated what is done with their own waste stream of sewage and solid wastes and the costs of processing, they probably would be much more open to discussion with agricultural planners about how to create a system that better serves all of us to manage and recycle waste nutrients sustainably.

The value of commercial fertilizer N sales represents a market opportunity for urban taxpayers to recover resource value in urban wastes if they can be processed to be a safe and effective fertilizer and offered at a price attractive to fertilizer users. This value also represents an opportunity to livestock producers that have excess waste nutrients on-farm and must transport excess nutrients from their farms; however, they too will need to produce a marketable fertilizer product. If urban dwellers and farmers can reduce their net expenses in processing wastes and market fertilizer products economically, all will benefit.

SUMMARY

Manure management systems should account for the fate of excreted nutrients that are of environmental concern. Amounts of manure nutrients—e.g., N, P, and K—originally excreted are predicted accurately with a nutritionally based input-output model, where input equals the amount consumed in feed and output equals amount in products produced—e.g., milk, eggs, meat, or offspring. Currently, N and P are most monitored to assure water quality. Land application of manure at acceptable levels to fertilize crops recycles manure nutrients and is the basis of nutrient management in most systems. Farms with insufficient crop production potential to recycle manure nutrients produced on-farm need to reduce nutrient excretions, if possible, and develop affordable systems to concentrate manure nutrients; this reduces hauling costs and possibly produces a salable product. Composting selected manure solids reduces volume and encourages combination with other organic wastes. Solutions to odor problems are needed. Generating energy from manure organic matter via anaerobic digestion reduces atmospheric emissions of methane and odors but, thus far, has not been economical. Reducing atmospheric emissions of ammonia from manures may become critical in the future.

REFERENCES

American Society for Agricultural Engineers (ASAE). 1994. Manure production and characteristics. *ASAE Standards D384.1.*

Apsimon, H.M., and M. Kruse-Plass. 1991. The role of ammonia as an atmospheric pollutant. In *Odour and Ammonia Emissions from Livestock Farming.* V.C. Nielsen, J.H. Voorburg, and P.L. L'Hermite, eds. London, England: Elsevier Applied Science Publications.

Arndt, D.L., D.L. Day, and E.E. Hatfield. 1979. Processing and handling of animal excreta for refeeding. *Journal of Animal Science* 48:157–162.

Arvanitidou, M., A. Tsakris, D. Sofianou, and V. Katsouyannopoulos. 1998. Antimicrobial resistance and R-factor transfer of salmonellae isolated from chicken carcasses in Greek hospitals. *International Journal of Food. Microbiology* 40:197–201.

Badger, P.C., J.K. Lindsey, and J.D. Veitch. 1995. Energy production from animal wastes. In *Animal Waste and the Land-Water Interface.* K. Steele, ed. Boca Raton, FL: Lewis Publishers. pp. 475–484.

Barrow, J.T., H.H. Van Horn, D.L. Anderson, and R.A. Nordstedt. 1997. Effects of Fe and Ca additions to dairy wastewaters on solids and nutrient removal by sedimentation. *Journal of Applied Engineering in Agriculture* 13(2):259–267.

Carter, S.D., G.L. Cromwell, M.D. Lindemann, L.W. Turner, and T.C. Bridges. 1996. Reducing N and P excretion by dietary manipulation in growing and finishing pigs. *Journal of Animal Science* 74 (Supplement 1):59.

Dagnall, S.P. 1993. Poultry litter as fuel. *World's Poultry Science Journal* 49:175–177.

D'Aoust, J.Y., A.M. Sewell, E. Daley, and P. Greco. 1992. Antibiotic resistance of agricultural and foodborne salmonella isolates in Canada: 1986–1989. *Journal of Food Protection* 55(6):428–434.

Elliott, H.A., R.C. Brandt, and K.S. Martin. 1990. *Atmospheric Disposal of Manure Nitrogen.* Technical Note No. 8. Harrisburg, PA: Pennsylvania Bay Education Office.

EPA (Environmental Protection Agency). 1989. Reducing methane emissions from livestock: opportunities and issues. *EPA 400/1–89/002.* Washington. D.C.: U.S. Environmental Protection Agency, Office of Air Radiation.

EPA (Environmental Protection Agency). 1998. National air pollutant emission trends, 1990–1998. *EPA 400/1–89/002.* Washington. D.C.: U.S. Environmental Protection Agency, Office of Air Radiation.

Fabian, E.E. 1989. Fundamentals and issues relative to anaerobic digestion on dairy farms. Part I. Production of biogas. In *Proceedings of Dairy Manure Management Symposium, Publication No. 31,* Northeastern Region of Agricultural Engineering Service. Ithaca, NY: Cornell University. pp. 133–144.

Finlay-Moore, O., P.G. Hartel, and M.L. Cabrera. 2000. 17 beta estradiol and testosterone in soil and runoff from grasslands amended with broiler litter. *Journal of Environmental Quality* 29(5):1604–1611.

Flachowsky, G., and A. Hennig. 1990. Composition and digestibility of untreated and chemically treated animal excreta for ruminants—A review. *Biological Wastes* 31:17–36.

Fontenot, J.P. 1991. Recycling animal wastes by feeding to enhance environmental quality. *The Professional Animal Scientist* 7(4):1–8.

Frink, C.R. 1971. Plant nutrients and water quality. *CSRS/USDA Agricultural Science Review* 9(2):11–31.

Fulhage, C.D. 1980. Performance of anaerobic lagoons as swine waste storage and treatment facilities in Missouri. Livestock waste: A renewable source. In *Proceedings, 4th International Symposium on Livestock Wastes—1980, April 15–17, 1980.* Amarillo Civic Center, Amarillo, TX. St. Joseph, MI: American Society of Agricultural Engineers. pp. 225–227.

Hill, D.T., G.L. Newton, R.A. Nordstedt, V.W.E. Payne, D.S. Ramsey, L.M. Safley, A.L. Sutton, and P.W. Westerman. 1990. Parameters for the efficient biological treatment of animal wastes. In *Proceedings of the 6th International Symposium on Agricultural and Food Processing Wastes.* St. Joseph, MI: American Society of Agricultural Engineering. pp. 515–524.

Hobson, P.N. 1984. Anaerobic digestion of agricultural wastes. *Journal of the Water Pollution Control Federation.* 1984(2):507–512.

Hobson, P.N. 1990. The treatment of agricultural wastes. In *Anaerobic Digestion: A Waste Treatment Technology.* A. Wheatley, ed. London, England: Elsevier Applied Science. pp. 93–137.

Holmberg, R.D., D.T. Hill, T.J. Prince, and N.J. Van Dyke. 1983. Potential of solid-liquid separation of swine wastes for methane production. *Transactions of the American Society of Agricultural Engineering* 26(6):1803–1807.

Janzen, R.A., J.J.R. Feddes, S.R. Jeffrey, N.G. Juma, J.J. Kennelly, G.R. Khoasani, J.L. Leonard, W.B. McGill, and E.K. Okine. 1994. Total nutrient management: A framework for sustainable management of livestock manure. *Proceedings of the Great Plains Animal Waste Conference on Confined Animal Production And Water Quality. Balancing Animal Production and the Environment.* Great Plains Agricultural Council (GPAC) Publication No. 151. pp 21–26. Fort Collins, CO: Great Plains Agric. Council.

Johnson, D.E., T.M. Hill, and G.M. Ward. 1992. Methane emissions from cattle: Global warming and management issues. In *Proceedings of the Minnesota Nutrition Conference,* Minnesota Extension Service. St. Paul: University of Minnesota.

Klopfenstein, T., R. Angel, G.L. Cromwell, G.E. Erickson, D.G. Fox, C. Parsons, L.D. Satter, and A.L. Sutton. 2002. Animal diet modification to decrease the potential for nitrogen and phosphorus pollution. *Issue Paper No. 21,* Council for Agricultural Science and Technology (CAST), Ames, IA.

Kornegay, E.T., D.M. Denbow, Z. Yi, and V. Ravindran. 1996. Response of broilers to graded levels of microbial phytase added to maize-soyabean meal-based diets containing three levels of non-phytate phosphorus. *British Journal of Nutrition* 75:839–852.

Lo, K.V., A.J. Whitehead, P.H. Liao, and N.R. Bulley. 1984. Methane production from screened dairy manure using a fixed-film reactor. *Agricultural Wastes* 9:175–188.

Mackie, R.I., and M.P. Bryant. 1981. Metabolic activity of fatty acid-oxidizing bacteria and the contribution of acetate, propionate, butyrate, and CO2 to methanogenesis in cattle waste at 40 and 60C. *Applied Environmental Micro.* 41(6):1363–1373.

Manie, T., V.S. Brozel, W.J. Veith, and P.A. Gouws. 1999. Antimicrobial resistance of bacterial flora associated with bovine products in South Africa. *Journal of Food Protection* 62(6):615–618.

Moore, J.A., and M.J. Gamroth. 1995. Selecting Nitrogen-Consuming Crops for Dairy Waste Utilization. Oregon State University, Corvallis: Bioresource Engineering Department.

Moore, P.A., T.C. Daniel, D.R. Edwards, and T.C. Sauer. 1997. Using aluminum sulfate to reduce ammonia emissions and decrease phosphorus runoff from poultry litter. *Proceedings of the Southeastern Sustainable Animal Waste Management Workshop,* Feb 11–13, 1997, Tifton, GA, Biological and Agricultural Engineering Department, University of Georgia, Athens, GA 30602. pp 97–102.

Morse, D., H.H. Head, and C.J. Wilcox. 1992. Disappearance of phosphorus in phytate from concentrates in vitro and from rations fed to lactating dairy cows. *Journal of Dairy Science* 75:1979–1986.

MWPS-18. 1993. *Livestock Waste Facilities Handbook,* 3rd Ed. Ames, IA: Midwest Plan Service, Iowa State University.

Newton, G.L., J.C. Johnson, J.G. Davis, G. Vellidis, R.K. Hubbard, and R. Lowrance. 1995. Nutrient recoveries from varied year-round application of liquid dairy manure on sprayfields. In *Proceedings of the Florida Dairy Production Conf. Dairy and Poultry Science Department,* University of Florida, Gainesville.

NRC (National Research Council). 2001. *Nutrient Requirements of Dairy,* 7th Ed. Washington, D.C.: National Academy Press.

Pain, B.F., V.R. Phillips, and R. West. 1988. Mesophilic anaerobic digestion of dairy cow slurry on a farm scale: Energy considerations. *Journal of Agricultural Engineering.* Research 39:123–135.

Patterson, P.H., and E.S. Lorenz. 1996. Manure nutrient production from commercial White Leghorn hens. *Journal of Applied Poultry Research* 5:260–268.

Powers, W.J., R.E. Montoya, H.H. Van Horn, R.A. Nordstedt, and R.A. Bucklin. 1995. Separation of manure solids from simulated flushed manures by screening or sedimentation. *Transactions of the American Society of Agricultural Engineers* 11(3)431–436.

Powers, W.J., and H.H. Van Horn. 2001. Nutritional implications for manure nutrient management planning. *Journal of Applied Engineering in Agriculture* 17(1):27–39.

Powers, W.J., H.H. Van Horn, A.C. Wilkie, C.J. Wilcox, and R.A. Nordstedt. 1999. Effects of anaerobic digestion and additives to effluent or cattle feed on odor and odorant concentrations. *Journal of Animal Science* 77:1412–1421.

Powers, W.J., A.C. Wilkie, H.H. Van Horn, and R.A. Nordstedt. 1997. Effects of hydraulic retention time on performance and effluent odor of conventional and fixed-film anaerobic digesters fed dairy manure wastewaters. *Transactions of the American Society of Agricultural Engineers* 40(5): 1449–1455.

Rorick, M.B., S.L. Spahr, and M.P. Bryant. 1980. Methane production from cattle waste in laboratory reactors at 40 and 60 C after solid-liquid separation. *Journal of Dairy Science* 63:1953–1956.

Sanchez, E.P., P. Weiland, and T. Travieso. 1992. Effect of hydraulic retention time on the anaerobic biofilm reactor efficiency applied to screened cattle waste treatment. *Biotechnology Letters* 14(7):635–638.

Sherman, J.J., H.H. Van Horn, and R.A. Nordstedt. 2000. Use of flocculants in dairy wastewaters to remove phosphorus. *Applied Engineering in Agriculture* 16(4):445–452.

Smith, L.W., and W.E. Wheeler. 1979. Nutritional and economic value of animal excreta. *Journal of Animal Science* 48:144–156.

Summers, R., and S. Bousfield. 1980. A detailed study of piggery-waste anaerobic digestion. *Agricultural Wastes.* 2:61–78.

Sweeten, J. 1988. Odor measurement and control for the swine industry. *Journal of Environmental Health* 50(5):282–287.

Sweeten, J.M., and J.R. Miner. 1993. Odor intensities at cattle feedlots in nuisance litigation. *Bioresource Technology* 45:177–188.

Tomlinson, A.P., W.J. Powers, H.H. Van Horn, R.A. Nordstedt, and C.J. Wilcox. 1996. Dietary protein effects on nitrogen excretion and manure characteristics of lactating cows. *Transactions of the American Society of Agricultural Engineers* 39(4):1441–1448.

USDA-SCS. 1992. Agricultural waste characteristics, Chapter 4, Table 4–5. In *Agricultural Waste Management Field Handbook*. Washington, D.C.: USDA-SCS.

Van Horn, H.H., G.L. Newton, G. Kidder, E.C. French, and R.A. Nordstedt. 1998. *Managing Dairy Manure Accountably: Worksheets for Nutrient Budgeting*. Florida Cooperative Extension Service, Institute of Food and Agricultural Services, University of Florida, Gainesville, Circular No. 1196.

Van Horn, H.H., G.L. Newton, and W.E. Kunkle. 1996. Ruminant nutrition from an environmental perspective: Factors affecting whole-farm nutrient balance. *Journal of Animal Science* 74:3082–3102.

Van Horn, H.H., G.L. Newton, R.A. Nordstedt, E.C. French, G. Kidder, D.A. Graetz, and C.F. Chambliss. 1997. *Dairy Manure Management: Strategies for Recycling Nutrients to Recover Fertilizer Value and Avoid Environmental Pollution*. Florida Cooperative Extension Service, Institute of Food and Agricultural Services, University of Florida, Gainesville, Circular No. 1016 (revised June 1997).

Van Horn, H.H., A.C. Wilkie, W.J. Powers, and R.A. Nordstedt. 1994. Components of dairy manure management systems. *Journal of Dairy Science* 77:2008–2030.

Varel, V.H., A.G. Hashimoto, and Y.R. Chen. 1980. Effect of temperature and retention time on methane production from beef cattle waste. *Applied Environmental Microbiology* 40(2):217–222.

Varel, V.H., H.R. Isaacson, and M.P. Bryant. 1977. Thermophilic methane production from cattle waste. *Applied Environmental Microbiology* 33(2):298–307.

Waggoner, P.E., R.L. Baldwin, P.R. Crosson, M.R. Drabenstott, D.N. Duvick, R.F. Follett, M.E. Jensen, G. Marland, R.M. Peart, N.J. Rosenberg, and V.W. Ruttan. 1992. *Preparing U. S. Agriculture for Global Climate Change.* CAST Task Force Report 119, Council on Agricultural Science and Technology, Ames, IA.

Watts, P.J., E.A. Gardner, R.W. Tucker, and K.D. Casey. 1994. Mass-balance approach to design nutrient management systems at cattle feedlots. *Proceedings of the Great Plains Animal Waste Conference on Confined Animal Production and Water Quality. Balancing Animal Production and the Environment.* Great Plains Agricultural Council (GPAC) Publication No. 151. Fort Collins, CO: Great Plains Agricultural Council. p. 27.

Yi, Z., E.T. Kornegay, V. Ravindran, M.D. Lindemann, and J.H. Wilson. 1996. Effectiveness of Natuphos® phytase in improving the bioavailabilities of phosphorus and other nutrients in soybean meal-based semipurified diets for young pigs. *Journal of Animal Science* 74(7):1601–1611.

Zahn., J.A., J. Anhalt, and E. Boyd. 2001. Evidence of transfer of tylosin and tylosin-resistant bacteria in air from swine production facilities using sub-therapeutic concentrations of tylan in feed. *Journal of Animal Science* 79 (Supplement 1):783.

NOTES

1. Department of Dairy and Poultry Sciences, University of Florida, Gainesville, FL 32611–0920.
2. Department of Animal Science, Iowa State University, Ames,IA 50011–3150.
3. Mcal = megacalorie; 1 Mcal = 4.184 MJ (megajoule).

11

THE EQUINE INDUSTRY –
ECONOMIC AND SOCIETAL IMPACT
Karyn Malinowski and Norman Luba

One of the largest industries in the United States revolves around horses, which make a daily impact on the lives of one in every 35 American citizens. The "Force of the Horse" as stated by the American Horse Council means more than the economic importance and $112 billion value of the United States Horse Industry. Horses and the industry that surround the animal are rich in the history and development of humankind, and millions of people worldwide actively participate in equine-related activities through occupations, recreation, and sport. Horses have a vast societal impact and are responsible for improving the quality of life for millions of Americans by preserving open space, providing outdoor sport and recreation, building a solid foundation for youth development, and providing mental and physical therapy to adjudicated youth and handicapped persons. Today the horse's role in American society is different than it was in the early 19th and 20th centuries. Because of the American horse owners' dedicated commitment to their horses, horses today enjoy a higher standard of living, increasing their longevity, productivity, and quality of life.

ECONOMIC IMPACT

More than 7.1 million (one in every 35) Americans are involved in the horse industry. The industry is vast and highly diverse, combining the rural activities of breeding, training, and housing horses with more urban activities such as racetracks, horse show and competition stadiums and grounds, and public stables. The United States horse industry is a $25.3 billion business associated with 6.9 million horses and 7.1 million Americans (Table 1.1).

The horse industry's contribution to the U.S. GDP of $112.1 billion is greater than the motion picture services, railroad transportation, fur-

TABLE 11.1. Size of the U.S. horse industry

Number of horses	6.9 million
Number of industry participants	7.1 million
Value of goods and services produced	$25.3 billion
Full-time equivalent jobs provided	338,500

TABLE 11.2. Value of horse-related goods and services by industrial subsector

Industrial Sector or Subsector	Value of Goods and Services ($ billions)
Agriculture, forestry & fishing	$5.3
Horse and other equine farms	$4.1
Boarding/training of horses (nonracing)	$1.2
Amusement and recreation services	$4.7
Riding instruction and horse leasing	$0.2
Horse shows	$0.6
Racetracks and OTB facilities	$2.2
Racehorse owners	$1.2
Boarding/training of racehorses	$0.5
Noncash value of horse services	$15.3
TOTAL	$25.3

niture and fixtures manufacturing, and tobacco product manufacturing industries (Tables 11.2 and 11.3).

Furthermore, the industry generates over 1.4 million FTE jobs across the country and directly employs more people than railroads, radio and television broadcasting, petroleum and coal products manufacturing, and tobacco product manufacturing (see Table 11.4, American Horse Council Foundation 1996).

More recent economic impact studies of the horse industry have demonstrated the importance of the equine industry (New Jersey Equine Industry 1996; New York Horse Industry 2000) in both New Jersey and New York, one of the smallest states, and one of the most diverse states, respectively, in the U.S.

The New Jersey equine industry (where the horse is the state animal) is invaluable as a major thrust in retaining agricultural acreage as open

TABLE 11.3. Horse industry GDP contribution compared to other representatives industries

Industry	GDP Contribution ($ billions)
Primary metal industries	$44.2
Radio and television broadcasting	$39.7
Petroleum and coal products manufacturing	$29.7
Apparel and other textile products manufacturing	$27.8
Horse production and entertainment services	$25.3
Motion picture services	$24.8
Railroad transportation	$24.3
Furniture and fixtures manufacturing	$19.0
Tobacco product manufacturing	$16.6
ALL INDUSTRIES	$6,931.4

TABLE 11.4. Horse industry employment compared to other representative industries

Industry	Employment (thousands)
Apparel and other textile products manufacturing	974
Primary metal industries	698
Furniture and fixtures manufacturing	505
Motion picture services	441
Horse production and entertainment services	339
Railroad transportation	241
Radio and television broadcasting	207
Petroleum and coal products manufacturing	149
Tobacco product manufacturing	43
ALL INDUSTRIES	117,581

space in the most densely populated state in the nation. Although horse owners do not market their product by the bushel, pound, or cubic foot, horses are bred, raised, bought, and sold in the Garden State like any other agricultural commodity. The horse industry in New Jersey is represented on the State Board of Agriculture, where policymaking decisions affecting agriculture are made.

Of the $698 million spent in New Jersey in 1996, $224 million was by racetracks, $407 million by New Jersey owners and operators, and $67 million by out-of-state equine owners and operators. In New York, the value of horses in 2000 was $1.7 billion, with total equine-related assets valued at $6.15 billion. New York equine owners and operators spent $704 million during 2000 for operating and capital expenses (Table 11.5).

RACING

More than 200 racetracks offer flat and harness racing in the United States. Thoroughbreds, Standardbreds, Quarter Horses, Arabians, Paints, Appaloosa, and mules participate in equine racing at everything from major racetracks, such as Churchill Downs and the Meadowlands, to small country fairs. The racing industry provides more than 472,800 full-time jobs and has an economic effect of $34 billion.

THOROUGHBRED RACING IN NORTH AMERICA

There are records of horse racing on Long Island as far back as 1665. However, the introduction of organized Thoroughbred racing in the United States is traditionally credited to Gov. Samuel Ogle of Maryland, at whose behest racing "between pedigreed horses in the English style" was first staged at Annapolis in 1745. But it was during the Industrial Revolution of the first half of the 19th century that breeding and racing began to flourish. The Thoroughbred played a key role in the country's westward expansion, prompting creation of racetracks and breeding farms along the way.

As racing proliferated through the fast expanding continent, the need for a pedigree registry of American-bred Thoroughbreds, similar to the General Stud Book in England, became apparent. The first volume of The American Stud Book was published in 1873 by Col. Sanders D. Bruce, who produced six volumes of the register until 1896 when the project was sold to The Jockey Club. The Stud Book ensures the correct identification of every Thoroughbred and, as such, is essential to the in-

TABLE 11.5. Impact of the horse industry on the U.S. economy

Total impact on U.S. gross domestic product	$112.1 billion
Total FTE jobs generated	1,404,400
Total taxes and fees paid	$1.9 billion

tegrity of Thoroughbred racing. When The Jockey Club published its first volume of the Stud Book, the annual foal crop was about 3,000. The foal crop peaked in 1986 at more than 51,000 and today averages about 37,000 foals annually.

In addition to its responsibilities as the North American Thorough-bred breed registry, The Jockey Club is the parent company of a hand-ful of subsidiary and affiliate organizations that serve the many and diverse needs of the Thoroughbred racing and breeding industries. The Jockey Club is also a founding member of, and continues to work closely with, the National Thoroughbred Racing Association (NTRA), racing's central league office charged with increasing the popularity and economics of the sport.

More than 70,000 Thoroughbreds race in North America each year and compete for ever-increasing prize money, which topped $1.2 billion in 2001. Likewise, the money wagered on North American Thorough-bred racing continues to increase and reached a record $15.2 billion in 2001. Approximately 85 percent of that total was wagered off-track at simulcast facilities, by phone, or over the Internet.

Additional statistical measures of industry performance, as well as a comprehensive directory of Thoroughbred organizations, are available on the Internet through The Jockey Club Fact Book at http://home.jockeyclub.com/factbook.

Standardbred Racing in North America

Standardbred horses and the sport of harness racing have been on the American scene as an organized sport for well over 150 years. The breed springs from the crossing of other breeds, most notably between Thor-oughbreds and Morgan Horses. The former contributed speed and the latter contributed stamina and driveability. The first Stud Book was com-piled in 1870s. The one-mile distance at which the overwhelming ma-jority of races are contested help give the name to the breed, as originally a horse could not be admitted into the Stud Book unless it had trotted or paced a mile in "standard time," which at the outset was three minutes.

Standardbreds compete on two gaits, the trot and the pace. The trot is familiar to any horse enthusiast, but pacing is seen in few other breeds. Pacers move both their left legs back while their right legs move forward which gives them the nickname "side-wheelers."

In the early years, trotters were so highly prized that they dominated the breed to the near exclusion of pacers, but the latter gait became pop-ular when hobbles (or "hopples" as they were sometimes called) were used to help pacers maintain their gait. This aid, which is a set of leather loops placed around the upper legs, helps keep the pacer on-stride. This meant young horses could learn and maintain their gait much earlier in their careers, thus bringing race purses to their owners much earlier in their careers.

Pacers now dominate the sport in North America, where about 80 percent of Standardbred racehorses are pacers. Pacers never compete against trotters and are, on the average, about three seconds faster over the mile distance. While pacing also dominates in Australia and New Zealand, two of about 20 countries where the sport exists, it is virtually nonexistent in Europe, except for Great Britain.

Today the sport in North America features purses as high as $1 million for a few major events, and is contested in 20 U.S. States and all the provinces of Canada. North American bloodstock and racehorses are eagerly sought by breeders and trainers all around the world.

There are approximately 300 harness racetracks in the U.S., including about three dozen big-city commercial tracks. The balance are state and county fairs, many of which have offered the sport for 150 years or more. The major commercial tracks are a relatively recent phenomenon; the first was built in New York City (Roosevelt Raceway) in 1940.

Annually, over $6 billion is wagered on the sport in North America each year, the value of yearlings sold at auction is more than $74 million each year, and horses in the U.S. race for total purse money of more than $334 million. About 11,000 new foals are registered each year, adding to the more than 800,000 entries in the United States Trotting Association Stud Book. Standardbreds are bred almost exclusively using artificial insemination, and have been since about 1970.

The Association has nearly 30,000 members worldwide, and not only maintains the Stud Book but keeps all breeding and racing statistics; licenses trainers, drivers, and officials; publicizes the sport; administers the rules on breeding and racing; and promotes the breed for not only racing but for the many other wonderful disciplines for which these powerful, intelligent, and tractable horses show an aptitude. You can learn more at http://www.ustrotting.com.

Quarter Horse Racing in North America

American Quarter Horse racing blends the beauty and athleticism of the American Quarter Horse, the fun of handicapping, and our western heritage into a popular sport that has thrived for hundreds of years. English colonists began importing horses to the East Coast in the early 1600s. By 1665, the "Quarter Horse" was considered an established breed. Mares were crossed with various other breeds, including Arabians, Barbs, Turks, and Spanish horses to produce this heavily muscled and compact horse. The horse proved to be the perfect size for racing down shallow streets and narrow fields that were common along the East Coast during the formation of the country.

Modern-day American Quarter Horse racing got its start in the 1940s, and today American Quarter Horse racing is conducted at more than 100 tracks throughout North America with purses reaching nearly $75 million and pari-mutuel handle exceeding $315 million. Thanks to the introduction of simulcast wagering, American Quarter Horse racing has a worldwide audience of race fans in the millions.

The first million-dollar race in all of horse racing was the All American Futurity for two-year-old American Quarter Horses, run each Labor Day at Ruidoso Downs in New Mexico. Today, the race purse exceeds $2 million and annually awards $1 million to the winner. First run in 1959, the race broke the $1 million barrier in 1978.

Debuting in 1993, the MBNA America Quarter Horse Racing Challenge is a $5 million series of races run in 10 regions at tracks across the United States, Canada, and Mexico. The winners of these regional races earn a starting position in a season-ending championship day where they vie for the title of "Best in North America."

While the sport of American Quarter Horse racing is primarily a regional sport in the west and southwest areas of the United States and Canada, the sprint-style of racing has strong roots in Mexico, where one-on-one match racing is extremely popular. In addition, South American countries such as Brazil and Argentina have established racing programs where short racing is preferred to distance racing. Unquestionably, the greatest potential for growth lies within South America, given Latin America's penchant for American Quarter Horse racing.

However, in recent years American Quarter Horse racing has started attracting the attention of people in Europe and in Australia. In Germany, small pockets of the sprint-style of racing can be enjoyed, and throughout Italy, special American Quarter Horse races are run at some of the country's premier Thoroughbred tracks. While still in its infancy, the sprinters' speed is proving popular and the American Quarter Horse Association expects slow but continued growth in other European countries as the sport is exposed to more people. Down under, thanks to the

American Quarter Horse's overall popularity, racing organizations have started fledgling programs for sprint racing. For more information on Quarter Horses and Quarter Horse Racing, go to www.aqha.org.

American Paint Horse Racing in North America

One of the hottest tickets at tracks across the country is American Paint Horse racing. In 1966, nearly 20 years after Paints first took to the track, the American Paint Horse Association (APHA) officially recognized the sport. Today, several hundred starters are racing at tracks throughout the United States and Canada, competing for millions of dollars in purse money. Paint horse racing has yet to peak. Newcomers have the opportunity to impact the industry and be a part of this successful sport. Each year, as breed improvement continues, purse structures grow, numbers of horses increase, and more races are offered, the sport becomes more rewarding for those involved.

Paints commonly race at distances of 220–870 yards. In addition to running in all-Paint stakes races, Paint Horses also run against Appaloosas, Quarter Horses, and Thoroughbreds at approved all-breed races across the nation. For more information on American Paint Horse racing, visit http://www.apha.com/racing or call the American Paint Horse Association at 817–834–2742.

Appaloosa Racing in North America

Appaloosa racing dates back to the original adoption of their rules and regulations in 1960. Gaining official recognition in 1962, the first parimutuel Appaloosa races were held in Albuquerque, New Mexico, with four official races and 23 starters. As it stands today, Appaloosa racing is small by comparison. Most recent statistics (2001) show there were 308 total horses competing in 446 races for total purse monies of $2,579,913. This equates to average earnings per starter of $8,376.

Non-Racing Sport and Recreation

Nationally, recreational activities on horseback have an economic effect of $28.3 billion and generate full-time employment for some 317,000 people. The 6.9 million horses living in the U.S. represent 100 breeds, which are registered in 133 horse and pony organizations (American Horse Council Foundation 1996) that are devoted to promoting horses. More than 14,000 sanctioned horse shows and thousands of unrecognized events are held annually. There are also more than 250,000 youth involved in horse programs through 4-H and Pony Club. Showing horses has an economic impact of $34.8 billion and supports 441,000 full-time jobs.

Many people enjoy horses simply for pleasure and trail riding. Horses provide a leisurely respite for the hectic daily life most Americans face. There is an increase in trail riding in the U.S., as well as an increased number of organizations devoted to the preservation of trails.

Recreational activities have an economic impact of $28.3 billion and generate 317,000 full-time jobs.

Therapeutic Aspects

Horses provide wonderful opportunities for an enhanced quality of life for people with physical and mental disabilities. Therapeutic riding has been used since the early 1950s in Europe as a tool for improving the lives of individuals with physical disabilities. NARHA was founded in 1969 to promote and support therapeutic riding in the United States and Canada.

The North American Riding for the Handicapped Association (NARHA), founded in 1969, promotes equine-facilitated activity programs in the United States and Canada. "NARHA is a membership organization that fosters safe, professional, ethical and therapeutic equine activities through education, communication, standards and research for people with and without disabilities" (NARHA 2001).

Currently, more than 650 NARHA program centers serve some 30,000 individuals with disabilities. Each year, dozens of new centers initiate new programs and thousands of individuals benefit from these activities.

Since 1969, NARHA has ensured that therapeutic riding is both safe for, and accessible to, those in need. In that time, the field of therapeutic riding has expanded along with the numbers of individuals profiting from involvement with horses. Today NARHA represents a growing number of equine-facilitated therapies and activities, including recreational riding for individuals with disabilities, classical hippotherapy, equine-facilitated psychotherapy, driving, vaulting, competition, and other therapeutic and educational interactions with horses.

NARHA is proud to provide opportunities for people with varying ability levels to challenge themselves physically and emotionally and to set goals to improve their quality of life via horses.

Individuals with almost any cognitive, physical, and/or emotional disability can benefit from therapeutic riding, driving, vaulting, competition, or other purposeful, safe, and supervised interaction with equines:

- Because horseback riding gently and rhythmically moves the rider's body in a manner similar to a human gait, riders with physical disabilities often show improvement in flexibility, balance, and muscle strength.
- For individuals with mental or emotional disabilities, the unique relationship formed with the horse can lead to increased confidence, patience, and self-esteem.
- Local, State, National and even International Competition often is the logical progression for individuals with disabilities who have mastered equestrian skills and seek to further challenge themselves.

The sense of independence and acceptance found through these activities with the horse benefits all. Individuals with the following disabilities commonly participate and benefit from equine-facilitated therapy and activities:

> Muscular dystrophy
> Cerebral palsy
> Visual impairment
> Down syndrome
> Mental retardation
> Autism
> Multiple sclerosis
> Spina bifida
> Emotional disabilities
> Brain injuries
> Spinal cord injuries
> Amputations
> Learning disabilities
> Attention deficit disorder
> Deafness
> Cardiovascular accident/stroke

Presently 30,000 Americans with disabilities benefit from equine-assisted activities at NARHA centers throughout the nation. In five years, 50,000 to 75,000 will be enrolled. As NARHA continues its remarkable growth, and as more individuals become aware of our significant mission, there will be an ever-increasing call for its services (NARHA 2001).

Rehabilitation of Adjudicated Youth

Horses, which require constant management and care, play an important role in the personal development of both children and adults. Management of horses teaches responsibility and appreciation for humane care of living things. The highly successful program entitled "Careers in the Green Industries" is a program that exposes high school-aged adjudicated youth, using horses, to an educational atmosphere that they would not normally experience (Malinowski 1995). Hands-on instruction in proper handling, grooming, tacking, and general management of horses is given to adjudicated youth with no prior horse management experience. Positive impacts of the program include increased communication between the participants, improved self-esteem, and career exploration and development.

Equine Ranching in North America

Since 1942, the estrogens in pregnant mares' urine (PMU) have been used to produce a leading estrogen replacement therapy prescribed to millions of women for the treatment of menopausal symptoms. Equine ranches that produce PMU are located throughout western Canada and North Dakota. These ranches are family owned and operated and are independently contracted to provide the raw material needed to manufacture the Premarin family of products.

For many years, scientists have known that as a woman ages, her body chemistry changes due to diminishing amounts of estrogen; this condition is called menopause. In searching for a way to alleviate this condition, scientists found that by replacing estrogen in women during menopause, uncomfortable symptoms were alleviated. The scientists tried obtaining estrogen from humans along with animal models, but soon discovered they were not able to collect sufficient quantities. They then found that horses produced the best known grouping of the critical estrogens required, and in a quantity able to meet the demands of women into the foreseeable future. Mares that are confirmed pregnant produce PMU from which the estrogens are extracted through a 125-step process—one of the most sophisticated in the pharmaceutical industry. This estrogen is used to produce Premarin.

The PMU collection season runs annually from October to March. During this time, mares are housed in temperature-controlled barns, which are approximately 40°Fahrenheit, shielding them from the outside temperature, which may reach minus 40°Fahrenheit throughout the harsh winter. Each horse is stabled in its own stall, the size of which varies depending on the size of the horse, with adequate bedding—typically straw, wood shavings, or sawdust.

The urine collection system consists of a lightweight, flexible pouch that is suspended from the ceiling by rubber suspension lines. The mares adapt very well to this system, which provides a full range of motion and allows the horses to lie down comfortably (Figures 11.1–11.3).

Ranchers provide feed and water for the horses in accordance with the nutritional standards provided by the National Research Council's Nutrient Requirements of Horses. In most barns, hay is available con-

FIGURE 11.1 The pregnant mare urine (PMU) collection system. This has also been utilized by several universities for nutrition research.

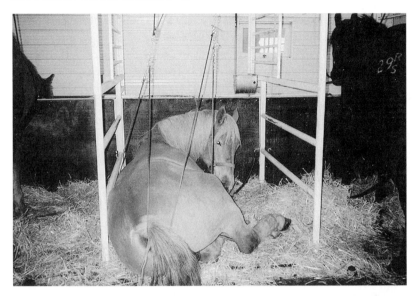

FIGURE 11.2 Mare used for collection of pregnant mare urine (PMU) lying down.

FIGURE 11.3 Mares used for collection of pregnant mare urine (PMU) in barn.

tinuously, while water is delivered multiple times per day by automated watering systems, designed to keep stables dry. Mares are given water at least five times per day ad libitum.

During each collection season, veterinarians visit the ranches at least three times to complete full-herd health reviews as part of a formal program, with oversight from veterinary representatives of the Alberta, Saskatchewan, and Manitoba Veterinary Medical Associations, as well as the North Dakota Board of Animal Health.

The ranchers that have a contract to provide PMU for this purpose are multigenerational horse breeders, and have been so long before they became involved in PMU ranching. They have had quality breeding and husbandry programs in place for generations, producing prize winning Quarter Horses, Paint Horses, Appaloosas, Belgians, Percherons, and Clydesdales, and supply these high-quality horses to a variety of markets, including show, recreation, workhorses, replacement stock, and rodeos. The North American Equine Ranching Information Council (NAERIC) is an association of equine ranching members that promotes several programs that reward buyers of these horses and some of the activities they are involved in, such as the NAERIC Incentive Program.

Equine ranchers adhere to the "Recommended Code of Practice for the Care and Handling of Horses in PMU Operations," which specifies standards for nutrition, watering, exercise, and barn environment among other standards. Equine ranching is the most highly regulated horse ranching activity in North America.

During its development, equine ranching has become a model of self-regulation in the agricultural industry, utilizing a system of extensive checks and balances that ensure that ranchers strive for the highest standards of practice rather than simply abiding by baseline laws and regulations. At the same time, ranchers have continually sought to increase the body of equine care knowledge so that ranching practices can be state of the art.

Equine ranchers have also endeavored to identify the best veterinary minds and equine experts available, in order to objectively and fairly review the industry and its management practices. Organizations, including the American Association of Equine Practitioners (AAEP), the Canadian Veterinary Medical Association (CVMA) and the International League for the Protection of Horses (ILPH), were invited to assess the industry and provide input and guidance for the future health and welfare of the horses. These organizations support the programs and the system of checks and balances that govern equine ranching.

Virtually every knowledgeable equine professional that has been involved with PMU ranching has recognized the outstanding care given to these horses. From a management standpoint, the model of self-regulation and incentive-based improvement directives that are employed in this industry is one that has been remarkably successful and can be replicated in a number of other agricultural and commercial enterprises (see www.naeric.org).

IMPORT/EXPORT OF HORSES

The international transport of horses has become more routine with the increased frequency of purchases, competitive events, and business and personal relocations in a global economy. With the international movement of horses comes the possibility of diseases and parasites being transmitted from country to country. Because of the possibility of such occurrences, most countries impose animal health recommendations on the importation of horses.

In most cases, import certificates that are obtained, in advance, from the receiving country's animal health or animal agriculture department must accompany horses coming into a particular country. Crucial to these important certificates will be appropriate health certification papers from the country of origin.

Countries with the objective of keeping diseases and parasites out monitor several types of equine diseases. Some of the more common diseases monitored are Glanders, Dourine, Equine Infectious Anemia, Equine Piroplasmosis, Venezuelan Equine Encephalitis, African Horse Sickness, and Contagious Equine Metritis. In addition, horses are typically checked for various types of external parasites.

In many countries, specific animal import centers have been established through which horses must pass in order to enter a particular country. In some cases, these centers specialize in the types of diseases of the originating country, and horses must pass designated quarantine times of specific duration at those centers.

Because some diseases are specific to the age or use of horses (breeding animal versus riding animal) there may be very specific and complex rules that involve the country of origin health certification, prescribed tests, and treatments and post-entry treatments under local and federal laws.

Since requirements can change on short notice, it is recommended that information be sought from attending veterinarians in concert with the governmental veterinarian from the country of origin. The changes in the world animal health situation can typically be monitored via the official web sites of the animal health or animal agriculture departments of the importing and exporting countries.

MEAT FOR HUMAN AND ANIMAL CONSUMPTION

Originally, the horse was a source of food for prehistoric man. Today, the United States Department of Agriculture recognizes the consumption of horsemeat, and the Federal Wholesome Meat Act, passed in 1967, includes horses as livestock. The same holds true for the Canadian Food Inspection Agency. While the majority of horsemeat is exported to Europe, horsemeat can legally be sold and consumed in the United States and Canada. In limited cases, some state and/or provincial and/or local ordinances may prevent its sale.

The decision to send a horse (or any animal) to a processing facility to be slaughtered for human consumption is a personal one, which should not be mandated by law. This is an option of disposing of up to 81,000 horses in the U.S. annually, when horses can no longer be provided for or have any viable use. There are presently laws providing for the humane treatment of horses on their way to slaughter. To take away the option of slaughter could make conditions worse for the many unwanted horses in the United States.

EDUCATIONAL AND SOCIETAL IMPACT

Horses play an important role in the personal development of children and adults. Horses also improve the quality of life by helping to retain rural agricultural character and scenic beauty in urban environments (Malinowski 1999). Departments of Animal Science around the country are challenged with changing demographics and demands of undergraduate and graduate students. Enrollment of undergraduates majoring in livestock production is shrinking. Today, undergraduate students enter departments of animal science because of their interest in domestic animals and veterinary medicine. A recent student profile at Oregon State University showed that 77–88 percent of total students were female and from an urban environment (32–62 percent). The students' main reasons for choosing animal science were love of animals (89–93 percent) and desire to be a veterinarian (69–77 percent). Interestingly, the animal species these students were most interested in were horses

(42–53 percent) and pets (9–38 percent) (Cheeke 1999). Horses as livestock and companion animals are shaping the way animal science departments of the future will look.

ANIMAL WELFARE

The care and regulation of horses and their related activities come under the auspices of the United States Department of Agriculture (USDA). Part of the responsibility of the USDA regarding the horse industry is to provide technical expertise and monetary support for research into the prevention of equine diseases, the enforcement of the Horse Protection Act, and the development and enforcement of the Safe Commercial Transportation of Equine to Slaughter Act.

HORSES: CLASSIFICATION AS LIVESTOCK OR COMPANION ANIMALS

Traditionally, and legally, horses have been considered livestock in Canada and the United States. Even today, horses are still kept and raised on a farm or ranch and are used in commercial enterprises. The Canadian and United States horse industries are major business sectors that make significant contributions to the economic well-being of both countries.

The care and regulation of horses and horse-related activities come under the purview of the Canadian Food Inspection Agency (CFIA) and the USDA on national levels. The Provincial Departments of Agriculture in Canada and the state Departments of Agriculture in the United States are charged with the regulation of horse-related activities on provincial and state levels, respectively.

Livestock is traditionally defined as animals kept or raised in a farm or ranch setting and used in a commercial enterprise. Economically, the horse industry has a huge impact, as previously described. While horses have long been classified as livestock in the U.S., individuals may still enjoy horses as companion animals.

There are many advantages to having horses classified as livestock. If livestock status is taken away from horses, there is a possibility of losing already limited support from the USDA for research, regulation, and disaster relief. All 50 states have animal anti-cruelty laws. Livestock anti-cruelty laws are written to ensure humane treatment for these animals while still providing for their use as an agricultural product. If horses were not classified as livestock, these anti-cruelty laws would no longer apply. Many states now have passed legislation known as "limited liability laws." These recognize horses as potentially dangerous animals and provide protection against lawsuits for stable owners, equine event organizer, and trail ride concessions. Many state laws are extended to livestock. If horses were no longer considered livestock, many horse and stable owners would no longer be protected by this critical legislation affecting every horse owner.

Tax ramifications to a reclassification of horses from livestock to companion animals are vast. Currently, under federal law, commercial horse owners, breeders, and in some states, trainers and boarding stable owners are considered farmers for tax purposes. This not only means property tax benefits but benefits from excise and sales taxes as well. If horse owners are considered farmers, their land can also be classified and sheltered as farmland preserved. This ensures that the land is kept open space for many years. (American Horse Council, American Association of Equine Practitioners 1999).

REFERENCES

American Horse Council, American Association of Equine Practitioners. 1999. *Legal Status of Horses as Livestock.* White Paper from committees of AHC and AAEP.

American Horse Council Foundation. 1996. White Paper from committee of AHC.

Cheeke, P.R. 1999. Shrinking membership in the American Society of Animal Science: Does the discipline of poultry science give us some clues?. *Journal of Animal Science* 77:2031–2038.

Malinowski, K. 1995. Careers in the green industry: Youth sow seeds for their future. In *Proceedings of the 14th Equine Nutrition and Physiology Society.* Ontario, CA. pp. 201–202.

Malinowski, K. 1999 Revised. The economic impact of the horse industry in New Jersey and the United States. *Rutgers Cooperative Extension Fact Sheet 517.*

NARHA (North America Riding for the Handicapped). 2001. *Annual Report.*

New Jersey Equine Industry. 1996. *New Jersey Department of Agriculture Circular #549.* Rod DeSmet, New Jersey Agricultural Statistics Service, New Jersey Department of Agriculture, New Jersey Equine Advisory Board.

New York Horse Industry. 2000. Policy Economics Practice by Barents Group LLC, (A KPMG Company) for AHC Foundation.

Policy Economics Practice, Barents Group LLC, (A KPMG Company). 1996. The Economic impact of the horse industry in the United States. *The American Horse Council Foundation Volume 1: National Summary.*

IV

FRONTIERS IN WATER

This section includes a series of papers on water that were presented at the 2002 World Food Prize symposium "From the Middle East to the Middle West: Managing Freshwater Shortages and Regional Water Security" held October, 2002. This critical subject is introduced by a chapter by Jacques Diouf, Director-General of the United Nations Food and Agriculture Organization. Overviews are provided in Chapters 13 by M. Falkenmark and J. Rockström and 14 by David Seckler and Upali Amarasinghe. Issues related to water in the Middle East are covered in Chapters 15–18, and water issues in India and Southeast Asia are considered in Chapters 19 and 20.

12

Managing Water Insecurity[1]
Jacques Diouf[2]

I am greatly honored to participate in this World Food Prize symposium this morning. It is also a great privilege that I have been asked to give a keynote speech on this auspicious occasion, and on a topic that is of vital importance to the Food and Agriculture Organization of the United Nations. In fact, as you may be aware, water for food security was the theme selected for World Food Day this year. What a fitting coincidence!

Let me start by asking a simple question: Are you aware that, to produce each of your breakfasts, probably more than 1,000 liters of water was necessary? For each and every one of us, the water required to produce our daily food intake varies from 2,000–5,000 liters per day, depending on one's dietary preferences.

We should never lose sight of the 800 million people who remain chronically undernourished—or food-insecure. Closing this gap in our world's food security has been my priority, and while we have seen overall improvement in satisfying the human demand for food in the last century—against very high rates of population growth—there is still a long way to go.

The focus of today's symposium is the Near East region, which has seen some of the highest population growth rates pushing at the limits of its scarce land and water resources. As a result, it has its share of undernourished people. We estimate that this regional figure will peak at some 40 million by 2015. The dependency upon water to sustain agricultural production has been fundamental for the region and will increase in the future, because improvement in yields have not been sufficient to match the growth of food needs. Furthermore, non-sustainable digging into strategic groundwater reserves is increasingly taking place.

Dependency on external resources of water, in the form of importing virtual water contained in food products, has largely increased over the last decades. It is estimated that the import of virtual water[3] to the countries of the Near East region corresponds to more than the actual runoff of the Nile River. These imports are likely to grow in the future. In some other cases, substitutes for water supply, such as desalinated seawater for cash crop irrigation, are used. Furthermore, nonsustainable digging into strategic groundwater reserves is increasingly taking place.

According to the FAO report "Agriculture towards 2030," agricultural water use is expected to grow by some 14 percent by 2030 at global level and at 10 percent only in the Near East region. These rates are much lower than those experienced in the last decade of the 20th century, but are nonetheless very high, considering the intensity of competition for the resource base. I should stress that this assessment is not normative; it therefore does not provide a judgment as to what is good or bad, but uses expert analysis within and outside FAO to project what is likely to happen in 2030. These projections offer a background against which to assess the performance of agricultural water use.

A normative approach would indicate that agriculture, accounting for 70 percent of total water use, is profligate in its use of water and should cap existing levels of use. It has been suggested that food production should grow on the basis of "zero growth" in water use. Water, it is argued, must be freed up for use in higher value and higher "utility" uses. I would argue that these sorts of normative statements are not particularly helpful. They do not help us examine comparative advantage in irrigated or rain-fed agriculture, they do not help us solve local and very real problems with food security, and they do not help us negotiate access to resources to deal with poverty alleviation and economic growth. The facts are that population growth and shifts in food preferences (from grain to animal proteins) will occur irrespective of what water professionals deem appropriate. What is pulling the bulk of our freshwater out of rivers, lakes, and aquifers is the demand for cheap food. This is a demand that will persist and grow.

Historically, improved access to water has been instrumental in buffering the inherent variability of rain-fed production. The Near East is no exception—it provides us with some of the oldest examples of gravity-controlled irrigation systems. Arguably, the advent of pumping technology in the last half of the 20th century has decoupled much agricultural production from the region's natural hydrological risk and has rapidly used up groundwater that has taken centuries and even millennia to build. While reducing the volatility of production patterns has been imperative for the Near East Region, as it has struggled to manage the inherent variability of its rainfall, perhaps nowhere else are the hydrological and hydro-geological systems stretched so tight. It is up to agriculture to respond—to adapt, to improve its water productivity, and to release the pressure before the resource base is exhausted and strategic reserves for vital water are lost.

How can this be done? Well, we will certainly see improvements in crop yields and some improvements in irrigation efficiency and post-harvest processing, but it may be the nontechnical interventions that will play a greater role. Setting up the socioeconomic tools for much more flexible allocation of scarce resources will be pivotal to address equity and efficiency concerns. For instance, improved systems of land tenure and water use rights can help enormously in spreading local water risk and improving food security by optimizing local food production patterns. Equally, the economic signals to food producers have to be clear and stable. Indeed, good agricultural policy can set a strategic balance between rain-fed and irrigated agriculture where there is effective demand and real comparative advantage in domestic resource costs. This is where there is real scope for savings in agricultural water demand.

Finally, let me go back to the issue of hunger. Managing demand for agricultural water *is* negotiable. Managing the demand for food is *not* negotiable for those who are undernourished—so while we are privileged to be enjoying this breakfast, let us think of those in the world who have none this morning, and in the process, do our bit to manage demand for our freshwater resources.

NOTES

1. Keynote speech at World Food Prize Symposium, Des Moines, IA, October 25, 2002.
2. Director-General, Food and Agriculture Organization of the United Nations.
3. Virtual water is the water used in producing tradable agricultural products, and which is therefore traded with the commodities.

13

NEITHER WATER NOR FOOD SECURITY WITHOUT A MAJOR SHIFT IN THINKING — A WATER-SCARCITY CLOSE-UP

M. Falkenmark[1,2] and J. Rockström[2,3,4]

WATER-DEPENDENT LIVELIHOOD SECURITY

The Johannesburg Summit, in line with the earlier Millennium declaration, formulated targets for a successive reduction of poverty, hunger, and unhealthiness. All these goals are in fact water-related. Alleviation of unhealthiness is closely linked to water security in the sense of the access to safe household water (and sanitation). Water security also includes water for irrigation for poverty alleviation through production of cash crops. This means that there is a fuzzy distinction between water security for humans and water security for food production, which one might rather speak of as water-dependent security. But food self-sufficiency in arid and semi-arid regions is not equivalent to enough irrigation water. The crops really do not mind what water the roots can get access to in the root zone, whether infiltrated rainwater or applied irrigation water. When considering the global scale, almost 2.5 times more water is consumed in rainfed crops than by irrigated crops (Rockström et al. 1999).

We may therefore conclude that the millennium goals involve

- Water security for households, employment, and income-raising activities such as cash crops and industrial production
- Food security, which involves enough water to meet the water requirements for crops, whether through irrigated or rainfed production

In the past, when addressing the issue of future food production and water, the approach has been predictions based on plausible assumptions of irrigation development and probable market responses—in other words, taking a forecasting approach. The problem with this approach is that it leaves uncovered a large hidden food gap by 2020 of

360 Mton/yr, which is twice as large as projected developing country imports (190 Mton/yr, Conway 1997). There are two geographical regions in the world particularly affected by the hidden food gap, namely sub-Saharan Africa and South Asia. These two zones can be characterized by their under-nutrition climatology in the sense that they are the two regions with the largest undernutrition and the most rapid population growth. At the same time, they share their location in a semi-arid zone with Savannah climate characterized by distinct dry and wet seasons—large rainfall variability both in terms of intra-annual variability resulting in frequent dry spells, and in terms of inter-annual variability through, e.g., recurrent El Niño events.

If we are, however, serious about the hunger alleviation goal, a major challenge is to try to close the hidden food gap, and if irrigation will not do the trick, other sources have to be found. This study therefore takes a backcasting approach. If we want the world population in its entirety to be nutritionally well-fed by, say, 2025 or 2050, what would that imply in terms of additional consumptive water use? What are the degrees of freedom? What are the main options? What are the challenges to be addressed in terms of trade-offs? The situation in the Savannah zone evidently represents a global hot spot, and the rapid population growth in large parts of that region—even when attention is paid to HIV/AIDS—shows that the situation is one of great urgency. For instance, the present drought situation combined with socioeconomic and political constraints in Southern Africa implies that some 14 million people are currently threatened by severe hunger.

In this study, the focus is on the additional water requirements to feed humanity one and two generations from now, and on the possible sources from which these requirements can be met. Therefore, both *blue water* and *green water* have to be considered; in both cases there will be trade-offs against ecosystems, which makes the balancing a fundamental issue to approach.

WATER REQUIREMENTS TO FEED HUMANITY

Crop water begins when photosynthesis splits the water molecule in the leaves; water is taken in by the roots, rising through plant vessels up to the leaves. When the plant takes up carbon dioxide—the other raw material—from the atmosphere through the stomata openings in the leaves, it releases large amounts of water as vapor. The amount of vapor depends on the evaporative demand and the photosynthetic pathway of the plant.

Let us now look at how much water that will be literally consumed —vaporized—in food production on an acceptable nutritional level as seen on a *per capita level*. Based on crop water requirements to produce different foodstuffs, the composition of different diets, and the food needs for a nutritionally acceptable diet, Rockström (2002) arrived at a per capita water requirement of 1,300 m^3/p yr in consumptive water use, irrespective of whether the roots get the water from infiltrated rainfall or from infiltrated irrigation water. This corresponds to almost 70 times the basic water needed on a household level, if taken as assessed by Gleick (1996) at 50 liters/day. Moreover, this water requirement to sustain human diets can be seen as more or less hydro-climatically generic, in the sense that similar amounts of water will be required to produce food irrespective of hydro-climate. Normally, it is assumed that much more consumptive water use is required to produce food in the tropics than in temperate regions. This is not necessarily the case, especially for grains, which constitute the bulk of the vegetative component of human diets. The reason is that the high evaporative demand in the tropics (resulting in a higher vapor flux from tropical crops) is largely compensated for by a more efficient photosynthetic pathway among tropical crops (C4 crops instead of C3 crops in the temperate-zone). This results in twice as high a carbon assimilation per unit of productive water flow (transpiration). The result is similar water use efficiencies for both temperate and tropical cereals in the order of 1,500 m^3/ton grain (Rockström and Falkenmark 2000).

In this forward-looking approach, we are interested in the *gross amount of water* that will be consumed in producing enough food to feed tomorrow's population. We have, therefore, to assume 1,300 m^3/year for each additional world inhabitant, but we also have to include the additional food needed to raise the nutritional level of the 800 million undernourished individuals in today's world, arriving at the following global amounts of additional consumptive water needs:

by 2025   + 3800 km^3/yr
by 2050   + 5600 km^3/yr

The 3800 km³/yr required by 2025 is a huge amount, in fact close to *all* the water withdrawals at present, sustaining irrigated agriculture, industrial water needs, drinking, sanitation, and other domestic uses. Today, irrigated agriculture represents some 70 percent of the overall water withdrawals of 3900 km³/year, out of which 2/3 is literally consumed. Rainfed agriculture consumes more than twice as much. Moreover, huge amounts of rainwater—in fact 2/3 of all the continental rainfall—are consumed in plant production in natural and anthropogenic ecosystems (forests, grasslands, croplands).

On the regional level, Rockström (2002) identified the following increases of consumptive water needs to 2025 to properly feed the expanding population:

- Sub-Saharan Africa, 3.1 times the present (from 460 to 1450 km³/yr)
- Asia (except former Soviet Union), 2.2 times the present (from 2830 to 6210 km³/yr)

These are not only large additional consumptive water needs, but they will be required in the two regions of the world characterized by the following:

- Largest proportion of water-scarcity–prone agricultural lands
- Highest levels of poverty
- A high degree of present human-induced land degradation, further deteriorating the capacity of the land to produce food

Meeting the Water Requirement

Food production involves a consumptive use of the per capita amounts given above. Altered plant mass production tends to be equivalent to altered runoff generation: a land-use decision is also a water decision. It may therefore influence downstream flow and also aquatic ecosystems.

Irrigated Versus Rain-Fed Production

The engineering approach to crop production distinguishes between irrigated and rainfed agriculture. It pays more attention to the former in view of the evident competition with other water uses and users. Most of the global food production, however, originates from rainfed agriculture. Seen in the drainage basin or catchment context (as in Figure 13.1), rainwater respresents the ultimate water resource, part of which

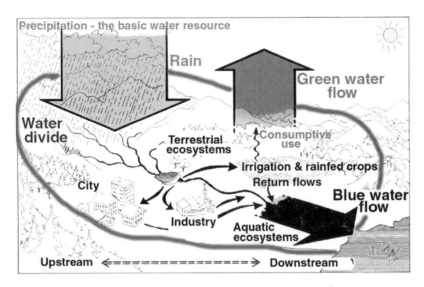

FIGURE 13.1 In a drainage basin perspective, the precipitation over the area represents the proper water resource, part of which is consumed in plant production and evaporation from moist surfaces (green water flow) while the surplus goes to recharge aquifers and rivers (blue water flow), available for societal use.

vaporizes as consumptive water use in plant production (green water flow), while the rest forms runoff (blue water flow).

There is, also, a broad gray area between the two modes of agricultural water use. From the perspective of the crops, the key is the amount of water available in the root zone, not how the water got there—whether it is infiltrated rainwater or applied irrigation water. The green revolution had its focus on irrigated agriculture. Yield levels of primary staple grains such as rice, wheat, and maize, more than doubled as a result of investments in new hybrid seed and fertilizers, which made sense because water supply was secured through the following:

- Adequate access to water
- Adequate purchasing power and human know-how to invest and operate small fuel pumps and irrigation schemes

Now, the challenge of a *new* green revolution stands in upgrading rainfed agriculture in water-scarce tropical environments, where present yield levels, due to frequent water stress and poor land management, oscillate in the region of 0.5–1.5 tons/ha. Dry spell occurrence is a key constraint and an entry point for upgrading. Not only because of the yield response to bridging water stress, but also because the high risk of dry spells affects farmers risk perceptions. The high risk of losing the crop due to water scarcity means a low incentive to invest in much-needed soil fertilization, hybrid seed, pest management, and weeding. But with dry spell mitigation efforts, rainfed agriculture can be upgraded in the tropical regions, doubling or even tripling the yields (small scale, short-term protective irrigation based on rainwater harvesting (Rockström and Falkenmark 2000).

THE GLOBAL PERSPECTIVE: POTENTIAL WATER SOURCES TO MEET FUTURE NEEDS OF ADDITIONAL GREEN WATER

Again, from a global perspective, the result suggests that huge additional amounts of green water flow will have to be appropriated for feeding humanity on an acceptable nutritional level. The crucial question is from where will this water originate? There are three basic sources:

- *Irrigation*, redirecting even more blue water for meeting green water needs—an alternative that is, however, strongly opposed by environmentalists who feel the need to conserve most of the remaining stream-flow for the benefit of aquatic ecosystems (IUCN 2000)
- *Increased "crop-per-drop" efficiency*, by which losses in current agricultural water use could be put to productive use, transforming pure evaporation losses from wet surfaces into productive transpiration through the plant—a solution strongly advocated in the international water community debate
- *Horizontal expansion*, by which green water now used for plant production by natural ecosystems (forests, grasslands) would instead be used for production of crops

Rockström (2002) has analyzed the potential contribution from the first sources, resulting in the following possibilities to meet the water needs two generations from now, by 2050:

Irrigation	maximum 800 km^3/yr;
Crop-per-drop improvements	maximum 1500 km^3/yr;
Horizontal expansion	minimum 3300 km^3/yr.

ATTENTION NEEDED TO TRADE-OFFS INVOLVED

As just indicated, there exists a strong opposition from ecological circles against both large-scale increase of irrigation (due to negative effects on aquatic ecosystems), but also of horisontal expansion (due to effects on terrestrial ecosystems). It is evident from the sheer scale of these assessments, however, that informed trade-offs will have to be made. What problems will have to be addressed? What sort of balancing between man and nature will be needed? And what would the criteria for priority setting be? Let us look closer at the three alternatives.

IRRIGATION

Irrigation involves redirecting blue water during the growing season, turning it into consumptive green water flow. During the wet season, the effect will basically be reduced flood flow. During the dry season, the resulting reduction of dry season flow may be more problematic. Current examples of river depletion are offered by the following:

- The Yellow river, which in 1997 went dry in the downstream stretch seven months a year
- The Aral Sea region where the river inflow has decreased to 10 percent of the natural flow, causing the lake evaporation to take over and the lake to shrink dramatically

Through water storage, wet season flow can be stored for use during the dry season.

Improved water productivity (crop-per-drop) can be secured in different ways in both rainfed and irrigated agriculture. On the one hand, infiltration possibilities can be improved by soil conservation measures so that more rainwater can infiltrate. This will also reduce the destructive overland flows that tend to cause severe erosion damage in large parts of the tropics. On the other hand, evaporation losses between plants can be reduced by increased foliage—for instance, by protecting the plants from dry spell damage to the roots (protective irrigation during dry spells with locally harvested overland flow). Depending on where the harvested blue water was heading—on its way to a local stream or on its way to evaporate—downstream effect may or may not happen. In irrigation systems, losses may be reduced by covering the canal or by lining the canal. In the latter case, however, groundwater recharge is reduced with possible downstream effects on groundwater-fed wetlands, or wells used for local water supply.

Interestingly, the largest and most immediate water productivity improvement can be achieved by increasing yield levels. Contrary to common assumptions, the relationship between water productivity and yield levels is dynamic, in the sense that every incremental increase in yield will improve the ratio of productive green water to total green water flow (the ratio of transpiration to total evaporation). This in turn will improve the water productivity, especially in the low yield range where nonproductive evaporation still constitutes a large share of total vapor flow from cropped land. Figure 13.2 illustrates this dynamic relationship between water productivity and yield for tropical grains grown in Savannah agroecosystems. As seen from Figure 13.2, doubling yield levels from 1 ton/ha (the present average grain yield in, e.g., sub-Saharan Africa) to 2 tons/ha would result in a consumptive water saving of approximately 1000 m³/ton.

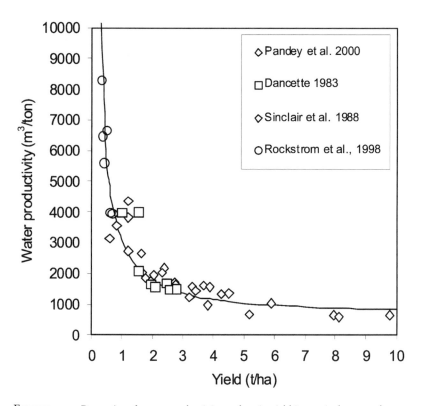

FIGURE 13.2 Dynamics of water productivity and grain yield in tropical savannahs.

HORIZONTAL EXPANSION

Horizontal expansion, turns forested land or grasslands into croplands, involves a land cover change, and may have effects on rainwater partitioning and therefore on local runoff generation. In cases where a year-round green water flow from a forest is replaced by a seasonal one from an annual crop, groundwater recharge and/or runoff production may increase. In Australia, where immigrants from Europe cleared the woodlands for croplands, the outcome was a disastrous, regional scale waterlogging and salinization (so-called dry-land salinization). The hydrological consequences of replacing grasslands for croplands are more complex, however, and difficult to generalize.

The balancing needed between water for existing ecosystems and water for feeding a growing human population will evidently be a difficult one. The International Water Management Institute (IWMI) has joined a large number of other international organizations, among them the International Union for Conservation of Nature (IUCN), initiating a broad dialogue on water, food, and environment. The aim is to find the way out of this considerable dilemma, which will need large-scale international attention in the next few decades.

REGIONAL CONTRASTS

OPTION 1: IRRIGATION

We may now try to find out how much of the huge water requirements needed to feed rapidly the growing regional populations can be met by irrigation from blue water sources. This is the same method that made the green revolution possible in regions with easy access to blue water in rivers and groundwater aquifers. The regional differences of population pressure on blue water availability are quite large, both in terms of technical scarcity (the possibility to mobilize more blue water to meet increasing demands) and in terms of demographic scarcity (population pressure on blue water availability or "water crowding"). The latter refers to number of individuals sharing each flow unit of blue water (see Figure 13.3). Evidently, population growth increases water crowding, and therefore also proneness to disputes.

Some irrigated regions are close to the "blue water ceiling" due to high per capita water use and have limited possibilities to mobilize even more. Other regions with mainly rainfed agriculture and therefore low per capita demands are very low on the technical axis but subject to

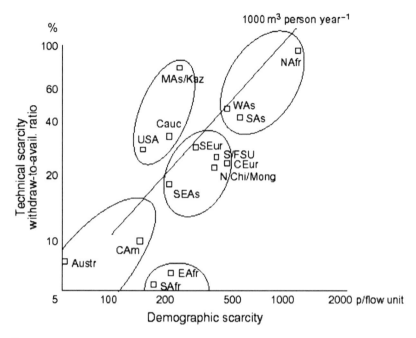

FIGURE 13.3 Regional differences between five different region clusters of the world in terms of population pressure on blue water availability (demographic scarcity, horizontal axis, people per flow unit of one million cubic meters per year) and withdrawal ratio (technical scarcity, vertical axis, withdrawal-to-availability ratio in percent). The diagonal line shows per capita water withdrawal of 1000 m³/p yr, needed.

rapid population growth increasing both food needs, water crowding, and dispute proneness. The diagram suggests that the two regions in the diagram denoted SAs (South Asia) and SEAs (Southeast Asia) are already high on the technical stress dimension but might be able to mobilize limited amounts of more water for irrigation purposes. The opposite is true for the sub-Saharan African regions (Southern Africa, East Africa, West Africa—the latter occurring just below the 5 percent axis). Most agriculture in these regions is rainfed. The population is pushed very rapidly toward higher water crowding levels. These regions will have grave difficulties mobilizing the water needed to support their food production needs by irrigation since they are poor in coping capability (expertise, data, financing sources).

In a river basin perspective, the situation may vary quite a lot, especially in the sense of whether the basin is open or closed (Keller et al.

1996). The former means that more blue water can be mobilized without serious effects on aquatic ecosystems, while the latter means that there is no more blue water that can be mobilized and put to additional productive consumptive use without serious effect on the aquatic ecosystems.

OPTION 2: UPGRADING RAIN-FED AGRICULTURE

Recall the earlier observation that the Savannah zone is the most critical hot spot globally. Falkenmark and Rockström (1993) have shown that the Savannah zone suffers from multiple bases for water scarcity:

- Unreliable rainfall with heavy downpours but also frequent dry spells even during wet years
- Frequent drought years due to El Niño effect
- Vulnerable soils that easily form crusts impeding infiltration
- Low runoff generation so that small rivers go empty except during the rainy period

Therefore, semi-arid regions with mainly rain-fed agriculture have particular food production challenges due to the combination of water scarcities A, B, and C and how they influence the water available to the plants in the root zone.

Rockström and Falkenmark (2000) also discussed the farmer's field dilemma, finding that the water problems can be considered under three categories (see Figure 13.4):

- Climatic deficiency: less rainfall than crop water requirements
- Soil deficiencies: infiltration problems so that part of the rainwater forms overland flow and low water holding capacity so that part of the infiltrated rainwater proceeds below the root zone, recharging groundwater
- Plant deficiencies: plants damaged by dry spells with poor capacity to absorb water available in the root zone

They also showed that, even in the Savannah zone, the potential crop yields—if soil and plant deficiencies could be mitigated—are around 6 tons/ha, whereas actual yields on the farmer's field are 0.5–1 ton/ha. How could this be possible, given the water-related constraints? The reason is that poor distribution of rainfall over time often constitutes a larger water problem than lack of water even in dry regions. There is enough total rainfall to produce food; it is just not accessible to the roots in the right time. Wise water management can assist in mitigating water

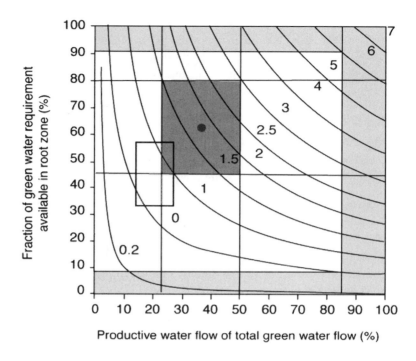

FIGURE 13.4 Analysis of the effects of rainfall partitioning and plant water uptake capacity on maize grain yields under semi-arid conditions. The larger shaded area shows the range of yields experienced on average in farming systems in sub-Saharan Africa; the smaller shaded area shows the yield range on degraded farmer's fields (Rockström and Falkenmark 2000).

scarcity, by leveling out the periods of excess and scarcity of water so characteristic for drought-prone tropical agroecosystems. There is a *huge window of opportunity* by combining soil/water management and dry spell mitigation based on local water harvesting. Experience from India shows that supplemental irrigation of rainfed grain crops with 50 mm/ha per season, resulted in average yield increases of 90 percent (Sivanappan 1997). Research in Burkina Faso on sorghum and on maize in Kenya shows similar results where supplemental irrigation from small water harvesting systems increased yield levels with 30–60 percent (Fox and Rockström 2000; Barron et al. 1999). Combining water harvesting with soil fertility management in these experiments, resulted in yield increases of 60–200 percent, indicating the critical need to address water and soil management together. Interestingly, in these on-farm experiments, even in the water-scarce Sahelian location in Northern Burkina Faso, soil nutrients were proven to be more limiting for crop growth than water alone. The critical role of plant nutrients, even in dry lands, has been pointed out by several authors (Breman et al. 2001; Klaij and Vachaud 1992).

Farmers in arid and semi-arid environments have been collecting surface runoff in water harvesting systems to upgrade rainfed agriculture for millennia. However, in many parts of the world, water-harvesting systems have been abandoned as a result of modernization of agriculture, especially in terms of irrigation development. In other parts of the world, such as large parts of sub-Saharan Africa, water harvesting for supplemental irrigation still constitutes innovative technologies that need to be adaptively tested and co-managed with local communities. Recently there is a new interest in water harvesting as a result of the realization of the important role played by rainfed agriculture in efforts of achieving food security among rural poor in the future. There is at present very little hydrological knowledge on the potential of water harvesting in semi-arid tropical farming systems and their impact on water availability for ecosystem support downstream.

COMPARING THE HUNGER GAP REGIONS

The options open for increased food production are different between different world regions. The two regions where the dilemma is largest are the semi-arid regions in sub-Saharan Africa and South Asia, both regions with large undernutrition and rapid population growth.

In South Asia, horizontal expansion is highly limited: Most land is already used and there are only limited reserves of arable but still unused land. The main options are therefore "crop per drop" and irrigation. To the degree that this will not be enough to feed tomorrow's populations, food will have to be imported.

In sub-Saharan Africa, however, plenty of unused land remains, mainly under forests. Since 95 percent of the farmers are rainfed and there is only limited irrigation, crop-per-drop in the sense of increased irrigation efficiency will contribute only to a limited degree. There are, however, considerable possibilities for upgrading rainfed agriculture, provided that dry spell mitigation can be developed on a regional scale and be made attractive among the sub-Saharan farmers (Rockström and Falkenmark 2000). During the transient process of social change and changing farmer attitudes in risk assessment, food import/food aid will probably have to play a central role.

A new research area appears in this connection: the water perspective of food trade and the future flows of so-called virtual water. This is the water involved in the production of food transferred from one region, better endowed in terms of water needed for food production, to a water-deficient region with large food needs. Japan has recently assessed its dependence of virtual water flow to 103.5 km^3/yr, which is almost 20 percent more than all domestic withdrawals (89 km^3/yr, Oki 2002).

MAJOR SHIFTS IN THINKING NEEDED

This backcasting study of the water needs for feeding humanity in the next half century has shown that major changes can be foreseen. The reported failure of the Johannesburg discussions of future food production should therefore cause serious concern. Already in the next generation, an additional amount of green water will be needed that is equivalent in size to *all* blue water use by humanity today. In the second generation, another 60 percent will be needed.

The study has also shown that the past approach, limited to irrigated agriculture and blue water needs only, will be highly insufficient. Probably, only some 14 percent of the additional water requirements may be covered. There will be no food security without a major shift in thinking. A new approach will have to be taken to address crop water requirements and the possibilities of meeting those requirements. It is no longer the irrigation needs that will remain in focus, but the overall water requirements—whether met by infiltrated rainfall or supplied irrigation water. Plant production will have to be addressed by referring to both green and blue water flows.

The crop water requirements represent green water flows. But when these flows change as a consequence of land cover change, runoff generation will be influenced and therefore also blue water flow. Conventionally, such relations were covered by the concept "water balance changes" but did not attract much interest, probably since the evaporative demand in the temperate zone tends to be too low to generate distinguishable stream-flow changes. South Africa has, however, started to refer to forest plantations as a "stream-flow reducing activity" for which foresters will have to pay.

Although the problems are already acute, there remains a *conceptual retardation* to be overcome.

This paper has shown that the green water approach is clarifying and gives an idea about the scale of the dilemma of feeding humanity and living up to the millennium declaration. The largest immediate challenge will be to prepare conceptually for the necessary trade-offs between water for humans and water for nature.

In the new approach, agricultural engineering will have to be complemented with agro-ecohydrology. There will have to be an active bridge-building between ecology and hydrology so that the conceptual void between climate, plant production, and stream-flow can be filled. Finally, virtual water flows will have to be focused.

SUMMARY

Future world food security represents a massive challenge, with the solutions hidden behind inadequate concepts. Water is a major entry point. Water-scarce regions are those suffering most from undernutrition, poverty, and population growth, and therefore represent the largest needs. In terms of implications, the situation presents an urgent need to realize that nothing less than a "New Green Revolution" is required, which targets the poverty-stricken rainfed farming sector in water-scarce environments. Even if successful, the achievements of a new green revolution will still leave large expectations on food trade, and large-scale ecological consequences may be foreseen.

REFERENCES

Barron, J., J. Rockström, and F. Gickuki. 1999. Rain water management for dryspell mitigation in semi-arid Kenya. *East African Agriculture and Forestry Journal* 65(1):57–69.

Breman, H., J.J.R. Groot,and H. van Keulen. 2001. Resource limitations in Sahelian agriculture. *Global Environmental Change* 11:59–68.

Conway, G. 1997. *The Doubly Green Revolution*. Ithaca, NY: Cornell University Press.

Dancette, C., 1983. Estimation des besoins en eau des principales cultures pluviales en zone soudano-sahélienne. *L'Agronomie Tropicale* 38(4):281–294.

Falkenmark, M., and J. Rockström. 1993. Curbing rural exodus from tropical drylands. *Ambio* 22(7):427–437.

Fox, P., and J. Rockström. 2000. Water harvesting for supplemental irrigation of cereal crops to overcome intra-seasonal dry-spells in the Sahel. *Physics and Chemistry of the Earth, Part B Hydrology, Oceans and Atmosphere* 25(3):289–296.

Gleick, P. 1996. Basic water requirement for human activities: meeting basic needs. *Water International* 21:83–92.

IUCN 2000. *Vision for Water for Nature.* Gland, Switzerland: IUCN.

Keller, A., J. Keller, and D. Seckler. 1996. *Integrated Water Resource Systems: Theory and Policy Implications.* Research Report 3. Colombo, Sri Lanka: International Water Management Institute.

Klaij, M.C., and G. Vachaud. 1992. Seasonal water balance of a sandy soil in Niger cropped with pearl millet, based on profile moisture measurements. *Agricultural Water Management* 21:313–330.

Oki, T. 2002. World water resources and global climate change. *Frontier Newsletter* 19, July 2002. Institute of Industrial Science, Tokyo.

Pandey, R.K., J.W. Maraville, and A. Admou. 2000. Deficit irrigation and nitrogen effects on maize in Sahelian environment. I. Grain yield and yield components. *Agricultural Water Management* 46:1–13.

Rockström, J. 2002. *Water for Food and Nature in Savannahs—Vapour Shift in Rainfed Agriculture.* Manuscript.

Rockström, J., and M. Falkenmark. 2000. Semiarid crop production from a hydrological perspective: Gap between potential and actual yields. *Critical Reviews in Plant Sciences,* 19(4):319–346.

Rockstrom, J., L. Gordon, M. Falkenmark, C. Folke, and M. Engvall. 1999. Linkages among water vapor flows, food production, and terrestrial ecosystem services. *Conservation Ecology* 3(2):5:1–28. http//www.consecol.org/vol3/iss2/art5.

Rockström, J., P.-E. Jansson, and J. Barron. 1998. Estimates of on-farm rainfall partitioning in pearl millet field with run-on and runoff flow based on field measurements and modelling. *Journal of Hydrology* 210:68–92.

Sinclair, T.R., C.B. Tanner, and J.M. Bennett. 1984. Water-Use-Efficiency in crop production. *BioScience* 34(1):36–40.

Sivanappan, R.K. 1997. *State of the Art in the Area of Water Harvesting in Semi-Arid Parts of the World.* Paper presented at the international workshop on water harvesting for supplemental irrigation for staple food crops in rainfed agriculture. Stockholm University, Department of Systems Ecology, June 23–24, 1997, Stockholm, Sweden.

NOTES

1. Stockholm International Water Institute; Hantverkargatan 5, House 6; SE 112 21 Stockholm; malin.falkenmark
2. Department of Systems Ecology, Stockholm University.
3. Unesco-IHE; P.O. Box 3015; 2601 DA Delft; the Netherlands; rockstrom
4. WaterNet, University of Zimbabwe; P.O. Box MP 600; Harare, Zimbabwe.

14

MAJOR PROBLEMS
IN THE GLOBAL WATER-FOOD NEXUS
David Seckler and Upali Amarasinghe

Prophecy is a good line of business, but it is full of risks.
Mark Twain, *Following the Equator*

Water and food are two of the basic necessities of life—and food production depends crucially on water. In this chapter we provide some of the major results of research in the global water-food nexus, as we call it, by the International Water Management Institute (IWMI).

The paper is divided into two parts. Part I presents the state of global water scarcity by country in 1995 with projections up to 2025. It is found that fully one-third of the people of the world in 2025 will live in countries suffering absolute, physical water scarcity. That is, they do not have sufficient water resources to meet minimum domestic, agricultural, and environmental needs even with full development and most productive utilization of their water resources. Another 45 percent of the people in 2025 will live in countries that have sufficient water resources to meet their minimum needs, but which will have to embark on extremely expensive and possibly environmentally harmful water development projects to actually utilize these resources. In short, the water outlook for countries containing two-thirds of the world's 2025 population is grim.

Part II discusses some of the major issues underlying these estimates and projections, and examines various courses of action for alleviating the problems of water scarcity. Since agriculture consumes over 70 percent of the world's developed water supplies, it is clear that most of the progress in conserving water has to be in the agricultural sector. But this is not as easy as many people think. While the efficiency of water use in agriculture is not as high as it could be, it is much higher than is commonly thought. Certainly, the greatest gains in the productivity of water in agriculture to date originate from outside of the water field per se, in crop science. By increasing the yield per unit of land, shortening the growing season, developing drought-resistant varieties, and extending productive crop systems to cool areas, crop research has substantially increased the productivity of water used in agriculture. Much more research on this subject is needed, and IWMI is shortly convening a workshop of leading water and crop scientists to develop a plan of integrated research in this area. Several other areas of opportunity are discussed in the text. These range from the vast but generally illusive potential of rain-fed agriculture, through small dams and water harvesting technologies, to increased water use efficiency through sprinkler, drip, and other kinds of improved irrigation systems.

PART I: GLOBAL WATER SCARCITY 1995–2025

The estimates and projections of global water scarcity are done through IWMI's Policy Interactive Dialogue Model (PODIUM), which is designed to simulate alternative food and water scenarios of the future. The results presented here are based on what we call the *basic scenario*.

The basic scenario is rather optimistic. Within an overall framework of social, technical, and economic feasibility, it relies on substantial investments and changes in policies, institutions, and management systems intended to achieve four major objectives:

- Achieve an adequate level of per capita food consumption, partly through increased irrigation, to substantially reduce malnutrition and the most extreme forms of poverty.
- Provide sufficient water to the domestic and industrial sectors to meet basic needs and economic demands for water in 2025.
- Increase food security and rural income in countries where a large percentage of poor people depend on agriculture for their livelihoods through agricultural development and protection from excessive (and often highly subsidized) agricultural imports.
- Introduce and enforce strong policies and programs to increase water quality and support environmental uses of water.

Realizing these objectives requires three major actions in the field of water resources and irrigation management in water-scarce countries:

- Greatly increase the productivity of water resources use.
- After productivity is increased, there generally remains a need for substantial increases in the amount of developed water supplies.
- Water resources development must be done with substantially reduced social and environmental costs than in the past—and people

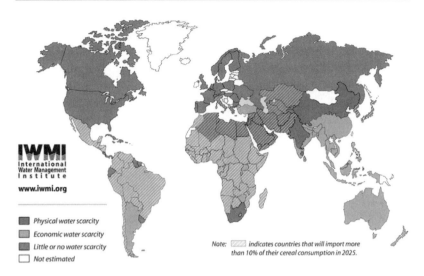

IWMI
International
Water Management
Institute
www.iwmi.org

- Physical water scarcity
- Economic water scarcity
- Little or no water scarcity
- Not estimated

Note: [ ] indicates countries that will import more than 10% of their cereal consumption in 2025.

FIGURE 14.1 Projected water scarcity in 2025.

must be willing to pay the increased financial costs this policy necessarily entails.

As shown in Figure 14.1, we have grouped the major countries of the world into three basic categories of water scarcity:[1]

- *Group I* represents countries that face *physical water scarcity* in 2025. This means that, even with the highest feasible efficiency and productivity of water use, these countries do *not have sufficient water resources* to meet their agricultural, domestic, industrial, and environmental needs in 2025. Indeed, many of these countries cannot even meet their present needs. The only options available for these countries are to invest in expensive desalinization plants and/or reduce the amount of water used in agriculture, transfer it to the other sectors, and import more food.
- *Group II* represents countries that face economic water scarcity in 2025. These countries have sufficient water resources to meet 2025 needs, but they will have to increase water supplies through additional storage, conveyance, and regulation systems by 25 percent or more over 1995 levels to meet their 2025 needs. Many of these countries face severe *financial and development capacity* problems in meeting their water needs.

- *Group III* consists of countries that have no physical water scarcity and that will need to develop less than 25 percent more water supplies to meet their 2025 needs. In most cases, this will not pose a substantial problem for them. In fact, several countries in this group could actually decrease their 2025 water supplies from 1995 levels because of increased water productivity.
- *Crosshatched* countries are projected to import over 10 percent of their cereal consumption by 2025. The correlation between the crosshatched set of countries and Group I countries is clear—but there are many reasons for food imports other than water scarcity.

PODIUM operates at the country level. Therefore, it generally ignores the substantial differences in water scarcity within countries at the levels of regions or river basins. For example, about one-half of the population of China lives in the wet region of southern China, mainly in the Yangtse basin, while the other one-half lives in the arid north, mainly in the Yellow river basin. This is also true of India, where about one-half of the population lives in the arid northwest and southeast, while the remaining one-half lives in fairly wet areas. Much the same is true of many other countries. A particularly vivid example is Mauritania, which is mostly desert but falls in Group II. The reason is that 90 percent of the total population lives along the southern border—along the Senegal River. This geographic issue needs to be addressed in the future; but for now, it is sufficient to say that we have ignored regional differences in the group classifications for all countries except India and China because of their huge size in terms of population and water use. IWMI plans to make further regional distinctions for countries, like Mexico, that have large regional disparities in water and agriculture (also see Alcamo et al. 1999).

It is also important to understand that we have used substantially lower population projections in preparing these estimates than are normally used. The United Nations presents *High, Medium,* and *Low* projections for 2025 (U.N. 1999). Almost no one now believes the High projections are relevant. Most people use the Medium projection. For reasons explained in Seckler and Rock 1995, we believe that the Low estimate is the best one. There are substantial differences between these projections. The Medium one projects a 38 percent increase in population over the period, to 7.8 billion people in 2025, whereas the Low one projects an increase of 28 percent, to 7.3 billion. However, in the spirit of compromise we have used the average of the Medium and the Low projections for 2025. A major consequence of these lower projections is that by 2040 population growth will have slowed virtually to zero. Thus, if the world can satisfy its food and water demand over the next 30 to 40 years, most of the problems will have been solved for the foreseeable future.

Summarizing, the following is projected by 2025:

- 33 percent of the population of 45 countries will be in Group I, with physical water scarcity.
- 45 percent of the population will have substantially underdeveloped water resources, requiring 25 percent or more development of additional water supplies.
- 22 percent of the population, mainly developed countries, will have little or no water scarcity.

Together, Groups I and II contain 78 percent of the population in 2025. Of course, this does not mean that everyone in these countries will directly be experiencing water scarcity. As usual, the economically better off members of most countries will have enough water and food, while poor and weak people will suffer the major part of the burden.

PART II: WATER AND FOOD—DEMAND AND SUPPLY

Food Demand

While most of this paper is concerned with the supply side of the water-food nexus, supply is meaningless without considering demand. Thus this section briefly outlines some of the major issues in world food—and, hence, water demand—and their implications for water supply.

FAO provides excellent data on food production and consumption in the world, conveniently entered on a CD-ROM (FAO 1998). These data are used extensively in PODIUM, and we are grateful to FAO for this and other data on agriculture and irrigation.

The IMPACT model of the International Food Policy Research Institute (IFPRI) provides projections of food demands for 16 major countries and 22 inter-country regions in 2025. PODIUM uses the food *demand* projections of the IMPACT model, adjusted for the population projection noted before (and much of the water part of PODIUM has been incorporated into the IMPACT model). However, the food *supply* projections are done independently in PODIUM. PODIUM also provides a means for policymakers to change the projections of food demand in order to *target* the nutritional standards they wish to achieve for their countries in 2025. Once these targets are set, the model provides a means of testing the feasibility of these targets in terms of agricultural and water constraints and the actions needed to achieve the targets.

The single most important component of nutrition is calorie consumption per capita. The average for developing countries is around 2,200 kcal/person/day. With reasonably varied diets, if people satisfy their calorie requirements, they will also satisfy their requirements for protein, minerals, and vitamins. A major exception to this rule is when a very high percentage of total calories is from rice, which is low in protein. Other exceptions occur with low vegetable consumption, which may cause vitamin and mineral deficiencies. But, on the whole, the principal target is adequate calorie consumption.

But even if the average calorie intake of a country is 2,200 kcal/capita/day, this is not enough to assure that everyone in the country is actually obtaining enough. People with relatively high incomes tend to overconsume calories, mainly from animal products. Therefore, it is necessary to get substantially higher average calorie consumption in a country to attempt to achieve the minimum for poor people. How much higher this amount must be is largely a function of the distribution of income in a country. As a rule of thumb, something in the range of 2,700–3,200 kcal/day is adequate for most countries to satisfy basic food needs, depending on the distribution of income and other factors in individual countries.

One of the most difficult issues in projecting the demand for food and related agricultural products in 2025 is consumption of animal products —meats, milk, cheese, etc. In most countries, the total calories consumed and the percentage of calories from animal products increase with income, even at high-income levels. However, because of a variety of causes, including urbanization, health concerns, and costs it is likely that there will be

- A reduction in excessive per capita calorie consumption by higher-income groups
- A rapid growth in consumption per capita of meat products in developing countries, such as China and India, as incomes increase,

combined with a tendency to plateau at lower levels of consumption than in the traditional meat-consuming countries of the west
- A shift toward more vegetarian, or "Mediterranean," diets, away from meats
- A shift from red meats, notably beef, to white meats, notably chicken

These changes on the food demand side will be accompanied by major changes on the food supply side. Traditional forms of animal husbandry produce most of the animal products in developing countries, where animal feeds are mainly from pastures and other lands not suitable for crops and from waste products. But the carrying capacity of these traditional feed resources is reaching its practical limit and most of the *additional* production of animal products will be from modern, commercial production units that depend on animal feeds (e.g., maize, barley, soybean meal, etc.). However, the carrying capacity of pastures in developing countries could be greatly increased with rotational grazing, better seeds, and application of inorganic fertilizers (and this would increase the supply of organic fertilizers for crops).

Given the propensity to consume more animal products and the conversion from traditional to commercial production of these products, it is reasonable to assume that the production of feedstuffs will have to increase dramatically by 2025. However, it is a remarkable fact that while world consumption of animal products has increased rapidly, consumption of feed cereals has increased very slowly, at an annual rate of only 0.5 percent since 1985. Somehow, the world has received a "free lunch" in the production of animal products. Part of the reason for this is shown in Figure 14.2. Developed countries barely increased consumption of feed cereals at all since 1985; developing countries nearly doubled their consumption from a comparatively small base, but this was offset by the decreased consumption in transitioning economies.

Underlying these data are changes in the production of animal products. Three important factors relating to the *conversion ratio*—the kg of feed required to produce one kg of animal products—have played an important role:

- The conversion ratios for red meats are about twice as high as those for white meats; thus, as consumption shifts from the former to the latter, feedstuffs are freed to produce more animal products.

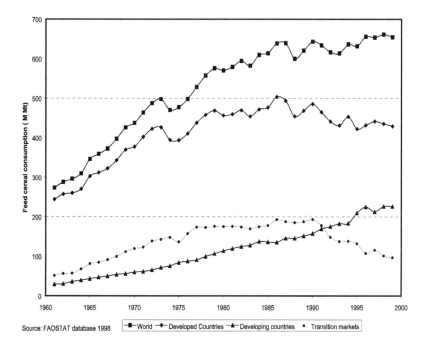

Source: FAOSTAT database 1998   ■ World ◆ Developed Countries ▲ Developing countries • Transition markets

FIGURE 14.2 Total domestic feed cereal consumption.

- The conversion ratios of all animal products have been decreasing rapidly due to technological change in terms of animal breeding, health, and nutrition. This frees up more feed to support additional consumption.
- There has been some substitution of feed cereals by other feeds, like oil meals and cassava.[2]

These factors all tend to decrease the conversion ratio for feed cereals. However, there is an offsetting factor. As developing countries move from traditional to commercial sources of feed for production, their incremental conversion ratios will *increase*. These effects have been incorporated in the analysis; conversion ratios decrease between 1995 and 2025 in most developed countries that are already under commercial forms of production of animal products, but they increase in most developing countries that have traditional forms of production.

It should be noted that the rapid growth in consumption of vegetables and fruits *increases* the demand for irrigation. The reason is that highly productive vegetable and fruit production is not possible without very good irrigation and drainage systems.

In sum, the demand for cereal grains is projected to increase by 37 percent in 2025; 49 percent of this increase is demand for feed grains. The demand for all food—including cereals, fruits, vegetables, etc.—is projected to increase by 40 percent. After 2025, the rapidly decreasing growth of the world's population will make the task of meeting food demands much easier.

WATER SUPPLY: IRRIGATED AND RAIN-FED AGRICULTURE

Irrigated agriculture has provided the base for the green revolution of the past 40 years and, hence, the source of most of the growth in food production over this period. As the World Bank observes (World Bank 1997):

> Irrigated farmland provides 60 percent of the world's grain production. Of the near doubling of world grain production that took place between 1966 and 1990, irrigated land (working synergistically with high-yielding seed varieties and fertilizer) was responsible for 92 percent of the total. Irrigation is the key to developing high-value cash crops. By helping guarantee consistent production, irrigation spawns agro-industry. Finally, irrigation creates significant rural employment. The Bank has been a major actor in the expansion of irrigation systems . . . More than 46 million farming families have benefited directly from the Bank's irrigation activities.

As shown in Figure 14.3, irrigated areas in the world have continued to expand at a fairly constant rate up to the present, with decreases in the growth rate in developed countries offset by increases in Asian and other developing countries. It should also be noted that the increase in gross irrigated area is probably even greater than indicated in Figure 14.3 since the extent of multiple cropping on irrigated areas has expanded greatly—largely through the use of tubewells in the dry season. This effect probably more than offsets the (uncounted) loss of irrigated areas due to urban sprawl, soil salinity, and other factors.

However, virtually since large-scale irrigation development began, it has been attacked by critics who contend that in terms of costs, equity, environmental quality, and even total food production, it would be better to invest in improved rain-fed agriculture. The total cultivated area of the world is about one billion hectares, of which only about one-third is irrigated. Thus, a 10 percent increase in the productivity of rain-fed agriculture would have twice the impact as the same increase in irrigated agriculture. Because the beneficial impact would be largely on poor farmers in marginal areas, this is an enormously attractive idea.

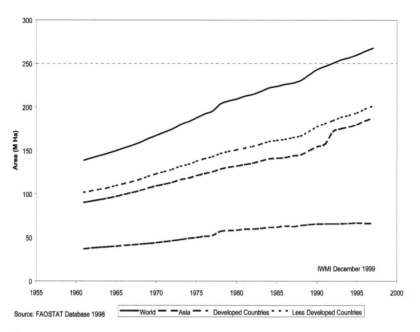

FIGURE 14.3 Net irrigated area of the world, Asia, DCs, and LDCs, 1961–1997.

This is by no means a new idea. The goal of increasing productivity of marginal rain-fed areas has been energetically pursued, using all the tools of agronomic science, for at least a century, with generally disappointing results—especially in developing countries. We believe that the sciences and technologies of agronomy and water management have now advanced to the point where there are grounds for optimism in this field—and, indeed, there are notable cases of success on the ground. But, before solutions can be found, the depth and extent of the problems must be thoroughly understood.

A major part of the problem is shown in Figure 14.4 (Hargreaves and Christiansen 1974). The vertical axis represents the relative yield; this is the actual yield obtained divided by the potential yield with all other factors, such as seeds and fertilizers at their physically optimum levels. The horizontal axis shows the relative water supply; this is the actual water supplied divided by the physically optimal water supply.

While the amount of water that needs to be supplied to crops differs enormously among agroclimatic regions, most crops have nearly the

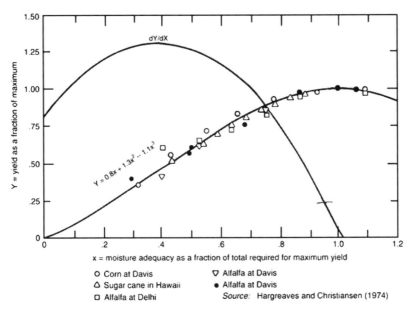

FIGURE 14.4 MOISTURE ADEQUACY AND YIELD FUNCTION.

same water requirement in the same agroclimatic region. This is because evaporation is by far the major determinant of crop water requirements. Thus the idea of conserving water by low–water-using crops is largely a myth. The exceptions are cool-weather crops—such as barley, sugarbeets, and winter wheat—and very high-yielding crops—such as maize and potatoes, which produce more per unit area and time, thus per unit of water (more "crop per drop").

Figure 14.4 helps to explain the great diversity of rain-fed yields in the world. On the one hand, there are vast areas of the more favored rain-fed areas—such as parts of the American Midwest and north-central Europe—which have reasonably adequate and reliable water supplies and thus are close to the optimum conditions for high yields. But most of these favored areas have already been fully exploited. The lower one-third of the relative water supply axis unfortunately, characterizes most of the underdeveloped rain-fed areas of the world, where yields are 25–35 percent of potential. There are also large areas of the world, indicated by the right side of the yield curve, that suffer from too much water. Many of these areas can grow only rice. The combination of high humidity and temperature in these areas also contribute to the growth of plant and animal pests and diseases, which further reduces agricultural productivity. Other researchers have found a significant positive correlation between agricultural productivity and the number of days of frost (which retards pests and diseases), up to a limit on days of frost.

These complications are considered in Droogers et al. (2001), which shows the maximum potential yield of crops in the world based on purely physical factors, such as soils, precipitation, radiation, and temperature. The maximum potential yield assumes ideal levels of inputs, such as fertilizers and farm management, and is thus unrealistic. But it provides an indicator of the physical constraints confronting farmers—and how, in some cases, farmers can overcome these constraints. The following are some more notable features of their analysis:

- The north-central regions of South America appear to have vast and underexploited potential for growth of agricultural production. The same is true of the central region of Africa. However, diseases and pests are a major problem in many areas of Africa.
- The highly favored regions of Indonesia and parts of India and China are so because of the water tolerance of rice.
- The low-to-moderate potential of the American Midwest, combined with its high productivity, reflects the triumph of good farm management.
- Many productive but low-potential areas, such as the western parts of the U.S. and Canada and the U.S.S.R., show the importance of winter wheat, fallowing, and irrigation.

Their results provide grounds for optimism on the potential for rain-fed agriculture. But before substantial progress can be made in this field, the problems must be clearly understood. This discussion concentrates on the dry end of the rain-fed spectrum, on marginal rain-fed areas. There are three central problems:

- Most of a farmer's costs are the fixed costs of cultivating land *area*, independently of yield. Thus as yields decrease, net returns to farmers decrease even faster. For example, if costs represent 2 Metric tons per hectare (MT/ha), the farmer earns a net of 3 MT/ha at an economic maximum yield of 5 MT/ha, with optimal water supply. But the farmer makes only 1 MT/ha if yield is reduced to 3 MT/ha due to deficient water supply.

- In most cases, rainfall is highly unreliable. Farmers rationally minimize their investments in labor, improved seeds, fertilizers, soil and water management, and the like to minimize losses due to drought. But this lack of investment in productive inputs means that even when good rainfall occurs, the yield is not as large as it should be.
- Since rainfall affects large areas, prices rise dramatically in times of drought, when there is nothing to sell, and collapse in periods of good rainfall, when harvests exceed subsistence needs and there is a lot to sell.

These problems have been partly overcome in marginal rain-fed areas of developed countries—such as the U.S., Canada, and Australia by large-scale, well-capitalized, and highly mechanized farming. With several hundreds, if not thousands, of hectares per farm unit, large tractors and other equipment, and sufficient capital to tide them over drought years, marginal rain-fed areas can be profitably farmed. Mechanization provides the ability to practice a variety of water and soil conservation practices—such as land leveling, terracing, fallowing, low-till agriculture, etc.—that are difficult and costly, if not altogether impossible, with only human and animal power. Because of their financial resources, these large farms can survive one out of three or four drought years. We believe that much of the future production of rain-fed farming in marginal areas will depend on the ability to bring these advantages of large-scale farming to small-scale producers through various methods of collective action (see Seckler 1992). But the history of such institutional innovations in developing countries has not been encouraging, to say the least.

It is hoped that advances in biotechnology will result in drought-resistant and more water-efficient crops. One problem with this idea is that, hitherto, drought-resistant crops and varieties are, for that very reason, low yielding. Such a crop may produce a more stable yield over varying climatic conditions but at such a low yield potential that it is uneconomical or unable to respond to favorable conditions.

However, *under specific agroclimatic conditions*, small-scale farming can be productive in marginal rain-fed areas through *supplemental irrigation*. Of course, all irrigation is supplemental irrigation because it is designed only to "top up" effective precipitation on the crops. But supplemental irrigation is a technique specifically designed for water-scarce regions, where scarce water is stored and used only in limited quantities at the critical growth stages of crops.

In many areas, for example, there is sufficient *average* rainfall over the crop season to obtain good yields, but yields are greatly reduced by short-term, 15–30-day, droughts at critical growth stages of the plant. Water stress at the flowering stage of maize, for example, will reduce yields by 60 percent, even if water is adequate during all the rest of the crop season. If there was a way to store *surplus water* before these critical stages and apply it if the rain fails in these critical stages, crop production would increase dramatically.

There are many ideas for water conservation and supplemental irrigation for small-holders. This is a long and complex subject that cannot be discussed at length here other than to say that often these ideas have failed in practice because of two important factors:

• They do not adequately consider the need to actually *have and store surplus water* before the drought episode.
• They fail to consider the economic costs, relative to benefits—which is all the farmer cares about.

One of the single most promising technologies in this field, which has gained wide adoption in India, is "percolation tanks." These are small reservoirs that capture runoff and hold the water for percolation into shallow water tables. The water is then pumped up onto fields when— and only when—it is most needed. Groundwater storage avoids the high evaporation losses of surface storage; with pumps, the water table provides a cost-free water distribution system to farms; and percolation losses from irrigation are automatically captured by the water table for re-use. These percolation tanks can be combined with highly efficient sprinkler and drip irrigation conveyance systems to provide just the right amount of water when it is needed most.

In sum, it is likely that an *increasing proportion* of the world's food supply will have to be from irrigation. An important need is supplemental irrigation, in marginal rain-fed areas such as in sub-Saharan Africa, using advanced irrigation technologies. In fact, this absolutely has to happen if sub-Saharan Africa is to produce enough food to feed its rapidly growing population without an unacceptably high level of food dependence and to provide remunerative rural employment.

Water Cycles and Water Use Efficiency

Water is the ultimately renewable resource. The amount of water on the planet has changed very little, if at all, since the earth formed some 20 billion years ago. This essentially fixed amount of water is in a perpetual state of cycling between the atmosphere and the surface and sub-surface areas of the globe. Water cycles between the atmosphere and the surface about seven times, so the actual quantum of water involved in total annual precipitation of the globe is only one-seventh of the total. The water cycle is extremely sensitive to temperature and such factors as plant cover, itself largely a product of the water cycle. Long-term changes in the distribution of water among its solid, liquid, and vapor states—between fresh and saline water—and the geographic and temporal distribution of rainfall have had profound consequences on plant and, therefore, animal and human life on the planet. Global warming would exacerbate these natural fluctuations, with important but as yet unknown consequences to food production. But degree and duration of global warming is itself a matter of serious scientific debate. For example, S. Fred Singer, formerly Director of the United States Weather Satellite Service, says, "Surface thermometers report a warming trend, but weather satellites, providing the only true global data, show no atmospheric warming" (Wall Street Journal 2001).

Hydrologic cycles have a direct bearing on one of the central issues in the field of water scarcity—the issue of water use efficiency (WUE) in agriculture and the other sectors. It is commonly thought that irrigation wastes enormous amounts of water. If we could just be more efficient with irrigation, more water would be available for all uses and we would not have to develop more water resources. Unfortunately, this perception is based on a misleading definition of water use efficiency.

WUE is broadly defined as the ratio of the amount of water required for a certain use (U) divided by the amount of water withdrawn or diverted (D) from a source—such as a river, aquifer, or reservoir—to serve that use: WUE = U/D. In irrigation, U is the amount of evapotranspiration (Eta) by crops minus the amount of water supplied by effective precipitation—or net evapotranspiration (NET). WUE can vary between 90 percent in the case of drip irrigation systems to as low as 20 percent in the case of paddy (rice) irrigation systems. But WUE is only a criterion of water delivery efficiency; it does not necessarily mean that water is lost, or wasted in low-efficiency systems. In order to determine this we have to know what happens to the drainage water from the system. Drainage water is mainly from the following:

- Seepage from conveyance systems
- Deep-percolation below the root-zone of plants in fields
- Surface runoff from fields

Drainage water may flow to saline areas or the oceans where it is effectively lost to further human uses. In this case, increasing WUE (reducing drainage) *can result in real water savings*. On the other hand, drainage water may flow to other surface and sub-surface areas where it can be beneficially re-used. This is the *return flow* of water, or water recycling. For example, surface flows of drainage from one field to another are characteristic means of irrigating paddy fields. It is obviously ridiculous to say that WUE is low based on water delivered to only one field. But this same effect occurs less obviously along river basins, with drainage water reentering the river at one point and being diverted and re-used downstream. Drainage water is also a major source in the recharge of aquifers, where because of groundwater storage and timing, it may actually have a higher value than the surface water. Thus one person's drainage becomes another person's water supply. Under these conditions, attempts to increase WUE, usually at large cost, can easily result in a zero-sum game.

The concept of *basin efficiency* includes WUE and all these recycling effects. For example in Egypt, the typical WUE on a farm is only 40 percent to 50 percent. But for the Egyptian irrigation system as a whole, basin efficiency is close to 80 percent—and much of the remaining 20 percent is beneficially used in other sectors. The logic of water recycling may also be illustrated in the case of domestic water supplies in Egypt. Nearly all the water diverted and domestically used in Cairo returns to the Nile, from whence it is re-used several times. But most of the water used in Alexandria flows to the sea, where it is lost to further uses.

While the concepts of both WUE and basin efficiency are valuable when properly understood and used, they are only physical concepts that must ultimately be used in a broader framework of analysis where the economic, environmental, and social *value* of water is considered. This framework is the concept of *water productivity*. Water productivity can be increased by increasing yields per unit of water or by allocating water from lower- to higher-valued crops—or, indeed, sometimes by allocating water from agriculture to other uses.

Esoteric as these concepts might appear to be, they are of enormous practical importance. In PODIUM, for example, we project an increase in food production of 38 percent from irrigated agriculture, with only a 17 percent increase in water diversions to agriculture. This effect is achieved by increases in WUE, in basin efficiency and in water productivity. And, as shown in a later section, recycling causes massive economic externalities in water management—making it very difficult to price water rationally.

GROUNDWATER DEPLETION

Another area of intensive competition for water and rapidly increasing water scarcity is the use of groundwater in irrigation. In terms of impact on food production, one of the greatest technical revolutions in irrigation has been the development of the small-scale pump. Tens of millions of small pumps are currently drawing water out of aquifers to irrigate crops. Over one-half of the irrigated area of India is now supplied by groundwater. Because pump irrigation provides water on demand, yields from pump irrigation can be two to three times those of canal irrigation. Since irrigation supplies about one-half of the total food production of India, one-third or more of India's food production depends on these humble devices and the aquifers that feed them.

Much the same is true in other arid countries. Yet, almost everywhere in the world, groundwater tables in areas that depend on irrigation from groundwater are falling at alarming rates. In many of the most pump-intensive areas of India and Pakistan, water tables are falling at rates of 2–3 meters per year. This is not surprising when you consider that the evaporation loss of a typical crop is around 0.5 m of depth and the yield of water in an aquifer is about 0.1 m per meter of depth. Without recharge, groundwater tables would fall by about 5 m per crop per year. Most of these areas receive sufficient average rainfall to recharge the aquifers, but most of the rainfall goes to runoff, not to recharge. We desperately need to change that relationship.

It is no exaggeration to say that the food security of India, Pakistan, China, and many other countries in 2025 will largely depend on how they manage this groundwater problem. Reducing the amount of pump irrigation is no answer; this simply reduces the most productive agriculture. The answer has to be in groundwater recharge. But this is not an easy solution. Indeed, to our knowledge, no one has devised a cost-effective way to do it on the large scale required. About the only idea that we in IWMI have been able to think of is to encourage, through subsidies if necessary, flooded paddy (rice) cultivation in lands above the most threatened aquifers in the wet season. Paddy irrigation has high percolation losses and is thus a very inefficient form of irrigation from a traditional point of view. But from the point of view of groundwater recharge, this is just what the doctor ordered.[3]

Of course, we have to be careful not to pollute groundwater through leaching nitrates and other chemicals. Restrictions on fertilizer, pesticides, and other chemicals in these recharge areas would be required. Ideally, the recharge areas would not be used for any other purpose except possibly fish production; but in densely populated areas, land is too valuable to simply be set aside for this single use. IWMI and others are conducting research on ways of maximizing the rate of recharge and controlling pollution effects in this exceptionally important area of water resources management.

COMPETITION FOR WATER AMONG THE AGRICULTURAL, URBAN, INDUSTRIAL, AND ENVIRONMENTAL SECTORS

By 2025, most of the world's population will live in urban and peri-urban areas. The people and industries in these areas will demand an increasingly large share of the total water available, and much of this will be taken from irrigated agriculture. Already, in India, the Philippines, and many other countries, large irrigation areas are literally shut down —either permanently or in times of drought—by cities taking water from farmers, with no compensation paid to them for loss of their livelihoods.

Urbanization is creating an enormous pollution load on freshwater supplies and estuaries. The amount of pollutants thrown into the waterways is increasing rapidly and, at the same time, the flows of freshwater are decreasing as more water is evaporated through intensive use. Thus the *concentration* of pollutants is increasing even more rapidly than growth of urban populations and industries would indicate. It is only recently that we have begun to appreciate the economic value of waterways as waste disposal systems. As this previously free service ends and we have to treat water discharges, the costs will run to tens of billions of dollars. But the human health costs of the alternative of doing nothing would be even greater. Already, most vegetables grown in developing countries are irrigated with untreated sewage water from the nearby market towns.

As if this were not bad enough, most of the urban population will be concentrated in coastal areas where sewage water, whether treated or not, is discharged into the seas. In addition to the pollution problem, this greatly increases the consumptive use of water and prevents water recycling, thereby contributing disproportionately to water scarcity.

One of the most important, although generally ignored, water-using sectors is the environmental sector. More water is allocated in California to wetlands, free flowing rivers, estuaries, and the like than to agriculture. The environmental sector has a strong impact on water scarcity because it can have high consumptive use. Exposed water surfaces evaporate rapidly and naturally flowing rivers generally end in the seas. This is a particularly acute problem in water-scarce developing countries. For example, the wetlands of the *Delta Central* on the Niger River in West Africa and the *Sud* on the White Nile in East Africa provide highly valuable wildlife sanctuaries and homes for migratory birds. But both of these wetlands evaporate around 50 percent of the water flow of their rivers. Both of these wetlands are under intensive pressure to redirect the water to lower evaporation losses and to provide water for human use downstream.

No one knows how large the water demands for the environmental sector actually are. Historically, water for this sector has been a naturally occurring free resource. But now that water is becoming more scarce, deliberate policies and water allocations to this sector have to be made. And it should be noted that a decision not to develop water supplies for the other sectors on environmental grounds is a de facto allocation of water to the environmental sector. Here is another important area for future research.

INTERNATIONAL TRADE IN FOOD

Tony Allan (1998) has coined the valuable term "trade in virtual water" to show how international trade can help alleviate water scarcity and other problems in many countries. Countries with plentiful water should export water-intensive crops, like rice, to water-scarce countries. This is a natural application of the principle of comparative advantage in international trade. It happens today with rice, which is exported mainly from wet countries such as Vietnam, Thailand, Myanmar, and the U.S. (which has excess irrigation and food production capacity). Wheat is exported from Canada, the U.S., and Europe, where it can be grown in cool seasons, with low-water requirements. Maize is exported from the U.S. largely because it can be grown without irrigation due to the exceptionally favorable agroclimatic conditions of the Corn Belt. This principle also pertains to trade within countries. Egypt could save nearly 10 percent of its scarce water supplies, for example, by replacing sugarcane production in the very hot south with cool season sugar beet production in the north.

Food imports are essential where countries cannot grow enough food because of water or other constraints, as in many countries of the Middle East and sub-Saharan Africa. This is also true for some countries in Southeast Asia, like Malaysia, where the expanding industrial and service sectors are creating severe labor resource constraints in agriculture. In some countries in sub-Saharan Africa, the costs of inland transportation make it better to feed coastal cities through imports than through domestic production—at least in the short term, until rural infrastructure can be created.

A major problem with trade, of course, is that food imports must be paid for in foreign exchange, earned from exports or by grants and loans. This fact is somewhat hidden by large amounts of donor assistance in hard currency and historically heavily subsidized exports from the U.S. and Europe. In the theory of comparative advantage, every country should be able to export enough to cover imports. But in practice, this does not happen. Many of the most needy countries, such as those of sub-Saharan Africa, do not have sufficient exports to pay for imports.

Economic consultants frequently have the revelation that water-scarce countries should devote their irrigation water only to high-valued crops, like flowers, fruits, and vegetables, export them and then buy the cereals they need on international markets. A team from a famous American university recently found that the Middle East North African (MENA) countries would not have a water problem if they did this! The problem, of course, is that high-valued crops constitute very narrow and highly competitive markets, where only a modest increase in supply drives prices virtually to zero. Even within India, there are times when apples and potatoes are given away free in the producing regions. Every country, developing and developed, is already trying its best to produce and export high-valued crops.

Another advantage of international trade is that imports help to build local markets, tastes, and skills that can result in new domestic industries through import substitution. For example, we expect a substantial shift toward import substitution in terms of domestic meat and feed-grain production in countries like India and China as local entrepreneurs catch up in these markets.

On the export side of the developed countries, it seems evident that there will be significant environmental and financial constraints on EU exports (we have heard that EU policy is to achieve self-sufficiency in food, but not to encourage food exports outside of the EU itself). In the U.S. and Canada, the ultimate results of the boom and bust cycles that the newly freed agricultural markets have been experiencing are not yet known, but they are currently encouraging an exodus from agriculture. Environmental pressures against irrigation and restoring water quality are also building in these countries.

The end result of these considerations is that we believe developing countries with a high percentage of their populations in rural areas will attempt to be as self-sufficient in agriculture as they reasonably can in order to conserve foreign exchange and provide rural livelihoods. They will gradually relax this objective over time as exports grow, the growth of the labor force slows, and employment opportunities in other sectors improve. Of course, many countries cannot achieve this objective because of water and other constraints and will need to import considerably more food by 2025.

It appears that the production potential of the exporting countries will be sufficient to meet needs for increased cereal imports without severe financial or environmental damage. While the trade positions of many countries will change, net cereal exports, as a percentage of total cereal consumption in the world as a whole, will decrease from about 3.3 percent in 1995 to 1.8 percent in 2025. This means that total cereal exports of the countries will increase from 187 M Mt in 1995 to 224 M Mt in 2025.

WATER PRICING AND INSTITUTIONS

It is one thing to estimate the potential for increased water productivity and quite another thing to achieve it. It is precisely because water is such an important economic good that, ironically, powerful forces do not want to treat it as such. As Mark Twain said, "Water flows uphill, toward power."

Some economists advocate pricing water at full marginal cost, both to achieve economic efficiency and to induce institutional change (see the discussion of these issues in Perry et al. 1997). But water resources management is subject to failure of not only the public sector but also the private sector: In economic terms, it is subject to "market failure." Technically, water-recycling effects create massive external benefits and costs that violate the optimizing conditions of free market systems. The *intensity of external effects in water use is perhaps greater than in any other sector of the economy*; that is why water resources have always been a publicly managed or regulated resource.

There are many advantages to pricing water, if it is properly regulated. First and foremost, it provides a means of financing water service agencies and, since they are being paid by their clients, of holding their feet to the fire of performance. Second, entitlements to water provide a means of forcing compensation to users who are harmed by unregulated public and private systems. In many countries, water is being arbitrarily reallocated from farmers to cities (India and Philippines) and for environmental purposes (U.S.) with no compensation for the loss of livelihoods this creates. Entitlements to return flows also would force payment of compensation to downstream losers created by upstream changes in use (as, it appears, happens under the unregulated market system in Chile). Of course, as economists point out, pricing water can induce water use efficiency and allocatable efficiency. But in many developing countries with hosts of small farmers to deal with, the transaction costs of marginal cost pricing are likely to be greater than the benefits.

Pricing water is a good way to regulate the external *costs* of water use —for example, in water pollution. This is because the higher the price, the less amount of water that will be used; thus, other things remaining equal, the lower the pollution. But it is very difficult to regulate the external *benefits* of water use through pricing. For example, the external benefits of field-to-field irrigation in paddy systems, or the recharge of aquifers from irrigation systems would require a *negative price*, or subsidy, to reach the optimum level of water use. While this can and is being done, it is not usually considered by the advocates of (positive) water pricing—and all one can say is that this omission is evidence of poor economic training and analysis.

Socially, a minimum supply of safe water is one of the essentials of life and most people would agree that everyone should be entitled to receive that minimal amount. Market systems, on their own, may not have sufficient incentives to achieve that social objective; it depends on technical conditions of the demand and supply curves. But free water supplies to poor people have sometimes resulted in bankrupt water supply agencies, massive subsidies, and preferred services to the rich.

Thus the introduction of water pricing and the need to manage water at the river-basin level means *more and better, not less, public management*. But a major problem in water management in developing countries is the large and growing disparity between the remuneration of public and private sector staff. Bloated bureaucracies can be supported only by low wages. This inevitably leads to corruption and brain drains to the private sector—just as the needs for well-trained and dedicated people in the public sector are becoming increasingly acute. The first step in beginning the revolution in water management is to provide generous redundancy payments to marginal staff in public agencies, as a onetime write-off and then use the future savings to upgrade the civil service. The lessons of Hong Kong, Singapore, and other countries where public servants are remunerated at rates comparable to the private sector indicate very high rates of return to such policies, if they are effectively implemented.

Given failure on both sides of the private-public table, it would appear that a partnership between the two is the only way out of the dilemma. There are innumerable experiments going on all over the world in designing and implementing these public-private sector partnerships. One of the most important research tasks for the future is to carefully and objectively monitor and evaluate these experiments so that everyone can learn from the experience. But until much more information is developed, it will remain exceptionally difficult to forecast the extent to which the potential gains in water productivity will become real gains.

CONCLUSION

We hope that this paper provides at least a glimpse at the many complex issues in the water-food nexus. To date, the world has had a comparatively easy ride on the back of generally ample water resources that had to be developed only to meet demands. But now, much of the world is simply running out of water—and much of the rest of the world is facing rapidly increasing financial and environmental costs of developing the water resources they have. The grounds for optimism are not in the supply side of the water-food nexus so much as in the demand side. Much of the world now consumes as much food as they need, or even want, and population growth is rapidly decelerating toward zero by the turn of the century. The world will require about 19 percent more water resource development by 2025. But after that big push is accomplished, normal improvements in water technology and management should carry us safely into the future.

REFERENCES

Alcamo, J., T. Henrichs, and T. Rosch. 1999. *World Water in 2025*. Global Modeling and Scenario Analysis for the World Commission on Water for the 21st Century. Unpublished Report.

Allan, T. 1998. Moving water to satisfy uneven global needs: "Trading" water as an alternative to engineering it. *ICID Journal* 47(2):1–8.

Droogers, P., David Seckler, and Ian Makin. 2001. *Estimating the Potential of Rain-Fed Agriculture*. Working Paper 21. Colombo, Sri Lanka: International Water Management Institute.

FAO. 1998. *FAOSTAT Statistical Database in CD-ROM*. Rome: FAO.

Hargreaves, G.H., and J.E. Christiansen. 1974. Production as a function of moisture availability. Association of Engineers and Architects in Israel. *ITCC Review vol. III* 1(9):179–189.

Perry, C.J. 1996. *Alternative Approaches to Cost Sharing for Water Services to Agriculture in Egypt*. Research Report 2. Colombo, Sri Lanka: International Irrigation Management Institute.

Perry, C.J., D. Seckler, and M. Rock. 1997. *Water as an Economic Good: A Solution, or a Problem?*, Research Report 14. Colombo, Sri Lanka: International Irrigation Management Institute.

Seckler, D. 1992. Irrigation policy, management and monitoring in developing countries. In *Roundtable on Egyptian Water Policy: Conference Proceedings*. M. Abu-Zeid and D. Seckler, eds. Cairo: Water Research Center, Ministry of Public Works and Water Resources.

Seckler, David, and Michael Rock. 1995. *World Population Growth and Food Demand to 2025. Water Resources and Irrigation Division*. Discussion Paper. Arlington, Virgina: Winrock International.

U.N. (United Nations). 1999. *World Population Prospects, 1998 Revision*. New York: U.N. Department for Policy Coordination and Sustainable Development.

Wall Street Journal. 2001. Letters to the Editor, September 10.

World Bank. 1997. *From Vision to Action*. Environmentally and Socially Sustainable Development Studies and Monograph Series 12. Washington, D.C.: World Bank.

NOTES

1. This grouping is based on the results of a detailed study of 45 countries representing major regions of the world and over 80 percent of the world population, and a less detailed study of another 80 countries. Group I countries include Saudi Arabia, Pakistan, Jordan, Iran, Syria, Tunisia, Egypt, Iraq, Israel, South Africa, and Algeria. Group II countries include Brazil, Turkey, Mexico, Philippines, Thailand, Ethiopia, Australia, Myanmar, Nigeria, Bangladesh, Argentina, Viet Nam, Sudan, and Morocco. Group III countries include Kyrgyzstan, Canada, U.S., Indonesia, Poland, Spain, France, U.K., Italy, Germany, Uzbekistan, Turkmenistan, Romania, Tajikistan, Japan, Kazakhstan, Ukraine, and Russian Federation. One-half of the population of China and India are included in Group I and the other half of their population in Group II.
2. We are grateful to Alexandros Nicharos of FAO for pointing this out.
3. After this statement was written, attracting some criticisms from colleagues, we found that India has been doing precisely this in a 180,000-hectare area for the past 10 years. We will do a study of the results of this important project soon.

15

WATER SCARCITY: FROM PROBLEMS TO OPPORTUNITIES IN THE MIDDLE EAST

Mona Mostafa El Kady[1]

History and wisdom tell us that inventions and discoveries are mostly challenge-driven outcome events (El Kady 1999). In fact, life-threatening challenges motivate and stimulate the creative thinking of the human brain. There is a popular Arabic proverb: *"The need is the mother of the invention."* Water scarcity problems in the arid regions, in general, and the Middle East in particular, have caused acute problems for human life. Conventional solutions, such as reducing controllable water losses and enhancing water resources development, have become unsuitable to address the increasing rate of water scarcity. Real breakthroughs of creative solutions are urgently needed. It should be remembered that chronic problems, if not resolved at the right time and with suitable means, will create conflicts, which in turn if not resolved, will develop into crisis, violence, and water wars. It has never been so important as it is now for politicians, water professionals, technologists, sociologists, economists, donors, and others to join forces to prevent water scarcity problems from becoming national, regional, and international conflicts. Hence, solutions must fully recognize, consider, and include measures to tackle the roots of problems based on water scarcity and not to focus only on the problem's impacts and effects. Food shortages, pollution, and low-quality water are a few examples of the problems based on water scarcity and severe aridity. Indeed, there are some human-caused problems—including poor management, a lack of vision, and absence of clear water policies.

Governments and international organizations have increasingly realized that water scarcity is not only the concern of every government, but it should also be the responsibility of the rich developed countries to work hand in hand with poor undeveloped countries to secure the minimum human right of clean water and sanitation. In view of this, the Rio convention Agenda 21, in 1992, resulted in a number of reports outlining a common understanding of the world water situation together with recommendations and an action plan. It was an alarming signal to the world's leaders to take seriously the problems of water scarcity and the environment. However, after ten years (2002), the Earth Summit Conference in Johannesburg revealed that very little had been achieved since Agenda 21. The obvious reason is that rich countries are not willing to commit toward poor countries. It was agreed in 1992 that rich countries would increase their financial contribution from 0.3 percent of their net economic return to 0.7 percent, as additional help for the 1.2 billion people without access to clean water and the more than 2.2 billion without sanitation in poor countries (Abu Zeid 2002a). Facts indicate the opposite. Even the 0.3 percent has been reduced in some cases. Problems relevant to water scarcity are numerous and varied in nature and priority from one arid poor country to another. Accordingly, the author presented to the world water council (WWC) in Turkey (2001) a proposal to develop a World Water Problems Map (El Kady and El Shibini 2002a). This is after several maps have been developed for the world showing aridity, population density, water potential, etc. It is time to set priorities for the world water problems.

Another important scope of research is renewable energy. This has been given very little attention from rich industrial countries but is of great importance to poor undeveloped arid countries and for the environment as well. The average number of hours per year that can feasibly produce solar energy in Europe is 100 hours. In contrast, in Arab countries and the Middle East, it is about 3000 hrs/year. Thus, solar energy has high potential in the arid regions, but less so in rich wet countries. In order to avoid the serious impacts that result from water scarcity, rich developed countries must recognize the human rights of the poor; hence they must dedicate budgets, efforts, and technologies to allow all people on earth to share a decent peaceful life without hunger, epidemic diseases, illiteracy, hatred, and envy. This is what all religious books call for—to address humanity—but man always forgets.

It may be appropriate to explore some background information and statistics about the water status in the Arab World before getting into the specific issue of this paper (also see El Kady 2002a).

PROBLEMS IDENTIFICATION AND STATEMENT

Deconstruction of complex problems and clear statements of their roots are effective ways to develop appropriate strategies and tools to face the challenges. In reality, water scarcity is among the factors leading to the rising terrorism movement witnessed in several parts of the world. The author believes that there are four main reasons causing the turbulence in the world: poverty, injustice, hopelessness, and inhumane environment.

POVERTY, INJUSTICE, HOPELESSNESS AND INHUMANE ENVIRONMENTS

Poverty, injustice, hopelessness, and inhumane environments are the main cornerstones causing the world troubles. Therefore, digging deep to the roots may lead to the basic reasons for the development of such problems, of which water scarcity might have one role. Accordingly, it is advisable to search for a road map of integrated solutions to the root problems rather than trimming the stem and neglecting the driving force generating problems. Diagnostic analysis of the predominant causes of the global problems may shed light on the proper integrated solution. Efficient tools and technologies and the required budget can be determined and allocated. The integrated solution may be implemented in phases, and Result Base Management (RBM) technique can be applied to follow the rate of progress according to specific performance indicators (National Water Quality and Availability Management Project Technical Report 2002).

METHODOLOGY AND ANALYSIS

Three analysis phases lead to an integrated solution:

- The first analysis phase is the branching tree formulation for each of the four reasons, leading to frustration and bitterness against the community.
- The second analysis phase deals with the relationship between the impacts of the water scarcity problem on each of the four main reasons and the degree of influence on each, using a judgmental causal scale.
- The third analysis phase is the interrelation between the joint water scarcity impacts on each of the four reasons, in order to identify the levels of importance for the integrated solution to relief pressures. The general ultimate objective is to set priorities and optimize efforts and techniques in an efficient and effective manner.

PHASE I OF ANALYSIS: HIERARCHY OF PROBLEMS

In the first phase of analysis, the hierarchy of problems can be expressed in a branching tree configuration (Duval et al. 1986), bottom up from the root level for each of the four main reasons (see Figure 15.1).

- **Poverty.** Poverty results from unemployment, low income, and insufficient food. These are the reasons for malnutrition and health problems. Indeed, poverty reflects the national economic strength of a country, providing democracy exists. Poverty is the inability to satisfy the basic human needs.
- **Injustice.** Injustice is normally due to discrimination between the powerful and the weak, the rich and poor, at different levels. It is judging with double standards. This causes the feeling of inferiority that results from the lack of equity in rights. This inequity is due to weak laws that do not protect the poor and the strong influence of capitalism on the policy at the national and international levels. Injustice is clear in shared river basins and groundwater aquifers, as far as water rights equity is concerned.

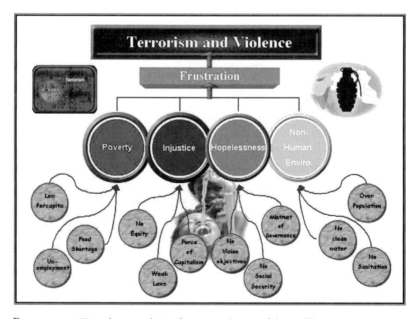

FIGURE 15.1 First phase, analysis of causes and roots of the problems.

- **Hopelessness.** Hopelessness develops due to little hope for the future —particularly among the younger generation. Other aspects of hopelessness include the feeling of insecurity due to a nonexisting social security system (health, unemployment, community, services, etc.) and the feeling of mistrust in government's performance. Out of an estimated Arab population of 300 million, about 20 million of the young generation are unemployed.
- **Inhumane environment.** The worst experience in life is to live without clean drinking water and surrounded by a sewage pond in slums with population densities exceeding $2000/km^2$. We might have *witnessed* that scene in several poor countries, but didn't *experience* such miserable inhumane life conditions. These conditions make the human inner feelings toward life worthless. The predominant factor of this dreadful problem is the lack of available funds and tangible plans for poverty alleviation.

After stating these causes leading to poverty, in-justice, hopelessness, and inhumane environment, one may imagine why people blow themselves up when pressure is exerted or imposed on them. In other words, frustration is most likely to happen. Consequently, terrorism and violence will find a fertile environment to emerge, grow, and develop. It should be emphasized that a single disciplinary effort will not be sufficient to address the problems, but it must be integrated, focused, and continuously monitored and measured.

PHASE II OF ANALYSIS: DEFINING THE RELATIONSHIPS

The second phase of analysis deals with defining the relationship between water scarcity issues, and the different problems that are the roots of the four main causes of terrorism and violent movements (see Figure 15.2).

What is the link between water scarcity and poverty? Water scarcity inhibits horizontal agriculture expansion, fish farming, animal husbandry, and other development projects for food security that are based on water availability. This leads to food shortage, unemployment, and consequently low per capita income in most arid poor countries: In arid countries, agriculture consumes more than 80 percent of the water, as well as intensive labor force employment, but is still incapable of securing self-sufficiency. Thus, poor countries remain food-dependent on rich

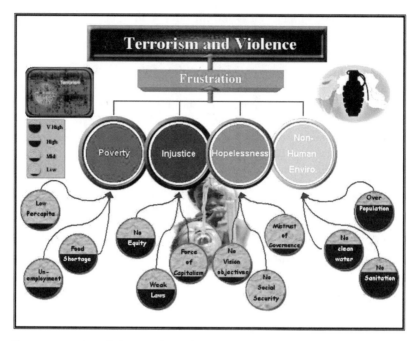

Figure 15.2 Second phase, analysis of water scarcity in relation to different problem causing troubles.

countries. The gap of food shortage will continue to increase and consequently more degraded living conditions will prevail. It is worth mentioning that water scarcity problems are directly linked to food shortage and indirectly linked to both unemployment and low per capita income. Low per capita income, food shortage, and unemployment are interrelated with water scarcity.

There is a relationship between injustice and water scarcity. This exists at two levels:

• At the national level, mismanagement and poor quantitative and qualitative allocation of water policy can occur among different users and create injustice. Also, pollution accumulates from upstream to downstream. There can be weak institutions and laws to secure water rights and insufficient government support for water users associations. Additional factors are lack of management participation by farmers, together with gender inequity (Water Quality Monitoring Project 1995).

- At the regional level, the same problems exist for some cases of shared river basins between neighboring countries and/or joint abstraction of groundwater aquifers. The following are examples from the Middle East:
  - The Israeli-Lebanon conflict regarding rivers Leetani, Al Wazani, and Al Hasabani in southern Lebanon. This problem started with the Israeli occupation of southern Lebanon and continued after liberation in 2000 until present. The main conflict is that Lebanon is unable to utilize its water right from the Al Hasabani River to provide some 30 villages with domestic water due to Israeli threats.
  - The Israeli–Jordanian-Syrian conflict regarding headwaters of the Jordan River, Golan Heights rain catchments, and Lake Tabaria.
  - The equity shares of Turkey, Syria, and Iraq from the Euphrates river basin.
  - The Israeli–Palestinian groundwater (quantity and quality) abstraction problems, including their joint management.
  - The Nile Water Initiative (NWI) is a unique model of initiating cooperation between ten Nile countries sharing its water potential, on the basis of "win-win" projects for the benefit of all. The general rule of thumb is: No country loses. The World Bank is supporting the initiative (Nile Water Initiative Report 2001), yet political turbulence in southern Sudan may impose some constraints.

It should be admitted that Israel is also witnessing water shortage; but a just water division policy through cooperation is the only way to reduce mistrust and build goodwill. Water should not be used as a strategic political issue. Donors' funds must not be allocated to achieve political goals; rather they should support water projects according to feasibility. This will prevent the vicious circle between politics and technicalities.

Hopelessness is a result of many factors, such as the mistrust in government's promises about a better future; degraded social, health, and unemployment security; the rising costs of living; housing; and the lack of transparency. The core of hopelessness is that poor countries in arid regions are usually self-centered because of their engagement in complex internal problems, such as the following:

- Scarcity of national tradable resources
- High rates of population growth
- Unbalanced import/exports—accumulation of debts
- Low manpower productivity

- Unstable political regime
- Other diverse social and economic problems

While under these harsh conditions, they watch their limited headwater resources being captured by military occupation. In the meantime, no international law can protect them, and no international organization can save their human and water rights. This causes loss of self-confidence, vision, and focused objectives, which leads to poor planning, which then adds more complexity to unemployment problems. Water scarcity in this context has a minor impact role, compared to other immense problems.

In this century, the non-human environment is a great tragedy to the poor, and should be a shame to the rich. The reason is obvious and well known to water professionals and experts: It is the world's population explosion. But if we imagine the amount of money spent during wars of the last century and what was spent on water treatment projects for the poor, one may notice the two extreme contradicting objectives. It is worth inquiring about the human rights of poor people who live without clean water and sanitation while wildlife in conservation areas are enjoying healthy environments and protected by man for ecological reasons. If we stop wars for one year and direct those costs to water treatment and solving water problems instead, the world will change to a much better state. The Gulf War cost in 1991, according to press media reports, was about $US65 billion, and the anticipated Iraqi war will cost about $US200 billion.

There are approaches to bridge the gap between demand and supply in poor arid counties that also suffer multiple victims of waterborne diseases and poor non-human environments. These include the following:

- Water reclamation projects
- Use of nonconventional water sources
- Desalination of brackish and seawater
- Water recycling
- Development of water resources at main river catchments
- Water harvesting (rainfall, fog, springs, flash floods, etc.)
- Groundwater protection and safe use

PHASE THREE OF ANALYSIS:
RELATIVE IMPACTS OF WATER SCARCITY

To illustrate the relative impact on the branching problems forming the four main reasons of terrorism, a conceptual judgment scale is used to differentiate between the mode and degree of water scarcity impacts (see

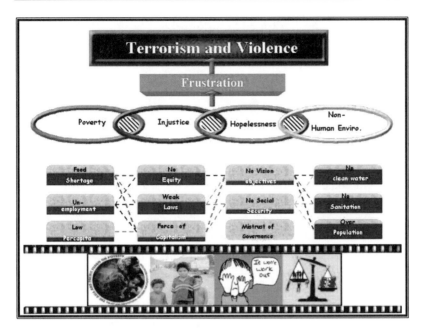

FIGURE 15.3   Third phase, relative impacts of water scarcity.

Figure 15.3). The greatest influence of water scarcity is on the problems causing nonacceptable human environment, and the least influences are on the problems leading to hopelessness. The main objectives for adjudicating these differences is to clarify for decision-makers in poor arid countries the relative weights of the problems leading to the main causes of frustration and ultimately terrorism, set priorities, and rank the problems before formulating policies and strategies for feasible solutions. Scenarios from possible alternative policies and strategies can be developed according to the targeted objectives for the solutions. The screening process of the feasible policies can be made to comply with the decision-makers' political, economic, and social visions.

It should be re-emphasized that the conceptual representation of water scarcity impacts is based on judgment estimates. However, the ranking and degree of water scarcity impacts may vary from one country to another. The objective is to clarify and explain a methodological procedure to visualize the compatible effects with the most likely accumulated problems leading to the four main causes. This approach also can estimate the degree of water scarcity that can enhance each problem. Furthermore, detailed analysis may be required to attain more accurate estimates and evaluation. This may suggest running a "Delphi" exercise for several rounds, involving a panel of experts, in order to improve results and obtain logic relationships between water scarcity and the general frustration problems.

CROSS RELATIONS OF THE PROBLEM'S IMPACTS

Having determined the branching tree of the rooted problems by the first analysis phase, followed by judgmental estimates of the impact of water scarcity on each problem, (second phase of analysis), cross relations is deemed important. It is critical to rank the priority problems from the national point of view prior to a suggested solution strategy. Also, common regional problems may be recognized and aggregated to optimize the use of the applied tools.

The cross relations between problems may be defined as enhancing mode relationship or neutral mode—that is, each problem occurs in separate mode from another. Therefore, the priority criteria for ranking consolidated problems must capture all the roots of the different problems. If the contribution of the water scarcity issues are estimated, their level of impact is evaluated (by a panel of experts), and the connecting mode and cross impacts are defined, appropriate remedial measures can be determined. Accordingly, the tools and mechanisms to deal with these problems can be applied and monitored. The targeted objective is to disintegrate the aggregate problems into fragments and maximize effective solutions results.

FROM PROBLEMS TO OPPORTUNITIES

A top priority is to plan strategic approaches and consider creative non-traditional solutions. It is important to focus our future vision on practical plans and tangible implementation mechanisms within a logical time frame. A monitoring and evaluation system is essential to ensure compatibility between plans and implementation. There are critical characteristics of the Middle East that have to be stated.

- Moslems are the majority of the population. The principal facets of the Islamic religion are: Mercy, Justice, Purification of the soul and spirit, Cooperation for the good of humanity, Discrimination never exists, Help the poor, Depend solely on the mighty God "Allah." If these are twisted or misinterpreted, it will be disaster for simple, poor, naïve people. They might be misused for non-human, anti-Islamic action against the community. The spirit of Islam is "Peace upon everyone."

- Arid counties have to do more with the little water available to sustain basic human needs and relieve the misery of malnutrition. There is always considerable water waste, but this can be avoided.
- Industrial development is slow due to the big gap between research and application. Bridging the gap is a must.
- Shortage of funds, rigidity in institutions, and poor management skills are severe inhibitory factors to the development of national economies and to poverty alleviation (El Kady 2002b).
- High pollution levels and deterioration of the quality of both surface and groundwater are developing environmental threats for the future.
- Insufficient trained manpower and awareness programs, and absence of appropriate technology adaptation policy exacerbate the problems.

In order to turn the problems to opportunities, there must be clear objectives to eradicate pandemic poverty, injustice, hopelessness, and non-human environment. This can be achieved if the rich developed world will be committed to provide the necessary elements to accelerate development toward the ultimate goals and objectives. This will consequently lead to future stability.

ACTION PLAN

The suggested action plan has a time horizon of 15 years. The strategy covers three successive phases:

1. A short-term action plan of five years for capacity building and establishing rural infrastructure at specific priority locations.
2. A medium-term action plan of about five years to demonstrate technology adaptation models (pilot projects) for different water activities, targeted to specific objectives (industrial wastewater treatment for recycling, sewage treated for agriculture, etc.). If primitive local technologies exist, they might be a good starting base for development.
3. A long-term action plan of five years for integrated water activities of water resources systems to secure future sustainability and open new opportunities linked to industry. At this stage, the road should be set to increase dramatically food production, full-capacity re-use

of marginal water in developments, and phasing out of dependence on virtual water.

The three planning phases complement each other; however, each phase should have its specific objectives, performance indicators, and action plan, including the tools and elements required for implementation. For each phase, an initiative mechanism will be developed to stimulate the human talent not only to solve problems, but also to invent local technologies and new development areas. The principal elements of sustainable development in undeveloped countries are the following:

- Optimum use of available resources
- Changing human behavior and attitudes toward water by changing habits
- Continuous upgrading of local technologies and adapting to new areas of development
- Enhancing regional cooperation
- Exchange of suitable technologies through pilot projects initiated by rich developed countries
- Development of team work spirit
- Provision of the appropriate training programs

The branching problems behind the four main suggested causes of frustration and terrorism can be grouped in three main categories:

- Human-driven problems include wrong or misguided beliefs, illiteracy, low productive outputs, rejecting the need to change toward improvement/progress, etc.
- Governance-driven problems include lack of institutional and legal reforms to comply with global changes, nonrealistic and over-ambitious plans, absence of a monitoring and evaluation system, lack of knowledge and information-dissemination mechanisms, inadequate attention to management training, misallocation of limited resources, and—in some cases—waste of resources.
- Non-governance–driven problems include shortage of funds, highly expensive technologies, lack of technical assistance for upgrading and modernization of facilities, worldwide political and economic instability impacts, etc.

The three successive action plans addressing water scarcity problems focus on mitigation measures, techniques and procedures, and use of lessons learned from past problems to turn its impacts to opportunities (providing the targeted objectives are achieved).

The first action plan phase lays the infrastructural foundation for solutions and opportunities. The fundamental needs to make it successful are the integrated efforts of the governments, the external technical and technological support and NGOs, in addition to the proper legal and institutional reform programs.

CAPACITY-BUILDING AND RURAL INFRASTRUCTURE

The role of governments (Abu Zeid 1998) should be to set a priority map for the following needs:

- Skilled manpower
- Non-human or capital resources, (equipment, funds, land, etc.)
- The techniques of production appropriate to the water scarcity issue —e.g., water treatment technologies, modern irrigation systems, monitoring and calibration instruments, maintenance and operation tools, groundwater development and protection techniques, and use of nonconventional energy sources in nonconventional water (desalination, etc.)

The first action plan phase aims at providing different project proposals to mitigate the scope of water scarcity and optimize water utilization. This should provide an appraisal of the available local technologies, possible policies to upgrade the technologies and/or replace it by appropriate technologies. It is, in fact, a self-study and evaluation of the prevailing conditions surrounding the water scarcity problems. This phase spots points of strength and weakness in the government's executive system, the national potential resources that are not efficiently used, and the stakeholders involved in the water scarcity problem.

The second action plan is the most crucial phase, since it deals with the answer to the "How?" question. It depends heavily on research and development, and the efficient capabilities of the role of a mediator between local technologies and external advanced technologies. The end result of this phase must be the implementation of a number of selected pilot projects that emerged from the first phase, according to national importance and regional priorities. The force driving this initiative should come from the advanced rich countries, which will provide technical assistance, technologies, and supplement funds.

The exemplar pilot projects must do the following:

• Educate, by hands on experience, the research teams.
• Provide appropriate technologies to solve problems associated with water scarcity that can be transplanted and developed in each country.
• Allocate funds to initiate and support national research organizations to take the role of continuation and dissemination.

It should be emphasized that assistance, from advanced countries to poor undeveloped countries, must focus on the tools, techniques, and HRD, rather than agriculture products and consumable goods (virtual water concept). This teaches the skills, gives the tools, and shows how to produce and secure food and water instead of being always dependent.

Biotechnology, tissue culture, genetic engineering, and membrane industrial technology are a few areas in which to develop food production and cope with water scarcity conditions, particularly as climate change is now considered a fact.

The solution programs need a suitable mechanism to sustain efficient links between technology transfer and adaptation. This should involve efficient networking between connecting research organizations concerned with water scarcity, in both developed and undeveloped countries. Pilot projects should be proposed and evaluated among developed and undeveloped research bodies to reach a regional consensus about a priority group of projects at different representative sites with various specific conditions. The funding of the agreed program may be covered jointly between regional and international sources. Results should belong to the network administrations, which, in turn, disseminate any information, conclusions, and/or advice from any of the experts and professionals forming the research team.

In this context, the author presented a proposal for an Arab Water Research and Learning Network in 2002 at an international conference in Kuwait (El Kady and El Shibini 2002b) and a ministerial meeting in U.A.E. (El Kady and El Shibini 2002c). The idea was highly appreciated and accepted.

The third action plan is the real target to turn problems into opportunities, since it lies within each country's own capability to change and use its potential to attain self sufficiency and cope with water scarcity (Abu Zeid 2002b), while maintaining active regional cooperation and optimizing resource allocations. However, since technology is dynamic, we should admit that reaching this target is a long-term process and needs a transition period during which developed countries have to support undeveloped countries. The role of the private sector is crucial to motivate and enhance continuous modernization of industries relevant to water activities. Regional trade-offs must be encouraged—in other words, political boundaries should not be a barrier to inhibit exchange of goods and services. There is no one country that could be totally independent in food production and trade, since trade requires producers and markets.

Monitoring and evaluation systems that analyze the progress should be applied. These follow the completion of the pilot projects and model the efficient use for technology adaptation. Periodic self-evaluation studies must be conducted (for example, on an annual basis), according to specific performance indicators, to assure the targeted rate and direction of change. A technical auditing team for evaluation progress will be of great importance after pilot projects are converted to real national projects.

Nonconventional water must be an integral part of the water policy. Also, both nonconventional water and nonconventional energy—developments and applications—must be areas of research supported by advanced developed countries. The expected achievements of the third action plan phase should lead to "endless and progressive" development with increased productivity and new opportunities.

CONCLUSIONS AND RECOMMENDATIONS

Some of the background information in this paper was gathered from the Arab media and reflects the general understanding about water scarcity problems by water experts and professionals in the region.

Analysis of the main motives for frustration and terrorism, reveal that poverty, injustice, hopelessness and non-human environment are the cornerstones. However, there are rooted problems behind each. These are diverse and at different levels of seriousness for each country in the Middle East.

Water scarcity is one of the enhancing causes. It has different levels of effect on each rooted problem, but it is not dominant. There are several other causes of the global problem that need to be analyzed by experts in other disciplines concerned with social, economic, educational, religious, national security, political, and other issues.

Regional cooperation in solving this common problem role of water scarcity mitigation is a *must*. An action plan is proposed to help poor communities lacking water and sanitation. This has three phases: infrastructure, pilot projects, and real-life national projects. To implement this, full assistance and support from advanced developed rich countries is needed. To achieve this, it is highly recommended that a special multinational contribution fund—a Water Scarcity Risk Management Fund—be implemented.

The essence of cooperation between developed counties and those undeveloped countries suffering water scarcity are knowledge, experiences, and technologies. Nonconventional energy (solar and wind) and its role in solving water scarcity problems are also essential areas of cooperation.

A strong emphasis on the importance of research and development and a wide area network of developed and undeveloped arid countries to work very closely in solving the complex water scarcity problems is imperative. Long experience tells us that solutions must emerge from tackling real conditions on the ground (not on paper). They cannot be imported or transplanted to different environments.

REFERENCES

Abu Zeid, M. 1998. *Water as a Main Issue of Tension in the 21st Century*. 1st Ed. Egypt: Al Ahram Publishing Center.

Abu Zeid, M. 2002a. Opening Speech, August 2002, Johannesburg.

Abu Zeid, M. 2002b. *Water in Africa and Future Challenges*. The 18th Congress and The 53rd International Executive Council Meeting International Commission on Irrigation and Drainage, July 21–28, 2002, Montreal, Canada.

Duval, E., A. Fontela, and A. Gabus. 1986. *A Handbook on Concepts and Applications, Report No. 1*. August 1986, Geneva.

El Kady, M. 1999. *History of Irrigation*. Egypt: NWRC.

El Kady, M. 2002a. *Vision for Water Security & Sustainability*. March 2002, Kuwait.

El Kady, M. 2002b. *Past, Present and Future*. Water Policy Meeting, May 2002, Alexandria.

El Kady, M., and F. El Shibini. 2002a. *Water Problems' Map*. Water Resources Management in Arid Regions Conference, March 2002, Kuwait.

El Kady, M., and F. El Shibini. 2002b. *A Vision for Water Security and Sustainability*. Water Resources Management in Arid Regions Conference, March 2002, Kuwait.

El Kady, M., and F. El Shibini. 2002c. *Coping with Water Scarcity*. Water Resources Management in Arid Regions Conference, March 2002, Kuwait.

National Water Quality and Availability Management Project Technical Report. 2002. *Result Based Management (RBM) Report*. Egypt: NAWQAM.

Nile Water Initiative Report. 2001. World Bank Document of the Nilotic Countries, Egypt.

Water Quality Monitoring Project. 1995. Fact sheets, Egypt.

NOTES

1. NWRC Chairperson.

16

THE ROLE OF BIOSALINE AGRICULTURE IN MANAGING FRESHWATER SHORTAGES AND IMPROVING WATER SECURITY[1]

Mohammad Al-Attar[2]

During the 20th century the world population tripled. As populations rise, so too does the demand for water. Yet we are aware that the total volume of water on earth is finite at about 1400 million cubic kilometers, of which only 2.5 percent is freshwater (UNEP 2002). Unfortunately most of the freshwater is locked up in the icecaps of the Antarctic and Arctic (see Figure 16.1).

Thus the usable portion of freshwater resources is less than 1 percent of all freshwater and only 0.01 percent of all water on earth (UNEP 2002). The shortage of this precious natural resource and the increasing demand for water as the world's population increases seriously limit development and threaten water security at local, regional, and global levels.

Yet the distribution of freshwater throughout the world is uneven. While some parts of the world, such as South America and Asia, are well endowed with freshwater resources, other parts of the world, such as the Near East, are not.

The Near East is one of the driest areas of the world, being poorly endowed with freshwater resources. The 29 countries in the region account for 14 percent of the world's land area and are home to 10 percent of the world's human population. Yet the region has only about 2 percent of the world's renewable water resources. Further, this is a region characterized by high temperatures throughout much of the year and where demand for water to meet human needs is correspondingly high (ICBA 2001). Drylands in the Middle East cover 99 percent of the surface area.

Many of the Near East countries have among the lowest renewable water resources per person in the world (Figure 16.2). The average for the Near East region is 1577 cubic meters of water per person per year,

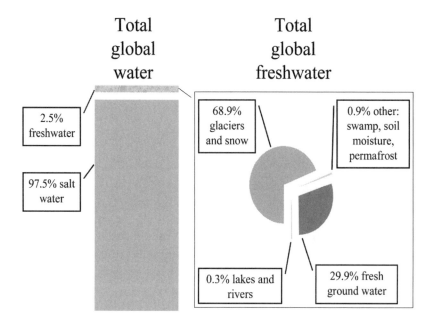

FIGURE 16.1 Global water resources (UNESCO/UNEP).

while the global average is 7000 cubic meters of water per person per year (ICBA 2001). Worse still is the situation in the following Middle Eastern countries: Jordan together with the six Gulf Cooperation Council countries of Bahrain, Kuwait, Oman, Qatar, Saudi Arabia, and the United Arab Emirates (Figure 16.3). In the Gulf Cooperation Council countries, there are only 170–200 cubic meters of renewable water resources are available per person per year (ICBA 2001). This is less than 3 percent of the global average.

In the Near East, demand for water has been increasing due to rapid population growth and an increase in per capita consumption. This has led to water rationing in some countries: Jordan restricts water supplies in Amman to 3 days a week; and in Syria, water for Damascus inhabitants is supplied for less than 12 hours a day.

Nevertheless, agriculture still consumes the largest proportion of the freshwater in the region. One estimation indicates that water withdrawal for irrigation is 70 percent of the total withdrawn for human

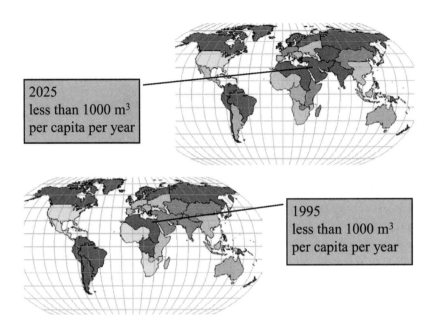

FIGURE 16.2 Water availability by region in 1995 and estimated for 2025 (1000 cubic meters per capita per annum) (UNESCO).

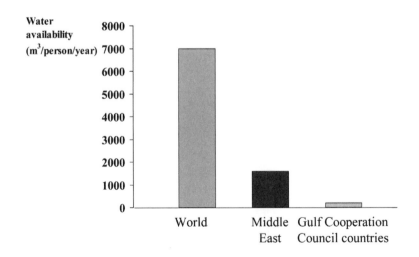

FIGURE 16.3 Availability of global water resources per person (ICBA 2001).

uses (Shiklomanov 1999), while the withdrawal for irrigation in the six Gulf Cooperation Council countries is 85 percent (Zubari 1997). Another study found that agriculture utilizes about 86 percent of the available water resources in the Arabian Peninsula and 80 percent in the Mashriq (Jordan, Syria, Iraq, Lebanon, and Palestine) countries (Khouri 2000). The proportion of water utilized for domestic and industrial use is much smaller.

Unfortunately the use of freshwater for agriculture per person is higher (Figure 16.4) in the Near East than the global use per person because of the aridity and high summer temperatures over much of the region. Thus irrigation in agriculture assumes a greater significance. Even in areas where the environmental conditions preclude intensive agriculture, there is a growing demand for water in horticulture and landscaping. In the six Gulf Cooperation Council countries, the area planted to horticultural crops increased by 12%–15 percent per annum from 1980 to 1999. Many countries in the region, particularly in the Middle East, have a policy of being as self-sufficient in food production as possible, but water limitations have kept them from achieving self-sufficiency (Cosgrove and Rijsberman 2000).

Globally, poor water management has resulted in the salinization of 20 percent of the world's irrigated land, with an additional 1.5 million hectares affected annually (CSD 1997), significantly reducing crop pro-

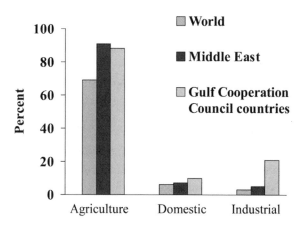

FIGURE 16.4 Use of freshwater by sector and region (ICBA 2001).

duction (WCD 2000). Other studies estimate the area affected annually to be much higher at 10 million hectares (Arnold et al. 1990). Evidently the true figure is somewhere between the two estimations. Soil salinity is most serious in arid and semi-arid regions where water is scarce, rainfall is erratic, and groundwater is often saline. Although some areas are saline because of the geology or geography, human activity is responsible for salinity of 77 million hectares worldwide. Of this, 45 million hectares is irrigated land (Tables 16.1 and 16.2).

It is not only irrigated land that is becoming saline but groundwater too is becoming increasingly brackish. More than 50 percent of groundwater in the region, it is estimated, is already saline and the proportion is increasing as the rate of extraction of water from aquifers exceeds recharge (ICBA 2001). In Saudi Arabia, water levels declined by more than 70 meters in the Umm Er Radhuma aquifer from 1978 to 1984 and this decline was accompanied by a salinity increase of more

TABLE 16.1. Area affected by soil salinity worldwide

Cause of Soil Salinity	Area (Million Hectares)
Geological and geographical	32
Manmade factors	77
Irrigation	45

Sources: IAEA, FAO.

TABLE 16.2. Area of irrigated land affected by salinity in Near East countries

Country	Percentage of Irrigated Land Affected by Salinity
Iraq	70
Pakistan	Over 40
Syria	50
Egypt	33
Iran	15
United Arab Emirates	25
Kuwait	85
Jordan	3.5
Central Asian countries	50

Sources: IAEA, FAO, UNEP.

than 1000 milligrams per liter (Al-Mahmood 1987). As water tables drop due to excessive extraction, intrusion of salt water increases in coastal areas. The aquifers of Bahrain, the Batenah Plains of Oman, and the United Arab Emirates are suffering severely from seawater intrusion (ICBA 2001). Groundwater salinity in most areas of the Syrian and Jordanian steppe has increased to several thousand milligrams per liter and overexploitation of coastal aquifers in Lebanon has caused seawater intrusion with a subsequent rise from 340 to 22000 milligrams per liter in some wells near Beirut (UNESCWA 1999).

ESTABLISHMENT OF THE INTERNATIONAL CENTER FOR BIOSALINE AGRICULTURE

The Arabian Peninsula is using up its water resources three times as fast as they are being renewed. Moreover, available water resources will be exhausted within the next 20 years unless consumption of freshwater is reduced. Based on these challenges, the Islamic Development Bank (IDB) took the bold initiative to set up the International Center for Biosaline Agriculture (ICBA) in Dubai, United Arab Emirates.

The primary objective of IDB in setting up ICBA is to promote the use of saline water in productive sustainable agriculture in 54 IDB member countries and elsewhere. Promoting the use of saline water and salt-tolerant species to increase food and feed production has the potential to free up scarce freshwater resources and bring into production marginal areas where agriculture is currently not viable because soils are saline or water is brackish.

The Center became operational in September 1999, with financial support from the Islamic Development Bank (IDB), the Arab Fund for Economic and Social Development (AFESD), and the OPEC Fund for International Development. The Center now has additional support from the International Fund for Agricultural Development (IFAD), the International Atomic Energy Agency (IAEA), the UAE Government, Dubai and Abu Dhabi Municipalities, and the private sector in Oman and Saudi Arabia.

This new center, now three years old, has already made strides in screening salt-tolerant species and developing techniques for using saline water irrigation in agricultural systems. ICBA recognizes the importance of scientific sustainable irrigation practices utilizing salty water and the importance of ensuring that salts do not accumulate in the fields implementing biosaline agriculture.

For example, in screening salt-tolerant plants, initial tests at ICBA have shown that elite pearl millet germplasm from the International Crops Research Institute for the Semi-Arid Tropics (ICRISAT), tested and proven to be heat and drought tolerant, has also proven to be salinity tolerant. Such salt-tolerant crops could form an integral component of a sustainable cropping system providing both food and feed.

As to developing saline water irrigation, in partnership with the private sector, ICBA is testing technologies to use low quality saline process water generated during oil drilling to grow food and feed in Oman. Petroleum Development Oman, a private company, sought ICBA's help in developing a technology to filter process water generated during oil drilling through reed beds and to test salt-tolerant species, which can be cultivated with this treated but still salty water. Such technologies could make economic use of saline water, which would otherwise be wasted, and barren land.

ROLE OF BIOSALINE AGRICULTURE IN MANAGING FRESHWATER SHORTAGES

What is "biosaline" agriculture and what role can it play in managing water shortages?

In many regions where freshwater is scarce, brackish or salty water is available and even abundant. Such water can be used to cultivate salt-tolerant plants. Salt-tolerant plants can also be grown where soils are saline, either because of geological or geographical factors or because of bad irrigation practices.

At present, little is known about the quantity, quality, and distribution of salty and brackish water. Assessment of water resources has focused almost exclusively on freshwater resources. Recent advances in biosaline agriculture now merit assessment of saline water resources and their potential for agricultural use. As a pilot study, an assessment of the saline water resources in five countries of the Near East, funded by the International Fund for Agricultural Development, is being undertaken.

Likewise, little is known about productive uses of salt-tolerant plants, although plants have been selected for salt-tolerance since humans first started to domesticate wild plants for agriculture. The interest in selection for salt tolerance, however increased dramatically in the late 20th century as concern about food security grew. Now there is wide acceptance that many salt-tolerant species can be used for food or fodder, timber production, horticulture, oil production, and other industrial uses. Some vegetables, such as asparagus, red beet, and zucchini squash, are salt-tolerant, as are some fruit trees, such as date palm and pomegranate.

Cereal crops vary tremendously at the species, cultivar, and genotype levels. Crop varieties developed in the Middle East have better salt tolerance than varieties developed elsewhere. Barley is the most salt-tolerant cereal, followed by sorghum and rice. Pearl millet is another crop that shows promise in salinity tolerance.

Forage crops hold good promise of being both salt-tolerant and suitable for cultivation on marginal lands. The most salt-tolerant grasses are wheat grass and Bermuda grass, followed by Rhodes grass and fescues.

However, the most promising salt-tolerant species for productive cultivation are halophytes, salt-tolerant wild plants that are now being included in pasture improvement programs in many salt-affected regions of the world. Many provide high quality fodder and use water very efficiently. Some of the important halophytes being investigated for productive use at ICBA are *Distichlis palmieri*, *Sporobolus airoides*, *Salicornia bigelovi*, *Atriplex lentiformis*, and *Salvadora persica*. While many of these species have potential as solely fodder crops, others, such as *Salicornia* may have multiple uses. *Salicornia* can be grown not only as an oil seed crop, but the plant tips can also be harvested as a high-value salad vegetable crop.

The conservation of plant genetic resources is critical to developing new plant varieties that are ecologically friendly and can flourish with less fertilizers, less pesticides, and in increasingly marginal lands where conditions are not ideal. While there are many genebanks for traditional agricultural plant species, ICBA has set up a genebank specifically to steward salt-tolerant plant genetic resources for future generations. The genebank houses over 200 salt-tolerant or potentially tolerant plant species and over 6,000 accessions of these species from all over the world.

To manage freshwater shortage effectively and improve water security for all, we will need to adapt and transfer technologies to farmers in the developing world, particularly regions such as the Near East where freshwater resources are dwindling and salty water is abundant. There have been rapid advances around the world in the use of saline water for irrigation, including development of irrigation systems, improved water management, and control of salinity within the root zone. Many tools are now available for monitoring irrigation systems and managing salinity. These systems have been successfully packaged and widely used in many countries such as the United States, Australia, South Africa, and some European countries

More than 50 percent of the land irrigated at present has been severely damaged by secondary salinization-alkalinization as well as waterlogging (Arnold et al. 1990). This is partly due to poor irrigation and drainage and partly due to improper selection of site, method, and design of irrigation systems. Every year, about 10 million hectares of irrigated land must be abandoned as a consequence of the process (Arnold et al. 1990). This hazard is particularly great in developing countries. A worldwide study by Dregne (1991) estimated that 145.5 million hectares are under irrigation, of which 38 million hectares (30 percent) are moderately or severely damaged by desertification/salinization. The principal cause is the imbalance between excessive irrigation and inefficient drainage (Kassas 1992).

Such imbalances can be redressed by appropriate technology. For example, ICBA was requested by officials in Abu Dhabi to help reclaim salt-affected agricultural lands by draining waterlogged soils. A drainage system was designed by ICBA and resulted in restoring productivity to over 100 hectares of agricultural land.

Hence, ICBA's role as an applied research center is not to duplicate the significant body of research relating to salinity in agriculture but to develop and refine available technologies that can be adopted by farmers to improve economic returns using saline water for irrigation. The Center will act as a focal point for gathering information on what has already been done and what is already known in the field. ICBA will provide this knowledge and information to farmers and landscape managers in the developing world through collaborative projects and its developing networks—the Global Salinity Network and the Inter-Islamic Network for Biosaline Agriculture.

Awareness is growing that global water security can be achieved only if all parties work together to develop Integrated Water Resource Management (IWRM) practices on a local, national and regional scale. Biosaline agriculture contributes to IWRM by developing the potential of salt-affected lands, saline water, and salt-tolerant plants to increase agricultural production, improve livelihoods, and conserve the environment.

Biosaline agriculture contributes to two of the primary objectives of global integrated water resource management:

- To produce more food and create more sustainable livelihoods per unit of water applied (more crops and jobs per drop) and ensure access for all to the food required for healthy and productive lives
- To manage human water use to conserve the quantity and quality of freshwater and terrestrial ecosystems that provide services to humans and all living things

Plant and environmental responses to water and salinity are highly dependent on climate, soils, topography, and management regime. Transferring irrigation and salinity technology from one environment to another requires an integrated approach. Systems must be developed to monitor the impact of the introduction of new technologies to ensure their maximum impact on sustainability and profitability. The role of ICBA is therefore not only to acquire, evaluate, assess, and transfer information or expertise from around the world but to work in partnership with those responsible for introducing new technologies into farming systems or greening programs in the region. Research and development projects are collaborative and, as needs are identified, collaborators' skills are developed in training courses, which bring participants from as far afield as Senegal and Bangladesh.

CONCLUSIONS

How will biosaline agriculture help in managing freshwater shortages and improving water security? Today, 1 kilogram of wheat needs at least 1000 liters of virtual freshwater, or water utilized to grow the wheat plants that produced the kilogram of grain. If a salt-tolerant crop were grown instead, the 1000 liters would have both freed up 1000 liters of freshwater for other uses and utilized brackish or saline water, an otherwise wasted resource.

CAN BIOSALINE AGRICULTURE MAKE A DIFFERENCE?

Biosaline agriculture technologies use salty water productively. Plants that tolerate salt in water and soil are being evaluated for productive use, perhaps replacing varieties that will grow only in sweet soil irrigated by freshwater. Biosaline agriculture focuses not on prime, productive areas but on bringing into production unused or little-used marginal lands where plants are difficult to grow and which support few livestock. If economically useful plants are grown with salty water and on saline land, more food and feed can be made available globally and land abandoned because the soil has become saline can be put to economic and sustainable agricultural use. An alternative supply of saline water that can be used in agriculture will relieve pressure on freshwater resources.

Freshwater we all know is scarce, shortages are commonplace and increasing. Let us conserve freshwater by using salty water instead to grow more food and feed. Biosaline agriculture can be part of the solution to integrated water resources management "linking social and economic development with protection of natural ecosystems and . . . land and water uses across the whole catchment area or aquifer" (GWP 2000). We cannot afford to waste water—even saline water—and neither can we afford to forsake salt-affected lands. Let us manage freshwater resources and improve water security where possible by making use of two unused resources—salty water and salt-affected land—and put them to productive use using the techniques of biosaline agriculture.

REFERENCES

Al-Mahmood, M.J. 1987. *Hydrogeology of Al-Hassa Oasis.* MSc Thesis, Geology Department, College of Graduate Studies, King Fahd University of Petroleum and Minerals, Saudi Arabia.

Arnold, R.W., I. Szabolcs, and V.O. Targulian. 1990. *Global Soil Change.* Laxenburg: IIASA.

Cosgrove, W.J., and F.R. Rijsberman. 2000. *Making Water Everybody's Business.* London, U.K.: Earthscan.

CSD. 1997. Comprehensive assessment of freshwater resources in the world. *Report of the Secretary General.* http://www.un.org/documents/ecosoc/cn17/1997/ecn171997–9.htm (GEO-2–117).

Dregne, H.E. 1991. *Desertification Costs: Land Damage and Rehabilitation.* Lubbock: Texas Tech University.

GWP (Global Water Partnership). 2000. *Towards Water Security: A Framework for Action.* Stockholm and London: GWP.

IAEA/FAO. *Productive Use of Saline Lands.* Information leaflet. Not dated.

ICBA. 2001. ICBA Strategic Plan 2000–2004. Dubai, UAE:ICBA.

Kassas, M. 1992. Desertification. In *Degradation and Restoration of Arid Lands.* H.E. Dregne, ed. International Center for Arid and Semiarid Land Studies, Lubbock: Texas Tech University.

Khouri, J. 2000. *Sustainable Management of Wadi Systems in the Arid and Semi-arid Zones of the Arab Region.* International Conference on Wadi Hydrology. Conference held in Sharm El-Sheikh, Egypt, November 21–23.

Shiklomanov, I.A. 1999. *World Water Resources and Water Use: Present Assessment and Outlook for 2025.* St. Petersburg, Russia: State Hydrological Institute.

UNEP. 2002. *Global Environmental Outlook 3. State of the Environment and Policy Retrospective 1972–2002.* Nairobi, Kenya: UNEP.

UNESCO. *Summary of the Monograph World Water Resources at the Beginning of the 21st Century.* Prepared in the Framework of IHP UNESCO. http://espejo.unesco.org.uy/summary/html/summary.html.

UNESCWA. 1999. *Updating the Assessment of Water Resources in ESCWA Member States. ESCWA/ENR/1999/WG.1/7.* Beirut: United Nations Economic and Social Commission for West Asia.

WCD. 2000. *Dams and Development: A New Framework for Decision Making. Report of the World Commission on Dams.* London, England: Earthscan.

Zubari, W.K. 1997. *Towards the Establishment of a Total Water Cycle Management and Re-use Program in the GCC Countries.* The 7th Regional Meeting of the Arab International Hydrological Programme Committee, September 8–12, 1997, Rabat, Morocco.

NOTES

1. Presented at the World Food Prize symposium, October, 2002.
2. Director-General, International Center for Biosaline Agriculture (ICBA).

17

COPING WITH WATER SCARCITY IN DRY AREAS: CHALLENGES AND OPTIONS[1]

Adel El-Beltagy[2] and Theib Oweis[3]

Over a billion people live in the dry areas, where about half of the work force earns its living from agriculture. Population growth rates here, at up to 3.6 percent, are among the highest in the world. An estimated 690 million people in these areas earn less than two dollars a day. Some 142 million people earn less than one dollar a day. Rural women and children suffer the most from poverty and its social and physical deprivations, which include malnutrition and high rates of infant mortality (Rodriguez and Thomas 1999). Natural resources mismanagement and degradation are among the most important contributors to poverty. Water is life. It is essential for human development and prosperity everywhere. But in the dry areas its importance becomes particularly high because poverty is very much linked to water scarcity.

Fresh water in the dry areas of the West Asia and North Africa (WANA) region is the scarcest in the world. In several countries of the Middle East, the available resources do not meet the basic needs of the people. Increasingly, water is being diverted from agriculture to other sectors. In the coming 25 years, agriculture's share of water in the region will drop from 80 percent to 50 percent, which will seriously threaten food security and affect adversely the already fragile environment. New fresh water supplies are currently limited by physical, economic, and political constraints. With growing water scarcity, higher water demand, inefficient use of available water resources and lack of sufficient regional cooperation, poverty and conflicts over water are likely to increase. However, improving the efficiency and effectiveness of water use can help meet the growing demand for water, alleviate poverty, and ease conflicts related to this precious natural resource.

Research at the International Center for Agricultural Research in the Dry Areas (ICARDA) and other institutions has shown that it is feasible to at least double the current productivity of water used in agriculture. This would be equivalent to doubling the amount of water resources. Improved water productivity could be achieved by implementing modern technologies, by adopting more efficient packages of water management, by improving cropping patterns and agribusiness practices, through use of improved germplasm, and through setting sound socioeconomic policies. The water saved could be applied to new lands to produce more food without damaging the environment. New technologies will impact the optimal use of available water, the development of nonconventional water resources—such as desalination—and the feasibility of water transportation within and across regions.

In water-scarce areas, such as WANA, water, not land, is the principal resource limiting agricultural development. Accordingly, the conventional objective of maximizing agricultural production per unit land (tonnes per hectare) needs to be adjusted to maximize the production per unit of water (kilograms per cubic meter of water). This new objective requires substantial changes in the way we think and conduct agricultural development in this region. A shift in the focus from land to water requires substantial changes in the management of water and the policies of using it in agriculture. These include changes in the following:

- Current water management guidelines to focus on water instead of land
- Current land use and cropping patterns to produce more water-efficient crops
- The way we value water to reflect actual scarcity
- Trade policies to encourage import, rather than domestic production, of those goods with high water demand
- The attitude toward enhancing regional cooperation on shared water resources
- The approaches to water management, from supply to demand and from disciplinary to integrated

These changes can be achieved only through adoption of national policies that provide increased support to research, encourage people to participate in the management of water, and promote regional cooperation in use of available water resources. Shortage of water will always be the scourge of the dry areas, but knowledge generated by science can address "thirst-driven unrest," where it exists, and keep water the essence of life and a vehicle for peace.

BACKGROUND

Degradation of land and other natural resources is both the cause and the effect of poverty. Environmental degradation sets in when the poor lose the capacity to support and sustain themselves from their natural resource base. Population pressure and a lack of adequate agricultural technologies are major forces driving the poor to make desperate choices. But the worst "enemies" of the poor in the region are water scarcity and drought. Drought is an inherent characteristic of climates with pronounced rainfall variability. In the last 10 years there has been a marked increase in the frequency of droughts in the dry areas.

Water is life and absolutely essential to human development and prosperity. In this region, water has shaped some of the greatest civilizations in history, along the Tigris, the Euphrates, and the Nile. Since then, water has always played a crucial role in the development and stability of this region. But this water is getting scarce as populations grow and demand increases. The average annual per capita renewable supply of water in WANA countries is now below 1500 m^3, well below the world average of about 7000 m^3 (World Resources 1999). This level has fallen from 3500 m^3 in 1960 and is expected to fall to less than 700 m^3 by the year 2025. In 1990, only 8 of the 23 WANA countries had per capita water availability of more than 1000 m^3, the threshold level for water poverty. In fact, 1000 m^3 looks ample for countries like Jordan, where the annual per capita share has dropped to less than 200 m^3 (Margat and Vallae 1999). Mining groundwater is now a common practice in the region, risking both water reserves and quality. In many countries, securing basic human water needs for domestic use is becoming an issue, not to mention the needs for agriculture, industry, and the environment.

The water scarcity situation in WANA is getting worse every day. It is projected that the vast majority of the 19 WANA countries will reach the severe water poverty level by the year 2025; 10 of them are already below that level (Seckler et al. 1999). Over the coming years this situation will worsen with increasing demand, given the fact that the possibility of new supplies is limited. The increasing pressure on this resource will, unless seriously tackled, escalate conflicts and seriously damage the already fragile environment in the region. This is particularly obvious between countries with shared water resources. In WANA, about one-third of the renewable water is provided by rivers flowing from outside the region. Two-thirds of the Arab people, forming the vast majority of the region, depend on water flowing from outside the Arab countries, and about one-fourth live in countries with no perennial water supplies (Ahmad 1996). Considering the importance attached to national sovereignty, and the fact that international laws on shared water resources are still inadequate, potential conflicts between two or more countries are a real concern.

Sufficient water is crucial for continued economic development and improved living conditions in the dry areas, especially in the Near East. The current water supplies will not be sufficient for economic growth of the countries in the region, except for Turkey and Iran. Water scarcity is already hampering development in the Arabian Peninsula, Jordan, Palestinian Territories, Egypt, Tunisia, and Morocco. Other countries, such as Syria, Iraq, Algeria and Lebanon, are increasingly affected as scarcity grows every year. It is, therefore, essential that substantial changes be made in the way water is managed to help alleviate poverty, promote economic growth, and overcome potential conflicts.

POTENTIAL NEW WATERS

The key to improved livelihoods, and consequently to alleviation of conflicts over water resources, is to secure adequate water supplies to meet basic human needs. Equitable distribution of water and protection of the environment are very much linked to sustainability of the solutions. Although limited, new waters can be secured from the following resources:

- Nonconventional water supplies (desalinization, marginal-quality water)
- Water transfers
- Rainwater harvesting

NONCONVENTIONAL WATER SUPPLIES

Nonconventional water supplies include desalinized water and marginal water, such as agricultural drainage, treated sewage effluent, and brackish water. These sources are gaining more importance as advances in water technologies are made and as scarcity increases.

DESALINIZATION

Desalinization is an expensive process, and hence is currently mainly used in areas where an affordable energy source is available, such as the Gulf countries, where part of the desalinized water is used for irrigation. The total production of fresh water might reach 913 billion m^3 annually for some 18 million inhabitants. Seawater desalinization costs $US1.00–1.80 per m^3. Some have reported lower costs, but with subsidized energy. Costs might become feasible for agricultural use, particularly with the development of new technologies, possibly based on natural gas as a source of energy. Breakthroughs in the cost of desalinization would open real opportunities for several countries of the region. However, our aspirations for a breakthrough in desalinization technology are hampered by lack of funds to support research in this field.

Marginal-Quality Water

Marginal-quality water offers some promise. Potential sources include natural brackish water, agricultural drainage water, and treated effluent. Research shows that substantial amounts of brackish water exist in dry areas. This water could either be used directly in agriculture or desalinated at low cost for human and industrial consumption. Treated effluent is an important source of water for agriculture in areas of extreme scarcity, such as Jordan and Tunisia, where it counts for about 25 percent of total water resources in the country. Egypt is currently producing about 1.2 billion m^3 per year of recycled water from the city of Cairo, and by 2010 it is expected that 4.9 billion m^3 per year will be produced in the country (El-Beltagy et al. 1997). There are, however, several health and environmental issues associated with use of marginal water for agriculture.

Agricultural drainage water is becoming an appealing option to many countries, not only to protect natural resources from deterioration, but also to make a new water resource available for agriculture. In the last two decades, the reuse of drainage water in agriculture and its impact on the environment have become the focus of research in many parts of the world, particularly in dry areas (El-Beltagy 1993). In Egypt, reuse of drainage water increased from 2.6 billion m^3 per year in the 1980s to about 4.2 billion m^3 per year in the early 1990s. Now, two new projects will bring the total reuse of drainage water to ~7.2 billion m^3 per year, some 12 percent of total water resources available to Egypt.

Water Transfers

Transfers between water basins and across national borders have been extensively discussed in the region over the last two decades (Kally 1994). Importation of water is under active consideration in the Near East. The two options most relevant are to involve transportation by pipeline (Turkey's proposed peace pipeline) and by ship or barrage (big tanks or "Medusa" bags). Both suggestions are subject to economical, political, and environmental considerations, which have yet to be examined (Lonergan and Brooks 1994). In the WANA region, attempts at transferring water by balloons and tankers have been made, but the cost is still high for agricultural purposes. The project to transfer water by pipelines from Turkey to other Near East countries was unsuccessful because of economical and political reasons. Potential for such projects can be realized only with good regional cooperation and trust between the parties concerned. As water scarcity in the region grows, the issues associated with cross-boundary water resources become urgent and require solutions. Internationally agreed on laws and codes of ethics need to be developed to insure water rights and to open the way for innovative projects in the region.

Rainwater Harvesting

Rainwater harvesting offers opportunities for decentralized community-based management of water resources. Hundreds of billions of cubic meters of rainwater in the drier environments are lost every year. This loss occurs mostly in the marginal lands, which occupy a major part of the dry areas. The development of water harvesting systems in these areas could save substantial amounts of water that would otherwise be lost. ICARDA has demonstrated that over 50 percent of this water could be captured and utilized for agricultural production if integrated on-farm water-use techniques were implemented properly (Oweis et al. 1999). To achieve success, special attention would have to be placed on policy and socioeconomic issues.

WATER AND AGRICULTURE

Agriculture is the major consumer of water in the dry areas. Currently about 80 percent of the total water resources are used to produce food. With fast-growing populations and improvement in living standards, more water is diverted to meet domestic and industrial needs, leaving less water for agriculture. Ironically, as water for agriculture is declining, more food is needed. Assuming that nonagricultural uses of water stay constant over the coming 50 years, agriculture's share of water is projected to drop from 80 percent to 50 percent during this period. However, if nonagricultural uses continue to grow at present rates, agriculture's share will drop to this level in half of this period. In several countries in WANA, such as Jordan, the only source of water for agriculture will be marginal-quality water.

Despite its scarcity, water continues to be misused. Farmers are able to extract water at rates far in excess of the recharge rate, thus depleting aquifers. Unless action is taken, the situation is likely to get worse. Water losses in irrigated agriculture in WANA amount to 25–40 percent. The productivity of water in the region is still low, but varies from one crop and country to another. A study was conducted by ICARDA and by the United Nations Economic and Social Commission for West Asia in farmers' fields in four countries of the region. This showed that water-use efficiency in producing major crops ranged from as low as 35 percent to as high as 75 percent, with great variations between crops and between countries. The study, however, shows that much improvement can be achieved. Water scarcity and mismanagement also have implications for the degradation of the environment, through soil erosion, soil and water salinization, and waterlogging. These are global problems, but they are especially severe in the dry areas.

Agricultural Water Demand Management

Research at ICARDA and other institutes has demonstrated that proper management can more than double the agricultural return from water (Oweis 1997). The green revolution, for instance, saw improved cultivars that produced twice the yield using the same amount of water as the old cultivars. Gains in production efficiency were also achieved through proper management of water and cropping systems (Drek 1994). In the foreseeable future, the potential for substantially increasing water supply is limited, but demand could be better managed, and this could contribute substantially to water availability for agriculture. The traditional strategy of responding to water shortages by increasing water supplies through capital-intensive water transfer or diversion projects has clearly reached its financial, legal, and environmental limits. Attention has, for the last decade, shifted from development to management. Three of the major areas contributing to water savings and improved demand management are discussed in the next sections.

Improving Water Productivity

Research at ICARDA has shown that a cubic meter of water can produce several times the current levels of agricultural produce through the use of efficient water management techniques. In supplemental irrigation, limited amounts of water are applied to rain-fed crops during critical stages, resulting in substantial improvement in yield and water-use efficiency. Wheat yields were increased from about 2 t/ha to more than 5 t/ha by applying less than 200 mm of irrigation water. Water-use efficiency of supplemental irrigation in producing wheat grains has reached 2.5 kg/m^3 compared to a maximum of 1.0 kg/m^3 under conventional irrigation. Water application based on deficit irrigation can maximize the return per unit of water rather than per unit of land (Oweis 1997). Application of water to satisfy less than full water requirement of crops was found to increase water productivity and spare water for irrigating new lands. Adopting a 50 percent deficit supplemental irrigation strategy for wheat in limited water resource areas of northern Syria resulted in 28 percent increase in total farm production compared to a full-irrigation strategy. Deficit and supplemental irrigation strategies are important in the dry areas because water, not land, is the principal factor limiting agricultural production (Oweis et al. 2000). Water is becoming scarcer, so guidelines for irrigation should be adjusted immediately to focus on water-use efficiency.

Optimizing farm cultural practices and inputs, such as selection of appropriate cropping patterns and fertilization regimes, can also increase water-use efficiency. Selection of crops should insure that water is used cost-effectively, in terms of social and economic considerations. It is, however, a dynamic process since the land-use in this area will be affected by globalization and the new world trade agreements. This issue is attracting substantial attention from advanced research institutions as well as decision-makers in the region.

Using Mendelian breeding techniques and modern genetic engineering, higher yielding and more water-efficient varieties could be developed. For example, ICARDA has developed winter chickpea to take advantage of winter rains, and drought-resistant barley varieties that use substantially less water to produce normal or higher yields. Winter varieties of chickpea can produce about twice that of spring varieties. More work is needed, however, to integrate all the above-mentioned approaches in practical packages to achieve the largest return from the limited water available (Erskine and Malhotra 1997).

Participation of all concerned in the management of scarce water resources is the key to successful implementation of more effective measures of water management. Players include the public and private sectors, but, most importantly, representatives of the users of water, particularly farmers and pastoralists, who should be involved in making decisions about water management issues. Users cannot, without appropriate policies, achieve the objectives of effective water management. It is widely agreed that lack of proper policies in this region is the main constraint to improved water use.

Recovering the Cost of Water

Although water is extremely valuable in this region, it is generally supplied free or at low and highly subsidized cost (Cosgrove and Rijsberman 2000). There is, therefore, little incentive for farmers to restrict their use of water or to spend money on new technologies to improve the use of available water. International agencies, donors, and research institutes are launching a major campaign to promote the adoption of a pricing scheme for water services based on total operational costs. Although it is widely accepted in the region that applying higher value for water would improve efficiency and insure better investment levels in water projects, the concept is seriously challenged in many countries of the region.

The reasons are mostly sociopolitical. Traditionally, water is considered in many countries of the region to be "God's gift," and, hence, should be free to everyone. Farmers' pressure for subsidized inputs for agriculture makes it difficult for decision-makers to implement water pricing. There is also a fear in many countries that once water is established as a market commodity, the market will determine prices, and the poor might be unable to buy water, even to meet their domestic needs. Downstream riparian fear that upstream riparian might use international waters as a market commodity in negotiations on water rights. This misunderstanding could lead to disruption to peace and stability in the region.

One cannot ignore these concerns, because they are real and derived from societies. With difficulties in pricing water in this region, innovative solutions are very much needed to put a real value on water in order to improve efficiency. At the same time, ways must be found from within the local culture to protect the right of people to access water for their basic needs. It can be seen that in countries with increasing water scarcity, there is a tendency to recover the operation and maintenance costs of irrigation supply systems.

Adopting Improved Technologies

It has been claimed that existing technologies, if applied in the field, might double the amount of food produced from the present levels of water used (Drek 1994). Precision irrigation, such as trickle and sprinkler systems, use of laser leveling, and other techniques, contributes to substantial water savings and improves water productivity. Along with the development of technologies to capture new water (such as water harvesting) or improve productivity of available resources, policies that promote the transfer of these technologies are vital. Farmers need to be provided with economic alternatives to the practices that lead to wasting water, and with incentives that can bring about the needed change.

THE CHALLENGE OF CHANGE

The world is passing through an exciting time, a time in which social, political, economic, and scientific realities are changing, in which a growing recognition of collective responsibility, facilitated by modern information technology, is driving the struggle for change. Fortunately, we now have cutting-edge science available to us to improve the pace and efficiency of our work in dealing with water scarcity and bridging the knowledge gap. The application of biotechnology has made it possible to develop crop varieties that produce more with less water. Remotely sensed data and geographic information systems have helped in devising more efficient water-capturing methods. Modeling tools can help in maximizing water-use efficiency. Advances in information technology have placed in our hands computer expert systems, important tools for technology dissemination (Rafea 1996). Scientific breakthroughs in renewable energy generation and economic desalinization, combined with low-cost means of transportation, can provide dependable water supply in the areas that are currently most stressed.

However, to benefit from advancements in science and from the recent world focus on water, there is an urgent need for action at the country and local levels. This involves the need for regulatory and legislative reforms in the water sector, protecting the most vulnerable strata of the population and attracting increased financial investment. It also involves supporting technological innovations, undertaking research and development, and building human and institutional capacity. Local policies are responsible for lack of adoption of much of the improved technologies developed by research institutions. Those policies need to be seriously revised to bring about a substantial change in the way we manage our precious water resources. Water determines the livelihoods of people and will be the major constraint to sustainable development in this region. "Business as usual" is no longer an option for water management in water-scarce areas. Unless we make the needed changes we will soon face a water crisis. The needed changes include the following:

- Change the emphasis from land to water. Traditionally strategies are developed to maximize yield per unit of land. This is good when land is the limiting resource for agriculture. With water becoming the limiting resource, strategies should be changed to focus on the return from a cubic meter of water rather than the return from a square meter of land. Maximizing water productivity can help alleviate water shortages.
- Change current land use and cropping patterns to make use of more water-efficient crops and cropping systems. Many crops that are grown in water-scarce areas are inefficient in their water use under certain conditions. Others could be much more efficient. New cropping patterns need to be studied to replace inefficient crops to reduce water demand and improve competitiveness. The comparative advantages of various agroecologies should be assessed, so they can be put to best use.
- Change the way water is valued to truly reflect the scarcity conditions. The value of water in water-scarce areas is very high and should be used as such. Since water is generally a common property, the equity of its use and sustainability require that it be given the real value when policies of development are drawn. Water demand management requires taking courageous steps in changing the current conventions of water allocation and pricing to promote sustainable farming systems and efficient water use and development.
- Change trade policies to import, rather than produce, goods that have a high water demand. Large amounts of water cross borders as virtual water. This needs to be adjusted to reduce water demand and support existing farming systems and associated socioeconomics.
- Change the attitude toward regional cooperation. Water use efficiency might be improved at the farm level, but will not be maximized unless it is tackled at the basin level. This requires regional cooperation, particularly among countries with river basins. It is also expected that increasing water scarcity will result in serious environmental problems across river basins. These need to be tackled collectively. Data exchange, transparency, and collective policies and decision-making are important for complementarity among neighbors.

- Change from disciplinary to integrated approach. A great deal of effort in the dry areas is wasted due to lack of integration. Maximum productivity will never be reached unless all production elements are optimized. Success lies in integrating natural resource management, including water management, with crop improvement, and in developing agricultural systems that would contribute to food security in the dry areas of the developing world.

CONCLUSIONS

Water scarcity is a serious problem in the dry areas of WANA. Scarcity is a threat to economic growth and a cause for possible conflict, particularly among countries with shared water resources or water basins. Possibilities for increase in renewable water resources are currently limited, either because the resources have been explored up to their safe yield potential or because of economic considerations. Breakthroughs in desalinization of seawater and brackish water might be achieved with substantial support to research. Major water transfers across basins and national boundaries are constrained by political and economic considerations and require substantial regional cooperation and active international efforts. Reducing water demand by changing water delivery schemes from a supply- to demand-driven basis, improving the efficiency of water use through advanced technologies, improved water management, appropriate cropping patterns, improved germplasm, and appropriate cultural practices could play a great role in alleviating the adverse effects of water scarcity in this region. Supporting research on the management of water under scarcity is vital to achieve this objective, but changing policies at the national and local level is vital to promote the adoption of improved technologies. Water, if properly managed, can be a vehicle for peace and regional cooperation and prosperity instead of a source of conflicts.

REFERENCES

Ahmad, M. 1996. *Sustainable Water Policies in the Arab Region.* Paper presented at the Symposium, Water and Arab Gulf Development: Problems and Policies. University of Exeter, England.

Cosgrove, W.J., and F. Rijsberman. 2000. *World Water Vision: Making Water Everybody's Business*. London, England: World Water Council, Earthscan Publications Ltd.

Drek, T.E. 1994. *Feeding and Greening the World, the Role of International Agricultural Research*. The Crawford Fund for International Agricultural Research, CAB International.

El-Beltagy, A. 1993. Capacity building for agricultural water management in Egypt. In *Integrated Rural Water management, Proceedings of Technical Consultation on Integrated Rural Water Management* Rome: FAO.

El-Beltagy, A., Y. Hamdi, M. El-Gindy, A. Hussein, and A. Abou-Hadid. 1997. Dryland farming research in Egypt: Strategies for developing a more sustainable agriculture. *American Journal of Alternative Agriculture* 12:3.

Erskine, W., and R.S. Malhotra. 1997. Progress in breeding, selecting and delivering production packages for winter sowing chickpea and lentil. In *Problems and prospects of Winter Sowing of Grain legumes in Europe*. AEP Workshop, December 1996, Dijon, pp. 43–50.

Kally, E. 1994. Cost of inter-regional conveyance of water and costs of seawater desalination. In *Water and Peace in the Middle East*. Issac and Shuval, eds. Amsterdam: Elsevier Science B.V.

Lee, E. 1990. Drainage water treatment and disposal options. In *Agriculture Salinity Assessment and Management*. K. Tanji, ed. New York: American Society of Civil Engineers.

Lonergan, S., and D.B. Brooks. 1994. *Watershed: The Role of Freshwater in the Israeli-Palestinian Conflict*. Ottawa: International Development Research Center (IDRC).

Margat, J., and D. Vallae. 1999. *Water Resources and Uses in the Mediterranean Countries: Figures and Facts*. Blue Plan, UNEP Regional Activity Center.

Oweis, T. 1997. *Supplemental Irrigation: A Highly Efficient Water-Use Practice*. Aleppo, Syria: ICARDA.

Oweis, T., A. Hachum, and J. Kijin. 1999. *Water Harvesting and Supplemental Irrigation for Improved Water Use Efficiency in Dry Areas*. Research Paper SWIM 7. IWMI, Sri Lanka.

Oweis, T., H. Zhang, and M. Pala. 2000. Water use efficiency of rainfed and irrigated bread wheat in Mediterranean environment. *Agronomy Journal* 92:231–233.

Rafea, A. 1996. *Natural Resources Conservation and Crop Management Expert Systems*. Workshop on Decision Support Systems for Sustainable Development, UNU/IIST, Macau. February 26–March 8, 1996.

Rodriguez, A., and N. Thomas. 1999. *Mapping Rural Poverty and Natural Resource Constraints in Dry Areas*. Social Sciences paper No 6. Natural Resource Management Program. Syria: ICARDA.

Seckler, D., D. Molden, and R. Barker. 1999. *Water Scarcity in the Twenty One Century*. IWMI, Water Brief 1.

World Resources Institute. 1999. *World Resource: A Guide to the Global Environment. Special Focus on Climate Change, Data on 146 Countries.* New York: Oxford University Press.

NOTES

1. Presented at the World Food Prize International Symposium *From the Middle East to the Middle West: Managing Fresh Water Shortages and Regional Water Scarcity* October 24–25, 2002, Des Moines, Iowa.
2. Director-General, International Center for Agricultural Research in the Dry Areas (ICARDA), P.O. Box 5466, Aleppo, Syria, a.elbeltagy@cgiar.org.
3. Senior Water Management Scientist, International Center for Agricultural Research in the Dry Areas (ICARDA), P.O. Box 5466, Aleppo, Syria, mailto:t.oweis@cgiar.org.

18

AN INTEGRATED APPROACH FOR EFFICIENT WATER USE – CASE STUDY: ISRAEL[1]

Saul Arlosoroff[2]

Since the early days of human settlements, food producers and nations have depended on irrigation to produce staple food supplies and meet the growing demand for agricultural products as population growth and standard of living increased in all countries of the world.

At the beginning of the 21st century, more than 40 percent of the global agricultural production and 60 percent of the world grains supply are being grown on irrigated lands. Irrigation accounts for approximately 70 percent of the global water use and close to 90 percent in the developing countries. The needs of the projected global population, estimated to exceed 9–10 billion, and the related demand for food and agricultural products, will be limited by water shortages unless water resources management policies are changed drastically. The past paradigm of expanding water supply as the demand increased can no longer be continued.

This paper will focus on one country in the Middle East, Israel, which since its establishment followed policies and implemented strategies that proved that increasing water use efficiencies in all sectors (especially in the irrigation sector) are feasible to sustain socioeconomic growth in a water-scarce environment.

Water demand management (WDM) is becoming a well-known phrase in the sector publications around the world. Moreover, several countries and regions have followed Israel's example. However, a drastic change in the water resources management is needed soon in many countries of the world, in order to prevent serious problems and disputes. Inter-boundary water conflicts can develop into military disputes —in the Middle East as well as other regions.

GENERAL WATER ECONOMICS ISSUES RELATED TO WATER DEMAND MANAGEMENT (WDM)

Water is getting scarce in many countries or regions. Its scarcity raises profound economic issues. Indeed, economics is the science of scarcity. Under regular conditions, air and ocean water are not scarce and do not pose economic issues, as long as they are not polluted. When population grows, the water it relies upon becomes scarce: There is not enough water to satisfy the needs and wants of everyone around. In a world of 6+ billion people, water is becoming scarce almost everywhere

When water is scarce, it is costly to develop and to use. Despite being valuable, even vital, fresh air is cost-less because it is not scarce. When one inhales air and consumes its oxygen, one does not deprive anybody from using as much air as he or she needs. However, when a farmer or an industry uses water in an area in which it is limited in amount, that quantity is not available to another farmer, a city, or an industry. The productive value of the water on the farm or the site where it is missing is the shadow cost of water.

Efficient use of water means that the contribution of water to human welfare is the maximum that may be achieved. Where people are poor and where food supply is not always assured, contribution to agricultural production is contribution to welfare. To a first approximation, where water is used in agriculture, it should be allocated to users so as to maximize agricultural production. Where water is used for human consumption, in urban or rural areas, it should be allocated to satisfy equally the needs of all.

Failure to realize that scarcity requires careful allocation of water and that such allocation is often not assured in a hands-off policy is one of the roots of inefficient use of water. The other root is the failure, or the absence of political courage, to realize that over-utilization of water destroys the resources—aquifers, rivers, soils, lakes, and habitats. This last failure is regarded sometimes as the creation of problems for future generations, but these generations are here already. Water resources are tolerant of our actions; despite abuses they have suffered and are suffering, they have continued to serve users for decades. But the "days of judgment" are coming; Aral Lake, Lake Chad, the Dead Sea are just three crying examples. It is time we face reality and insist on efficient allocation and use of water. Sustainability is but one aspect of efficiency; efficiency in the distribution of the benefit of water equally over long periods of time is the other aspect.

Efficient utilization means that water contributes equally wherever it is used, within a defined project, river, or area. Central allocation (county, state, or national) can be efficient. But as experience has taught, central allocation, lacking omniscience, is not efficient. Moreover, it breeds political struggle and corruption. Experience has also taught that wherever successfully applied, markets and prices are efficient instruments for allocation.

Water prices function in two ways:

- To provide information
- To transfer income

The first role of prices (to provide information) is the least understood. The right price will reflect the cost of water to society; that cost materializes only under scarcity. When water is scarce, the quantity used on one farm is missing to others in the same project or basin. The product not produced where the water is missing is its cost. The right—or shadow—prices will reflect the cost of the resource. This is the basis, for example, for the abstraction fees imposed in Israel since 2000.

When prices are right—reflecting its real economic value—farmers, industries, and households will utilize water only if its contribution to production and to welfare exceeds its cost. "Right prices," including prices paid for the extraction of water from common sources, assure efficient allocation of water. The right prices for irrigation are appropriate when water is scarce, because they maximize production of food and fibers to the satisfaction of human needs. The price system is basically clean; it encourages efficient allocation with no reliance on bureaucratic human interference. Prices are not intended to encourage people to use less water; their aim is to promote people to use the right quantity of water—on the farm, in industry, in the household, or in the urban sector.

Prices also transfer income. Failing to understand the information and allocation role of the prices, farmers and urban dwellers oppose the adequate price system. They see it only as a means of income transfer. Politicians oppose the price system because they strive to reflect public opinion and because they support the patronage potentials of central allocation. But prices will encourage efficient use of water; they will increase, not reduce, income in rural and urban communities alike, and will reduce the reliance of water users on the whims of the political "powers-that-be."

However, water is intended to satisfy the needs of all members of society and thus must be under public control, especially in periods, like droughts, when prices by themselves cannot maintain needed distribution.

Water is already scarce around the world; we have already over-utilized and damaged our resources. We must promote the sustainable and efficient utilization of the gift of nature. We have to turn to the management instruments like simple maxims of economics and start "the water revolution" by the introduction of adequate prices into the water sectors.

ISRAEL: A CASE STUDY IN IRRIGATION WATER DEMAND MANAGEMENT

This paper is not intended to advocate that Israel's level of water resources management (WRM) can or should be implemented universally. Water management experts might examine the model of Israel as a laboratory for others, or as a long-range objective. A number of countries have already followed and inserted water management into their regulations and codes, employing elements pioneered in Israel more than 50 years ago:

- Israel is a water-scarce, semi-arid country, small (20000+ sq.km/ ~8000 sq.miles), with a present population of 6.5 million. The total availability of average natural freshwater resources is ~1600–1650 million Cubic Meters (CM) per average year (~260 CM/capita/year). Moreover, irrigation is essential to maintain economic agricultural production, because rain occurs predominantly in the five months of winter and then mainly in the northern and central regions of the country.
- Since Israel's establishment in 1948 as an independent state, water resources (research, planning, development, control, and management) have national priorities. This was supported by legislation, the institutional setup, the needed budgetary investments, research and development in water-related technology and agronomy, and the creation of monitoring data and tools as a feedback to the continued research and development efforts.
- As was clear from the start, groundwater is the major resource of the country. The Water Drilling Law and Water Metering Law preceded

the General Water Code and declared all water resources of the country to be public property. This was followed by the Water Pollution Prevention Law and regulations.

Groundwater development started early in the history of Israel, with archeological findings from urban and irrigation projects by the occupants of the country in 2000–1000 B.C.E. In the 20th century, groundwater resources became the major initial tool of development with the aquifers developed (over 65 percent of the total water supply of the country is pumped from the aquifers). All wells are metered, as well as all other resources used. The total supply is metered to each farmer, apartment in the cities, house in villages, industry, etc.; water is allocated to all consumers and progressive water pricing has been imposed on all consumers, as well as pumping levies per CM pumped:

- There is constant monitoring of water tables in all aquifers, lakes, springs, etc. Water quality is determined by sampling and analysis. In addition, the operation of the water meters and any changes in use patterns are recorded and investigated.
- This communication is a retrospective look at the results of the strategy after 50 years of operation.

Development and water experts, who have an interest in the Middle East and in the economic development process of semi-arid countries, often pose the following question. How does Israel, a semi-arid country, prosper with less than 300 cubic meters of water per capita per year for all uses? International organizations define arid countries with less than 1000 CM/cap/year as high-stress countries, with low water availability a severe constraint to socioeconomic growth. Semi-arid countries with less than 500 CM/cap/year are considered under acute water conditions.

This section will try to clarify some of the policies, legislative basis, and selected economic issues that enabled Israel to achieve the following:

- Reach a GDP of $16,000 per capita per year
- Supply much of its agricultural needs (except grains)
- Export agricultural products
- Supply its population and industry
- Maintain a high standard of living, all with very limited freshwater resources

The basis of the strategy lies with a balanced combination of measures:

- Legislative, institutional, economic, and technological focusing on water demand management
- Increased efficiency of water use in agriculture, as well as the industrial and urban water use
- Re-use of most of its treated sewage effluents
- The economic and integrated use of all its total surface and groundwater resources

It is envisioned that potential future water markets (internal and possibly regional), continuous updating of its water pricing policies, and future large scale sea and brackish water desalination (as of 2005) will enable Israel and its immediate neighbors (as part of the peace process) to continue economic growth. This is despite the water-scarce conditions that all the Middle East is facing.

The policy of Israel to meet the growing demand for water focuses on combined supply and demand activities and investments, while the long range solution lies with the total re-use of its wastewater as well as brackish and seawater desalination. Past and present activities are aimed at delaying the high investments and the associated costs involved with the integration of large-scale seawater desalination, an expensive unlimited source of water, which will be a major supplementary source of freshwater as of 2005 and after.

In summary, the main instruments of the national water resources development and water demand management are the following:

- **Pricing and economic policies.** Progressive block rates are coupled with total metering system (for every farmer, house, apartment, and industry). Prices are updated automatically with a cost of living formula. Subsidies are minimized. Recently a water abstraction fee/levee has been approved by Parliament.
- **Re-use of sewage effluents.** Regulations have been promulgated to increase the quality of both sewage treatment plants and their effluents. This is to maximize the re-use potential of effluents and to minimize the health and environmental risks. This also enhances the trading instruments for the exchange of freshwater allocations, with treated effluents mainly used for irrigation purposes. The allocation policy concentrates on reduction of freshwater allocations to the farming

community and replacing it with treated wastewater effluents. (Total sewerage costs are borne by the city, while the re-use component costs are borne by the water sector.)

- **Water conservation/improved efficiency of water use.** Continued policies concentrate on mixed tools including the following:
  - Allocations, norms, and progressive block rates for each sector.
  - Research, development, and implementation of agronomic techniques. The most famous such technique is the large-scale implementation of "drip irrigation" and automated irrigation. In addition, technological means have been employed to improve water use efficiency and reduce water consumption in the domestic, commercial, and industrial sectors, and the irrigation of urban parks and gardens.
- **Agricultural sector water allocations system.** The irrigation water allocations are based on norms developed by agricultural researchers together with the farming community. It reflects the potential economic gains by introduction of new irrigation technologies and changes of cropping patterns, and moves away from crops where the product value per unit of water is relatively low—like grains, for example.
- **Virtual water policy.** Upon realizing that water resources would not meet demand, the Israeli authorities made the difficult decision (in the 1960s) to import the great majority of grain needs, instead of growing them. In today's terms, this means the "virtual import" of almost 3 billion cubic meters of water annually, almost twice the total availability of freshwater resources in Israel. It should be stressed that this was a courageous decision—creating the situation where essentials such as bread, beef, poultry, eggs, dairy products, etc., are imported at the risk of world grains markets and of potential political impacts.
- **Water markets (internal and possible external).** The authorities and Parliament have recently approved a change in the water code, enabling holders of water allocations to sell their permanent or temporary allocations to others. The actual transaction is via the national water carrier. This opens the sector for a market-like operation. The water commissioner has been doing it for years already by trading freshwater with treated sewage effluents. The market concept could well serve or even promote peaceful exchanges of water between the countries of the Middle East.

Modern hydro-geological studies started in the 1930s, when the Hydrological department was created and intensive research was initiated. Preliminary master planning of water resources and works to support forecasting of water supply and demand were mainly started after the formation of Israel, in its new boundaries and populations.

Israel was established as an independent state in May 1948, and had to struggle with waves of post–World War II immigrants; these waves tripled its population within a few years. The only feasible way was to resettle the immigrants in the rural areas, in villages. This was followed by rapid water resources development and distribution. This provided income as well as irrigated crops as agricultural products, for a developing nation with limited financial resources.

The situation forced the authorities to initiate the basic legislative tools, the institutional setup, and a rapid groundwater development. Most of the surface water resources were in the northern part of the country, but the population growth was mainly in the central and southern regions. (It took more than 10 years to complete the transfer project.)

Wide-scale drilling and pumping water from the various aquifers enabled the authorities to meet the water demand throughout the country, and allowed for over-pumping as the National Water Carrier (NWC) was planned and its construction started.

This planning concept has enabled the country to move surplus water from the north to the centers of population and development in the south. In 1964, the National Water Network was completed, thus creating a national integrated network of all groundwater and surface sources. The Sea of Galilee, in the Jordan basin, was linked with a closed piping (108" + 66", 70", and 84" lines) to the southern end of the country, interconnecting along its way all regional projects, which are based on the various aquifers.

THE NATIONAL WATER STRATEGY MODEL

A number of cornerstones of the national water strategy have been established:

- The Water Code is based on both the declaration of all water resources as public property and the promotion of water conservation and promotion of water use efficiency. The Code includes all the relevant and specific laws, including The Water Drilling Law, the Water Metering Law, the Water Pollution Prevention Law, and others.

- A Water Commissioner's office was established with complete responsibility for all water affairs and management.
- The vast majority of the water resources in Israel (ground and surface) are integrated in one network, the NWC. This enables maximum flexibility in the operation of the various water resources in the country. The aquifers and the groundwater reserves act as the most effective storage of water between seasons, dry and wet years, and weather fluctuations.
- Total water metering was started and completed within 10 years. Each well and water producer together with consumers in the rural, urban, and industrial sectors are legally obligated to install standard water meters, calibrated regularly, by certified laboratories. These meters form the basis for the water demand management strategy, as well as the basis for the ground and surface water resources management. Each well, each farmer, each industry, each apartment in the urban sector, and any other producer or water user are all metered, read routinely, and reported to the water authorities.
- The water annual licensing and allocation system was initiated in 1959, based on water production and consumption norms, mainly used in the agriculture and industrial sectors. The norms were developed, and are being changed along time, through an intensive research and development program, in order to establish the optimum water use by crops, regions, industrial products, and others.
- Water prices are based on the Progressive Block Rates principles, set by Parliament commission for over 70 percent of the total water in the country, mainly for water produced and distributed by the NWC. The water rates for the rest of the producers and consumers are based on costs in addition to abstraction levees/fees that reflect the average economic shadow values of water.

A number of other important issues—such as the prevention of water pollution, water quality control, water drainage regulations, and others —supplement the main features of the strategy as described above, but will not be detailed here. However, it is worthwhile to stress, that as a result of the recent dry spells and water demand beyond natural recharges, the government has decided to initiate and accelerate the construction of Reverse Osmosis Sea Water Desalination Plants (ROSWDP), adding in

2005/2006 (by about 25 percent) to the total freshwater availability of the country. This follows significant cost reductions of ROSWDP during the international tenders of Israel in 2001/2002.

The decision includes the following:

- The completion of a nationwide treatment and re-use of all treated wastewater, tertiary or secondary treated
- Allocating these new sources of water to the farmers in exchange for their freshwater allocations

This policy and investment will allow the country to continue, indefinitely, its socioeconomic growth despite the increase in population and standard of living, as well as opening the door for the potential solutions of water conflicts between Israel and its neighbors.

NATIONAL WATER DEMAND MANAGEMENT PROGRAM

The present population of Israel is approximately 6.5 million (700,000 in 1948) and is increasing at an approximate rate of 2.2–2.5 percent per year. Best estimates for the year 2020 indicate a potential population of 10–12 million Israeli citizens. (The variation is mainly due to unpredictable future immigration levels.)

Present average of urban water consumption (domestic, commercial, and industrial) is approximately 108 CM/capita/year. It would have been approximately 150 CM/cap/year today if it were not for past efforts that resulted in >30 percent savings. Present industrial forecasts, coupled with projections for urban water consumption per capita, converge at an estimate of 110–120 CM/capita/year by the year 2020. These figures assume a much higher standard of living coupled with the continued rigid and wide-scale implementation of demand management policies.

Inelastic agricultural demands for water to supply basic fresh food (dairy products, eggs, and vegetables) are estimated at 25–30 CM/capita; this adds an additional 220–330 MCM/year. Re-use by the farmers of treated effluents in Israel will reach 70–75 percent of the total DCI (domestic, commercial, industrial) use. This is ~100 percent of the total sewerage flows (the entire population will be connected to the sewerage by 2010). The estimated treated effluent flow by 2020 will be approximately 700 DCI (domestic, commercial, industrial) 1000 MCM/ year.

This endeavor includes continued efforts—technological, economic, and agronomic—to further reduce water demand and improve the efficiency of water use for the two production sectors. Incremental costs of water saved in these two range from $US0.05–0.40 per CM. The figures for irrigation assume increased production per unit of water in real terms. They reflect changes in the basic production cycle that is adapting to more economical cropping patterns. It assumes benefits of genetic engineering as well as modifying industrial processes.

The levels of "indirect" additional water production through savings and improved efficiency of water use are very important because they represent permanent reduction in demand. Israel has already gone a long way in its efforts in these two sectors. The term "effort" is much more complicated than it sounds. It means the large-scale application of appropriate irrigation technology (drip, sprinkler, and automation of irrigation) and changes in industrial water use and water processes (like "cascading" water uses and cooling methods).

Special funds, training, public education, and effective extension services have been active in the past and must continue as promotion and implementation instruments. Finally, the further impact of improved pricing mechanism and the application of a market or trading system can play a dominant role in the whole operation.

The significant achievements of Israel's agricultural sector have lead to 300 percent real-term increase in 50 years. A comparison of prevailing prices for irrigation water between most irrigating countries and Israel illustrates and partially explains the gap in the country's agricultural yield/unit of water, and the potential for reducing agricultural water demand in other countries when following comprehensive and rigid water demand policies.

In Israel, water prices for irrigation are one of the highest in the world. Farmers pay an average of 20–25 USC/CM (approximately 260–$290 per acre.foot). They still manage to compete due to training, public education, and effective extension services. Finally, the further impact of an improved pricing mechanism and the application of a market or trading system can play a dominant role in the whole operation.

Unaccounted for water (UFW) causes significant water and financial losses to urban utilities and municipalities. UFW has been substantially reduced in Israel (down to 11–12 percent on average from 25 percent,

15 years ago), but remains a serious problem in other Middle Eastern countries, where for example, UFW rates in some cities are over 50 percent and represent critical water and financial losses.

RE-USE OF HUMAN AND INDUSTRIAL TREATED WASTEWATER EFFLUENTS

Water effluents become an integrated water resource and are traded for freshwater. The price mechanism as well as effluent charges are gradually being enforced and are contributing their share to urban and industrial water demand management. Many of the industries are located in the urban sector and are subject to the additional utility prices. The industrial subsector has observed an increase of 250 percent of industrial production per unit of water (in real terms) following 20 years of demand management campaign.

Re-use of secondary effluents in Israel is restricted to industrial field crops (cotton, maize, etc.) (Figure 18.1). The use of tertiary treated effluents are for unlimited irrigation with subsurface drip irrigation in horticulture.

WATER MARKET—A TEMPORARY OR PERMANENT SOLUTION?

Water in Israel is used within a system of allocations (annual or multi-annual); in most other countries in the region it is user rights that de-

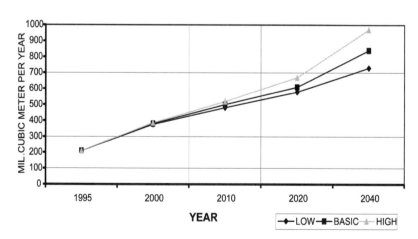

FIGURE 18.1 National forecast for the human wastes re-use program.

termines the demand. In many regions, a person who owns land (or cultivates it) has the right to the water flowing beside or under the plot. In other regions, various quota systems allocate the amounts of water on an annual, monthly, weekly, daily, or even hourly basis. Veteran users usually have the rights to continue to use the resource when shortages prevail.

In Israel, it has been shown that the efficiency of water resource allocation and use can be substantially improved through the increased use of price and trading mechanisms. Trading water on the margin or using a system in which urban/industrial demand is met by supply from farmers selling quotas reduces inefficiency of administrative allocations. Irrigation water in Israel was, and is today, partially subsidized when supplied by the NWC. The administrative allocation system creates a "rent seeking" operation for the development of new resources and higher demand. It leads to built-in inefficiency, which could be improved when water trading is active between consumers using the national water system as conduit and/or using the aquifers as common pools. This allows some to pump more and others less, with those others to be compensated according to the regulations.

THE MIDDLE EAST, EXCLUDING ISRAEL

Many of the Middle East and North Africa countries face an environmental crisis, much of it as a result of water scarcity and the existing and potential pollution of their water resources. It is estimated that the investment needed to deal with and solve the problem could reach $US70–80 billion in the period of 1995–2005 (World Bank).

The hydro-geological conditions are constantly deteriorating. As extraction from ground- and surface water resources increases, so do the problems associated with low water levels and decreased quality. Inadequate human and industrial waste discharges and inappropriate wastewater re-use programs lead to higher concentrations of chemicals and organic contaminants. The concentrations of heavy metals and toxic compounds have already reached alarming levels in various countries in the region, and the projected future cleaning costs could reach prohibitive levels unless urgent and strict measures are introduced.

The expected population growth in the region is likely to exacerbate the problems. The World Bank forecasts growth of ~40 percent in population (from 250 million in 1990 to 350 million by the end of the century). Some regional governments may be unable to generate the financial and human resources needed to provide adequate water and sanitation facilities to meet the future demand. Already, almost 20 percent of the total population in the region lack an adequate potable water supply, and almost 35 percent lack appropriate sanitation. Less than 20 percent of the urban water supplied in 1990 has been properly treated, while in the industrial world this figure is above 70 percent.

Most of the countries in the Middle East therefore face serious water scarcity and pollution problems already, while water shortages are reaching acute levels. During the last 20 years, the average water availability per capita has dropped from 3500 CM/capita and will fall to approximately 1500 CM/capita by the year 2020 for the whole region. Israel, the kingdom of Jordan, and the Palestinian Autonomy are in the most acute situation. All fall below 300 CM/capita.

SUMMARY

Recent reports claim that more than 40 percent of the world food and agricultural needs are produced on irrigated lands. As the world, and especially the urban populations, continue to grow at a rapid rate, the forecasted food and agricultural demand will increase the pressure on the dwindling water resources in many of the world countries, especially the developing ones.

As most of the feasible water resources in river basins and in the aquifers have already been connected and used in the various countries, one can not avoid asking these questions: "From where will the demand for more food and water come?" and "How will this demand be met?"

The largest and cheapest untapped water resource in the world, is hidden in the expression water demand management/water conservation/increase of efficient water use. The Israel water resources strategy will be used as a case study of adequate water management under highly water-scarce conditions. It is not intended to claim that what was followed in Israel has a universal application potential. Rather, one should observe it as a pilot project, or see that country as a laboratory. However, this national experiment could serve others, who may learn and possibly implement various elements of that comprehensive water resources management strategy, in semi-arid to arid conditions.

The main lesson may be the one that early realization and planning —as well as the need for courageous politicians and a relevant decision-making process—could minimize future water supply problems. As approximately 70 percent of all water resources, globally, are used for irrigation purposes, the experience of Israel could be seen as proof that the Earth water resources would be able to support the forecasted population's growth and needs—for a long period of time.

NOTES

1. Presented at the The World Food Prize International Symposium.
2. Chairman of Israel Water Engineers Association.

19

THE MEKONG RIVER BASIN – SEARCHING FOR TRANSBOUNDARY WATER ALLOCATION RULES
Claudia Ringler[1]

SUMMARY

The Mekong River is the dominant geohydrological structure in mainland Southeast Asia. Whereas water resources are more than adequate during the wet season, there are regional water shortages during the dry season, when only 1–2 percent of the annual flow reaches the Mekong Delta. Recent rapid agricultural and economic development in the basin has led to increasing competition among the riparian countries for Mekong waters. This development calls for a structured approach to the management of the basin, including efficient, equitable, and environmentally sustainable water allocation mechanisms that support the socioeconomic development in the region.

The Mekong Regime—comprising the lower basin countries of Cambodia, Laos, Thailand, and Vietnam—has, over more than four decades, continued to successfully adjust its mandate and objectives according to the needs and political situation given. However, at the onset of the 21st century, the Mekong River Commission (MRC) is at a turning point as it attempts to negotiate a series of transboundary water allocation rules to achieve a reasonable and equitable utilization of the Mekong River waters, envisioned in the 1995 Mekong Agreement.

This communication introduces the geopolitical situation of the Mekong River Basin and the general basin hydrology—water supply and demand across water-using sectors and countries. The organizations involved in water allocation in the Mekong River Basin will be introduced with a focus on the MRC. The challenges facing the MRC include the negotiation of protocols for water allocation.

NOTES

1. International Food Policy Research Institute; 2033 K Street, N.W.; Washington D.C. 20006; c.ringler@cgiar.org.

20

JAL SWARAJ: A WIN-WIN SITUATION FOR ALL
M. S. Swaminathan

The Water Emergency prevailing in India is due to both drought and the nonsustainable and inefficient use of most water resources, particularly of groundwater. This has brought home the urgent need for launching a Jal Swaraj movement based on the conservation of every drop of water and its sustainable and equitable use for domestic, agricultural, and industrial purposes. This was the vision of the late Anil Agarwal, founder of the Centre for Science and Environment, New Delhi, who coined the term "Jal Swaraj." In 1980, when I was in charge of Agriculture, Rural Development, Irrigation, and Science and Technology in the Union Planning Commission, a five-point strategy was included in the VI Plan (1980–85) for developing a sustainable water security system for the country:

- Rainwater harvesting and storage in a manner that evapo-transpiration losses are minimized
- Participatory watershed development and management and desilting and renovation of ponds, tanks, lakes, and reservoirs
- River water sharing and efficient use
- Waste water (including sewage water and industrial effluents) treatment and recycling
- Seawater use along the coast for raising mangrove and salicornia plantations together with aquaculture

All these water sources need to be used in a conjunctive manner. Also, in the case of dry farming areas, community conservation of rainwater will happen only if there is equity in water sharing. For this purpose, all farm families should agree to grow only low-water–requiring but high-value crops, such as pulses and oilseeds. Such a method of water conservation with improved livelihood security can be achieved through the organization of Pulses and Oilseeds Villages.

In addition to these immediate measures, steps were proposed in the VI Plan for developing a long-term plan for linking the major rivers of peninsular India—such as Krishna, Godavari, Mahanadi, and Cauvery—as well as for the desalination of seawater. Peninsular rivers are under our political control, unlike the Ganges, Indus, and Brahmaputra, which have international dimensions.

The suggestion for linking our major rivers is not new. Dr. K. L. Rao is famous for his advocacy of the Ganga-Cauvery tie-up. Various experts have also been writing books and papers from time to time urging steps for linking our major rivers. During the tenure of Bharat Ratna Morarji Desai as Prime Minister (1977–79), the proposal of Mr. Dastur for constructing "garland canals" in the Himalayas, central and peninsular India attracted his serious attention. He however abandoned the idea when it was pointed out that constructing a canal at a high altitude in the Himalayas was beset with serious ecological, political, and national security issues.

The revised National Water Policy (April 2002) of the Government of India contains the following statement:

> Non-conventional methods for utilisation of water such as through inter-basin transfers, artificial recharge of ground water and desalination of brackish or sea water, as well as traditional water conservation practices like rainwater harvesting, including roof-top rainwater harvesting, need to be practised to further increase the utilisable water resources. Water should be made available to water short areas by transfer from other areas including transfers from one river basin to another, based on a national perspective, after taking into account the requirements of the areas/basins.

Thus, sharing of waters is an idea whose time has come. The present time is particularly a propitious one, since there is scope now for launching a massive Food and Cloth for Jal Swaraj program, organized on the model of the Employment Guarantee Scheme of Maharashtra.

Based on the financial provision made in the VI Plan for initiating steps for linking the rivers of peninsular India, the Central Water Commission had undertaken detailed studies. Given the necessary political will and consensus, the linking of the rivers of Peninsular India and the equitable and efficient use of this gigantic water grid will help create a win-win situation for all the states involved. If there are winners and losers, there will be only conflict and confrontation, as is happening now.

It is high time we move in the direction of rainwater harvesting, watershed development, and conjunctive use of rain, river, ground, treated sewage, and seawater on the one hand, and river water linking on the other. China, which has taken a long-term view in the area of building a national water security system, has already made much progress in building the gigantic Three Gorges Dam, for taking the waters of the Yangtze River to the parched lands of the northern parts of that country. The Three Gorges Dam will confer multiple benefits in the areas of flood control, irrigation, drinking water supply, and power generation.

Our country unfortunately is one where there is often "paralysis by analysis," particularly in vital areas such as inter-basin sharing of water. For example, a National Water Policy was adopted in September, 1987, but it remained largely on paper. This policy was reviewed and updated in April 2002. However, implementation structures for converting the policy into reality do not exist. There has been no effort to mobilize the Panchayati Raj institutions for harvesting and sharing water at the local level. The Centre for Science and Environment and the Akash Ganga Trust have established in Chennai a Rain Centre, which is a single-stop information center on all aspects of rainwater harvesting, storage, and use. We need such rain centers in every village and town in the country. In each Panchayat and Nagarpalika, at least two women and two elected members should be trained as Water Security Managers, capable of looking at water security issues in their totality, namely the consumption of water to meet the needs of human settlements, agriculture, industry, and ecosystem maintenance.

While there is discussion and debate relating to the quantitative aspects of water needs, the same interest is not evident with reference to the qualitative aspects of water, with particular reference to domestic consumption. For example, the CPR Environment Education Centre in Chennai has observed the following position in relation to groundwater quality in the Chennai area.

Water samples from 25 different locations in Chennai, collected from wells and bore wells were analyzed for their potability in the Center's laboratory. The pH of the samples ranged from 6.50 to 8.81. The Total Dissolved Solids (TDS) exceeded 500 mg/l (the desired limit of TDS in drinking water) in 88 percent of the samples, and 64 percent of the samples contained more than 1000 mg/l of TDS. Twenty-eight percent of the samples contained more than 2000 mg/l of TDS, the maximum permissible limit as per IS 10500:1991. Twenty-four percent of the samples contained high levels of iron content. This clearly indicates the sad state of groundwater quality in Chennai.

The Ganga Action Plan, after huge investment and many years of work, could not succeed in making the water of this "holy" river potable. This is attributed to non-point pollution, caused by human beings, many of them pilgrims. The same fate is true of most of the rivers and tanks revered for their spiritual significance. It is clear that unless water becomes everybody's business, as Anil Agarwal put it, it will remain a source of conflict and contamination.

India has enough water resources to fulfill everybody's need. We live in a democracy. Let us designate this year as a Water Emergency Year and initiate during the year a Water Literacy Movement, which will stimulate State and Central Governments to launch an integrated water security strategy with the following short- and medium-term action plans. In the short term, emphasis may be given to the following:

- *Collect* every drop of water and use it for raising more crop per drop.
- *Treat* all wasted water and use it conjuctively with ground and surface water resources.
- *Ensure* that drinking water is not polluted.
- *Create* consciousness of the need to regard holy rivers, reservoirs, lakes, and tanks as gifts of God on earth by not polluting them.
- *Manage* ground and surface water resources in a sustainable and equitable manner through regulation, education, and social mobilization.

All these steps are politically doable and economically affordable. The short-term steps mostly need nonmonetary inputs, since they depend upon people's awareness and participation on account of enlightened self-interest.

In the medium term, we should initiate action to use conjunctively sea- and groundwater along the nearly 8000 km shoreline of India for raising plantations of mangroves, salicornia, casuarina, cashewnut, and coconut. Also, agro-aqua farms can be established all along the coast for promoting integrated systems of aquaculture and agroforestry.

Above all, we should not lose even a day more in fostering political consensus to link all peninsular Indian rivers in the form of a Dakshin Water Grid. When coupled with the Prime Minister's National Highways Project, the Dakshin Water Grid will usher in uncommon opportunities for strengthening both food and livelihood security, leading to a hunger- and poverty-free Dakshin.

Jal Swaraj will become a reality only if such a movement is built on the foundation of the twin pillars of equity and ethics. For this purpose, all political parties must agree to abide by the rule of law and respect the findings of institutional structures set up to settle disputes relating to sharing of water, particularly under conditions of scarcity. When Lal Bahadur Shastri coined the slogan "Jai Jawan, Jai Kisan," he wanted us to respect the contributions of every Indian Jawan and Kisan to national security. Since about 75 percent of our water resources go to agriculture, sharing the joys and sorrows of all Indian farmers in an equitable manner is vital for both agrarian prosperity and sustainable food security.

In these days of water peril, we should follow the advice given in 1955 by Bertrand Russell and Albert Einstein when the world was confronted with a grave nuclear peril.

> There lies before us, if we choose, continual progress in happiness, knowledge, and wisdom. Shall we, instead, choose death, because we cannot forget our quarrels? We appeal, as human beings, to human beings; remember your humanity, and forget the rest. If you can do so, the way lies open to a new Paradise; if you cannot, there lies before you the risk of universal death.

V

FRONTIERS IN
POLICY AND ETHICS

This section presents essays and reviews related to agricultural policy and ethics, placing agriculture into a societal and ethical context. Zuhui Huang and colleagues discuss the effect of China's accession to the World Trade Organization on its grain policies. Paul Thompson, one of the foremost philosophers addressing issues related to agriculture, contributes a seminal essay featuring the legacy of positivism. Radford Davis discusses the threat of agroterrorism. David Zilberman discusses very recent research on the economic impact of agricultural biotechnology in developing countries.

21

CHINA'S GRAIN SECURITY AND TRADE POLICIES AFTER ENTRY INTO THE WORLD TRADE ORGANIZATION (WTO): ISSUES AND OPTIONS

Zuhui Huang, Jianzhang You, and Jiaan Cheng[1]

As China's economy has grown rapidly during the past 20 years, the share of primary industry to GDP has declined steadily, from 28.1 percent in 1978 to 15.9 percent in 2000 (Figure 21.1). However, employment in agriculture is still high, being about 50 percent, and three quarters of the population live in rural areas.

With a population of more than 1.2 billion, China is both a major producer and a consumer of grain. Any change in its grain output will impact other developing countries. A percentage-point drop in China's grain output will result in an increase of 5 million tons of grain imports. China is an important trader in agricultural products in the world markets. China was an important rice and soybean exporter, but now maize has become one of the major agricultural exports. China's import and export of cereals and cereals flour totaled US$1.73 billion in 2001, a decrease of 24.4 percent comparing to 2000. Exports of cereals and cereals flour were US$1.10 billion, down 35.2 percent; and imports came to US$0.63 billion, up 6.8 percent. Imports of soybeans were US$28.1 billion, up 23.8 percent.

THE CURRENT SITUATION FOR SUPPLY AND DEMAND FOR GRAIN BY CHINA

China is now experiencing a surplus of grain, with production exceeding demand. China's gross grain output first exceeded 500 million tons in 1996 with the quantity of grain per capita being 400 kg. In 1998, the gross grain output reached a record of 512.3 million tons. In 2001, China overcame serious natural disasters and still produced over 452.6 million tons of grain. Its grain output in 2002 is expected to hit 500

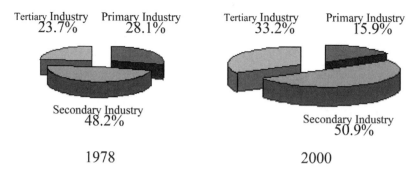

FIGURE 21.1 Composition of China's gross domestic product.

million tons. China's gross grain output has been maintained at about 500 million tons during recent years (see Figure 21.2).

There is considerable grain storage. This carryover increases by about 3.5 million tons every year and accounts for 50 percent of the quantity traded. However this does not include grain storage by the farmers. It is estimated that there are about 400 million tons of grain stored by farmers.

Commercial marketing of grain is improving. More and more farmers sell or buy grain from the market. It is well known that the commercial grain represents only about 30 percent of the total grain in China.

The price of grain has increased over the last two decades. It had gone up 8.4 percent per year from 1978 to 1997. Currently, prices for domestically produced wheat, maize, and soybeans in China are higher than those from abroad by 20–50 percent.

The grain-supplying behavior of China's farmers is characterized by *the spider web theorem*. There were eight years in which the grain elas-

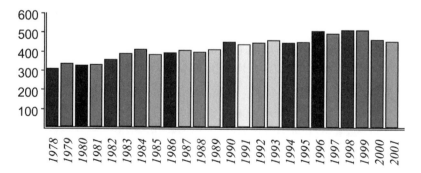

FIGURE 21.2 China's gross output of grain 1978–2001 (million ton) (China Statistical Yearbook 2001, Beijing, China Statistics Press).

TABLE 21.1. Elasticity of grain supply

Year	Elasticity
1979	0.6174246
1980	0.2443070
1981	1.3437149
1982	6.3745591
1983	3.3654237
1984	1.6182135
1985	5.8589143
1986	1.0447806
1987	1.5624719
1988	1.3328333
1989	0.1451780
1990	−1.3420833
1991	0.2793652
1992	−1.2970818

Source: Quanhai Dong, 2000, *China's Grain Market: Fluctuate and Govern*, Beijing, China's Price of Goods Press, p. 44.

ticity of supply was more than 1 and four years in which it was less than 1 in 1979–92 (Table 21.1).

There were five years in which the grain elasticity of demand was negative, and five years in which it was positive in 1989–98 (Table 21.2).

EQUILIBRIUM PLANNING IN THE LONG TERM

White Paper—The Grain Issue in China described the equilibrium planning of China's grain in the long run. This chapter summarizes it in the following sections.

PROSPECTS FOR CONSUMPTION/DEMAND FOR GRAIN BY CHINA

In the years to come, a moderate pattern of increased food consumption, which keeps pace with the national economic growth and conforms to the situation of the nation's agricultural resources, should be brought into being among both the urban and rural residents. The Chinese government will strive to avoid a rapid increase in grain demand beyond the supply capacity through guided grain consumption and tapping the grain-producing potential as well as the potential of non-grain

TABLE 21.2. Elasticity of demand for grain

Year	Elasticity
1989	−0.0396863
1990	0.1860753
1991	−0.0195864
1992	−0.2502563
1993	−0.2067681
1994	0.3175312
1995	0.1416344
1996	0.4704978
1997	0.3578417
1998	−0.4249558

Source: Quanhai Dong, 2000, *China's Grain Market: Fluctuate and Govern,* Beijing, China's Price of Goods Press, p. 49.

food production. China's consumption/demand for grain have been predicted in accordance with this scientific and moderate food consumption pattern. Predictions of China's consumption/demand for grain are shown in Table 21.3.

POSSIBILITY OF ACHIEVING SELF-SUFFICIENCY IN GRAIN

At present, China has basically achieved self-sufficiency in grain production. There are objectively many factors favoring her maintaining this in the course of future development:

- There is great potential to increase production due to natural agricultural resources, production conditions, technical knowledge, and some other conditions.

TABLE 21.3. Projection of China's consumption/demand for grain in 2010 and 2030

Year	Population (Billion)	Per Capita Demand (kg)	Aggregate Demand (Million Ton)
2010	1.4	390	550
2030	1.6	400	640

Source: *White Paper—The Grain Issue in China,* Information Office of the State Council Of the People's Republic of China, October 1996, Beijing.

- Stabilizing the grain-producing area at about 110 million hectares (ha), through the increase of the multiple crop index, will keep the area of cultivated land constant for a long period of time. China now has 35 million ha of wasteland that may be suitable for farming. Of this, about 14.7 million ha can be reclaimed.
- Given a relatively stable sown area, it is envisioned that an annual average increase in yield per unit area can be 1.0 percent (1996–2010) and 0.7 percent (2011–2030), which will achieve China's desired total grain output target. This can be compared to the annual yield increase of 3.1 percent per unit area for the past 46 years. It is, therefore, clear that 1.0 percent and 0.7 percent are low. Recently, it has been reported that pilot cultivation of "super hybrid rice" in south China was very successful with an average per hectare yield hitting 10.5 tons (Xinhua News Agency, 12/08/2001). The basic principle for solving the problem of grain supply and demand in China is to rely on domestic resources to achieve self-sufficiency in grain.
- There are three reasons for China to even up its grain supply to meet the demand. These are as follows:
  - To maintain social stability. Grain production plays an important role in maintaining social stability. China is a country with a population of more than 1.2 billion, which makes it imperative for the government to ensure a high rate of grain self-sufficiency as a necessary condition for stability. Otherwise, it will not be able to maintain its national economy's sustained, rapid, and healthy development.
  - To ensure the stability of the grain market. The quantity of grain consumed in China every year is one-fifth of the world's total. If China were to import a great deal of grain from other countries, the international grain market would be under severe pressure, and poorer countries would be unable to obtain enough supplies of cheap grain from it.
  - Employment of the rural work force. At present, China has more than 400 million laborers in the countryside, and the development of grain production is one of the main ways of stimulating the employment of the rural work force and increasing the income of the farmers. To import too much grain would have an unfavorable impact on grain production at home as well as on the employment of the rural work force.

China's striving to rely mainly on her own efforts to solve the grain problem will serve to improve the stability of the world grain market and to strengthen the stabilizing factors of the international grain trade. China will maintain its strategy of self-reliance for grain demand and supply in the 21st century with the rate of domestic grain supplied being no lower than 95 percent.

CHARACTER AND TREND OF GRAIN TRADE

While standing for balancing supply and demand for grain, China will not refuse to use international resources as a necessary complement. Nowadays, no country could develop solely on domestic resources. The entry of China into the World Trade Organization (WTO) has surely increased the chances for China to optimize its own resource consumption with global supply. However, we would like to point out here the continued importance of securing China's food supply by ensuring the basic farmland coverage in the country.

China has never relied too heavily on the international grain market. From the founding of New China to the eve of the 1960s, China was a net exporter of grain. After that, China began to import more than it exported. Since the introduction of the reform and opening policies at the end of the 1970s, net imports as a percentage of domestic grain production have been decreasing. It was 3.2 percent from 1978 to 1984, 1.2 percent from 1985 to 1990, and 0.4 percent from 1991 to 1995. Therefore, the small quantity of grain imported by China will not imperil the stability of the international grain market. (Figure 21.3).

China's recent imports and exports of grain (including cereals and cereals flour and soybeans only) totaled 7.92 million tons in the first quarter of 2002, an increase of 46.7 percent over the year earlier. Exports

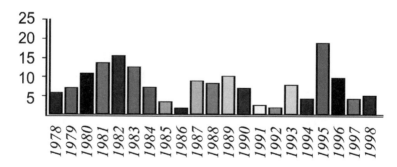

FIGURE 21.3 The net import of China's grain in 1978–98 (million ton) (China Statistical Yearbook 2001, Beijing, China Statistics Press).

reached 4.84 million tons, up 72.2 percent and imports came to 3.08 million tons, up 18.9 percent. Imports of cereals and cereals flour were 0.79 million tons, up 20.8 percent and exports, 2.74 million tons, down 0.2 percent. The net export of cereals and cereals flour came to 1.95 million tons. Imports of soybeans were 2.29 million tons, up 18.7 percent.

The trends of China's grain trade can be characterized as follows:

- China's imports and exports of grain play the role only of regulators in varieties, in case of crop failures and to support poor regions.
- The international grain trade is not a dominant role in China's food security, since its self-sufficiency rate under normal conditions is above 95 percent and the net import rate 5 percent or less (see Table 21.4).

TABLE 21.4. China's imports of wheat and its share of consumption and of wheat imports in the world

Year	Quantity of Import (Million Ton)	Share of the Domestic Wheat Output(%)	Share of the Import of Gross Grain (%)	Share of the Total Import of Wheat in the World (%)
1981	13.00	18.67	89.84	12.62
1982	13.80	20.15	85.45	13.92
1983	11.11	13.65	82.11	10.49
1984	9.87	11.24	94.81	9.33
1985	5.41	6.30	90.17	5.24
1986	6.11	6.79	79.04	6.93
1987	13.20	15.37	81.08	12.23
1988	14.55	17.03	94.91	12.45
1989	14.88	16.39	89.75	13.57
1990	12.53	12.76	91.33	11.82
1991	12.37	12.89	91.97	10.60
1992	10.58	10.41	90.04	8.71
1993	6.42	6.03	85.37	5.67
1994	7.18	7.23	78.04	4.30
1995	11.59	11.34	55.69	10.61
1996	8.25	7.48	76.18	8.07
1997	1.86	1.51	44.60	17.00
1998	1.49	1.36	38.40	1.41

Source: Chinese Academy of Sciences, *Two Sources and Two Markets,* People's Press of Tianjin, p. 147.

- The quantity of grain trade fluctuates heavily (see Figure 21.2).
- China's grain trading partners have become multipartite with more than one hundred countries and districts having traded grain with China. China exports cereals to more than forty-nine countries and districts. Tables 21.5 and 21.6 show the major trading partners.

EXPECTED IMPACT OF CHINA'S ENTRY INTO THE WTO

China's entry into the WTO has brought both opportunities and challenges to China's rural economy. Chinese are experiencing both gains and pressure arising from the nation's accession to the WTO.

As in many other developing countries, China's tariffs were set at relatively high levels. It was understood that these high levels were adopted to safeguard agricultural sectors from competing imports in the short

TABLE 21.5. The main countries and districts importing corn from China

1985		1999	
Country or District	Import Quantity (Million Ton)	Country or District	Import Quantity (Million Ton)
Japan	2.461	Malaysia	1.571
Poland	1.605	Korea South	1.385
China HK	1.046	Indonesia	1.314

Source: Chinese Academy of Sciences, *Two Sources and Two Markets,* People's Press of Tianjin, 2001, p. 153.

TABLE 21.6. Major countries exporting wheat to China

1985		1999	
Country or District	Import Quantity (Million Ton)	Country or District	Import Quantity (Million Ton)
Canada	2.290	U.S.A.	0.181
Australia	1.225	Canada	0.124
U.S.A.	1.154	Australia	0.111
Argentina	0.686		
France	0.227		

Source: Chinese Academy of Sciences, *Two Sources and Two Markets,* People's Press of Tianjin, 2001, p. 153.

run, particularly from possible disruption as non-tariff barriers (NTBs) were removed. China has promised to "slash" its import tariffs for farm products. For entry into the WTO, China has lowered its tariffs to 12 percent from 15.3 percent this year and canceled the quota license administration on grain and other materials. In 2002, the Chinese government unveiled the tariff quotas for imported grains (18.3 million tons). This includes the import quota for wheat at 8.468 million tons, the quota for corn at 5.85 million tons, and the quota for rice at over 3.9 million.

The import gates are "opening" for a large amount of imported grains. More and more imports of foreign agricultural products will affect the production and price of domestic items. This will lead to decrease in income for farmers if the quantity of imported grains increases too much. This would be disadvantageous for domestic grain production. Moreover, its impact would be especially hard for the rural work force where it is difficult to transfer employment from the agricultural sector to other sectors.

If export subsidies are abolished, the price of grain for export will go up. It will result in challenges for the export of China's grain and reduce the competitive edge of formerly advantaged domestic agricultural products. Soybeans face a particularly difficult situation: Potentially large volumes of soybeans imported as domestic edible oil plants are turning to imported soybeans because of their lower prices and higher oil content. Enterprises in some less competitive sectors, such as the State-owned Grain Enterprises, will sustain greater pressure from outside competitors after WTO accession. It will be harder for their grains to sell.

The entry of China into the WTO will cause two profound changes in China's grain trading policies:

- More nongovernmental enterprises will engage in the movement of agricultural products, especially in grain trading and distribution. In 2002, over 7.812 million tons of the quota for wheat was held by State-owned foreign trade companies (90 percent of the tariff quota). In contrast, only 68 percent of the tariff quota (3.978 million tons) for corn was given to State-owned applicants.
- Changes will occur in the pattern of grain production within China. Grain output in north and central China will be used to supply the west. Part of the grain demand of southeast China will be satisfied through imports.

The country's imports of 10 key agricultural products will be decided in accordance with both the demands of the Chinese market and the tariff quotas promised by Beijing during negotiations with the WTO. It is harder for the government to control the trade and prices of grains for national grain security.

OPTIONS

China should make full use of WTO rules, which favor developing countries, when formulating legislation to protect its agriculture, directly or indirectly. The most important measure is to apply an import tax quota to control imports by levying tax and setting quotas. Products within the quota limits enjoy lower or zero tax, while products outside the quota will attract an average or higher tax.

The agricultural subsidy that has been used mainly to market agricultural products should be given directly to producers of agricultural goods. What's more, China needs to improve the quality and the international competitiveness of domestic agricultural products.

In addition to trading barriers, China would need to address new issues on agriculture, the most important of which relate to multifunctionality and other non-trade concerns and the controversial trade in biotechnology products.

China needs to revise the commodity inspection law to stipulate a uniform national certification system that conforms with WTO rules. There is also a need to establish a legal system controlling grain exports by levying export taxes. This is in order to safeguard the domestic grain market sufficiently and avoid excess grain export, especially when the price of grain in the international market is much lower than that in the domestic market. A legal system needs to be established authorizing enterprises to participate in the import or export of grain.

REFERENCES

Analysis and Research Team on National Situation of the Chinese Academy of Science. 2000. *No. 8 Report of National Situation: Two Resources and Two Markets*. Tianjin: People's Press of Tianjin.
China Economy Information Net. 2002. *China's Famous Grain Base Preparing for WTO Challenges*. http://ce.cei.gov.cn/,04/24/2002.
China's State Development and Planning Commission. 2002. *China Published Import Tax Quota on Farm Products*. China Economy Information Net. http://ce.cei.gov.cn. 02/09/2002.

Commodities and Trade Division, FAO. 2000. *Agriculture, Trade and Food Se-curity Issues and Options in the WTO Negotiations from the Perspective of Developing Countries.* Report and papers of an FAO Symposium held at Geneva on September 23–24, 1999.

Information Office of the State Council Of the People's Republic of China. 1996. *White Paper—The Grain Issue in China.* October 1996, Beijing.

N.A. 2001. *Super Hybrid Rice Yields over 10 Tons per Hectare.* Guangzhou: Xinhua News Agency. http://ehostvgw2.epnet.com/. 12/08/2001.

N.A. 2002. *China's Grain Output in 2002 to Hit 500 Mln Tons This Year.* Asia Pulse. http://ehostvgw2.epnet.com/. 02/04/2002.

Peoples Daily. 2002. *China Adjusts Agriculture Policy for WTO Entry.* Beijing: Peoples Daily. http://english.peopledaily.com.cn/. 01/11/2002.

Peoples Daily. 2002. *China Releases Quotas for Imported Grains.* Beijing: Peoples Daily. http://english.peopledaily.com.cn/. 02/11/2002.

Peoples Daily. 2002. *China to Upgrade Agriculture Facing WTO Challenges.* Beijing: Peoples Daily. http://english.peopledaily.com.cn/. 07/03/2002.

Quanhai Dong. 2000. *China's Grain Market: Fluctuate and Govern.* Beijing: China's Price of Goods Press.

Ruizhen Yan and Shulan Cheng. 2001. *Economy Globalize and China's Grain Issue.* Beijing: People's University of China Press.

Xinhuanet. 2002. *Better Laws Needed to Help Rural Areas Cope with WTO Entry.* Beijing: Xinhuanet. http://news.xinhuanet.com/. 04/28/2002.

Xinhuanet. 2002. *China Revises Laws to Meet WTO Requirements.* Beijing: Xinhuanet. http://news.xinhuanet.com/. 05/02/2002.

Xinhuanet. 2002. *China to Revise Commodity Inspection Law.* Beijing: Xinhuanet. http://news.xinhuanet.com/. 02/27/2002.

Xinhuanet. 2002. *Chinese Experience Gains, Pressure from WTO Membership.* Beijing: Xinhuanet. http://news.xinhuanet.com/. 07/03/2002.

Yueyun Li, Naihua Jiang, and Zhongxing Guo. 2001. *Theory of China's Grain Fluctuate.* Beijing: China's Agriculture Press.

NOTES

1. Zhejiang University, Hangzhou 310029.

22

THE LEGACY OF POSITIVISM AND THE ROLE OF ETHICS IN THE AGRICULTURAL SCIENCES

Paul B. Thompson[1]

A slow but steady shift in perspective on the need to address ethical issues within the professional activity of agricultural scientists has been underway for almost 20 years. Increasingly, the issue is less *whether* they should be addressing ethical issues in their classes, at their professional meetings, and in their interactions with client groups, but *how* they should do so. This chapter will provide some philosophical support for this general trend. The basic argument is that ethical issues have been neglected due to the confluence of complex social forces and the philosophies of science that were especially influential during the period of the 20th century when programs in the plant sciences experienced their historic success with hybrid maize, and when livestock husbandry was reconstituting itself in the image of animal science. Philosophy's unfortunate contribution to this neglect should be countered. By the last half of the 20th century, philosophers had recognized some basic errors in the positivist viewpoints that dominated earlier.

The word *positivism* has been associated with many different intellectual movements over the last two centuries. It has been used broadly to indicate the tendency toward logical rigor and quantification in a number of academic disciplines, and more narrowly as the name for some specific doctrines in the philosophy of science that were particularly influential in the 19th and early 20th centuries. Glenn L. Johnson (1976, 1982) used the term *positivism* to describe the then-dominant philosophy in the agricultural sciences, arguing that the main weakness of this philosophy was the exclusion of research on normative issues. This theme was applied specifically to the animal sciences by Kunkel and Hagevoort (1994), who attributed an unjustified and unwarranted narrowing of the topics, perspectives, and methods within livestock research to the positivist philosophy of animal scientists. Bernard Rollin

(1990) discusses positivism as one source of the philosophical values that led to more than a century of inattention to the pain that animals endure in laboratory testing, and suggests that this general cast of mind may have contributed to the neglect of animal welfare in the development of livestock production systems after World War II. Jeffrey Burkhardt uses the term *positivism* to characterize the mindset of agricultural scientists in explaining why researchers fail to recognize the need for and legitimacy of research and teaching on ethics (Burkhardt 1997, 1999).

These criticisms of the agricultural sciences suggest two distinct kinds of inquiry. One is empirical and historical. It would attempt to discern what philosophical views agricultural scientists hold, both currently and over time, and would also take a critical look at whether these beliefs truly have had the types of influence that critics allege. A second type of inquiry—the one that I will undertake here—is philosophical. It attempts to characterize positivism as a set of beliefs about the nature and conduct of scientific inquiry, and queries as to whether these beliefs are justified. The beliefs in question provide a philosophical basis for distinguishing science from non-science. As such, they can be applied in making value judgments about what topics, perspectives, and methods belong in the agricultural sciences. Such judgments have their main practical importance in decisions about which research to undertake, fund, and reward, and in decisions about the curriculum. It is through decisions about who to hire and promote, which research projects to support financially, and what to teach that positivism might lead agricultural scientists to the various types of neglect that the critics have noted.

My argument entails that the term *positivism* has been used too broadly by many of the commentators who argue that agricultural scientists reflect a positivist orientation to the role and conduct of science. Positivism was a pervasive and influential philosophy of science during the first half of the 20th century, but even during this period many well-known figures in logic, epistemology, and philosophy of science were opposing its core doctrines. By 1960, virtually all of the philosophers who had advocated positivist views 30 years before had abandoned their position. The mainstream view in philosophy of science that had taken the place of positivism continued to emphasize rigor and logic, but it was particularly influenced by the views of Karl Popper (1902–94). In one of his papers lamenting the influence of positivism on the agricultural sciences, Johnson (1976) offers an accurate summary of

Popper's views on scientific discovery and explanation, and then characterizes Popper as a positivist. But Popper always dissociated himself from positivism and never held the philosophical views that led some positivists to impugn ethics and ethical debate. Untangling positivist and Popperian philosophy of science is crucial to understanding the legacy of positivism.

POSITIVISM AND POSITIVIST PHILOSOPHY OF SCIENCE

Looking back on positivism from the vantage point of the 21st century, we can see that it was, in fact, the vanguard of the philosophical movement we now call *postmodernism*. The sciences of the 17th and 18th centuries were committed to foundational methods of inquiry. Some, following Descartes, stressed logic and mathematics as the foundational basis of natural laws, and argued that the pursuit of knowledge must proceed through a rigorous process of deduction from basic definitions and self-evident principles of reason. Others, following Locke and the empiricists, took the foundational metaphor more literally and argued that knowledge can be built, brick by brick, from observations of the world. The debate between these philosophies quickly became embroiled in extremely complex arguments concerning the way that deductive and empirical knowledge was connected to the real world beyond human knowledge and experience. The core doctrine of positivism attempts to get around this debate by pinning a number of methodological principles on the observation that human beings cannot have knowledge of entities or existences beyond human experience. Rigorous methods for seeking and organizing knowledge should thus be limited to observations reported directly from human sensory experience, plus logical or mathematical relationships that may be deduced from these primary reports. Someone adhering to such methods strictly will construe knowledge in terms of structured relationships of data. The problems begin, so a positivist asserts, when we start making assertions that go further.

This core doctrine of positivism can be traced to the French philosopher Auguste Comte (1798–1857). It was developed and promoted throughout the 19th century by Ernst Mach (1838–1916) and Friedrich Nietzsche (1844–1900), among many others. Within the sciences, positivist philosophy laid stress on careful recording and interpretation of data, and upon the development of methods for analyzing quantitative relationships between variables of interests. Beyond their disciplinary work, the 19th century positivists were impressed by the way that positivism promised to liberate empirical science from endless philosophical and theological debates about the existence of God, the soul, and abstract entities in general. In confining oneself to the collection and organization of data, one had nothing to say about such things. It was, given the positivist conception of knowledge, impossible to have knowledge of God or religion. In the case of Comte and Nietzsche, in particular, the view extended to a broad social vision in which scientists and other similarly rigorous thinkers would take over social decision-making and political power.

By the 20th century, positivist philosophy no longer entailed any particular vision of social progress. Methods of observation and statistical analysis improved dramatically. Furthermore, logicians such as Gottlob Frege (1848–1925) and Bertrand Russell (1872–1970) stressed the need to distinguish carefully between meaningful symbolic representations (that is, theory and data) and sensory experiences. The construction of scientific theory should not be understood as the collection and organization of sensory experiences, which were, after all wholly private psychological states, but in terms of the sentences or propositions that report and organize data. This logical and linguistic term was pursued by the Vienna Circle positivists who were active primarily from 1924 until about 1938. Like 19th century positivism, Vienna Circle positivism attempted to avoid metaphysical disputes about the nature of ultimate reality. Questions about the existence or non-existence of entities were redefined as questions about whether propositions had meaning. Vienna Circle positivists proposed the Verification Principle as the criterion of meaning: Statements that cannot be verified by appeal to empirical observations or logical analysis are meaningless.

The key ideas of Vienna Circle positivism were popularized in A. J. Ayer's *Language, Truth and Logic* (1936). Ayer's book propounded the verification principle in support of a restatement of the 19th century positivist doctrine eschewing knowledge of religious topics. Statements about God, the soul, and traditional metaphysical disputes were now pronounced to be meaningless. Ayer also included an attack on statements expressing value judgments of all kinds. Hence ethics, as well as theology, became characterized as the pandering of meaningless propositions, incapable of being verified. Ayer's book was particularly strident in making the attack on ethics, though it had not really been a central dogma of the Vienna Circle. *Language, Truth and Logic* was widely read in introductory philosophy classes for the succeeding half century, in part because it combined clarity and brevity in defining a number of concepts and themes that analytic philosophers continue to find important, even if Ayer's positivist view on them has been largely abandoned (not the least by Ayer himself). The book may have served as a manifesto and a warrant for two generations of budding scientists to return to their empirical studies confident in the belief that not only could they safely ignore the ministrations of theologians and ethicists, but also that it was their duty as rigorous scientists to do so.

Deep problems in positivism were seen early on by Popper, who had participated in the original discussion groups that gave the Vienna Circle its name. The contradictions inherent in the positivist view can be seen in the way that budding positivists took rigorous abandonment of theology and ethics to be their duty as scientists. Strictly speaking positivism rules out the possibility that statements articulating duties are meaningful in virtue of their normative, hence unverifiable, content. The statement that a rigorous scientist should not indulge in theology and ethics expresses a norm. As such, it is meaningless, and if one is to eschew meaningless statements, one should have nothing to do with it. The statement that science should be value free thus involves a performative contradiction: In saying that science should be value free, a scientist expresses a value judgment. The act of saying that science should be value free is contradicted by the very norm that it expresses. Thus, positivism rules out ethics, on the one hand, but *is* an ethic for the practice of science, on the other. In fact, positivism involves several performative contradictions, not the least of which is that the verification principle itself turns out to be unverifiable, hence meaningless. Neither Russell nor Popper ever allied with positivism, because both saw that their own views were very far from the abandonment of metaphysics that the positivists advocated.

POPPERIAN PHILOSOPHY OF SCIENCE

Positivism originated in the 19th century as an account of knowledge that shifted the philosophical emphasis away from the manner in which knowledge is grounded or somehow foundationally tied to a reality that transcends human experience, and toward an account of the way that

knowledge is justified when it meets certain logical and empirical conditions. But many 19th and 20th century philosophers were committed
to this shift of emphasis without also being committed to the positivist
views on meaning, verification, and the impossibility of ethics. By the
mid–twentieth century, the problem was being articulated in terms of
the need for a principled way of distinguishing scientific knowledge
from forms of ordinary knowledge, such as how to change a flat tire,
where to get a good steak, and who is the reigning Queen of England.
This is "the demarcation problem." What distinguishes or demarcates
science from ordinary learning and recall of facts about the world? Positivism provided one answer to this question, but it was Popper who
provided what has come to be regarded as the most promising approach
to the demarcation problem, at least among scientists and philosophers
who still believe that science and ordinary knowledge can be sharply
distinguished. Popper's answer stresses the way that scientific claims are
falsifiable through processes of logic and observation (Popper 1959).

The present context is not an appropriate venue for discussing the details of Popper's views on falsification. Yet it is worth noting that the
failure of positivism need not amount to a failure of the core philosophical project that attracted budding scientists to its doctrines. Falsification and the deductive-nomological account of scientific explanation
developed by Carl Hempel (1965) provide a philosophical basis for
characterizing rigorous and objective empirical science that is very much
in the spirit of positivism. Yet this mature philosophy of science does not
entail that normative statements are simply expressions of emotion; nor
does it support the belief that debate over normative issues is futile or
without logical foundation. Popper and Hempel recognized that science
is committed to norms of logical rigor and experimental hypothesis testing, and accepted the task of arguing for these norms as a component of
the philosophy of science. In this their views are quite different from
those expressed in *Language, Truth and Logic*.

Popper's response to the demarcation problem owes a debt to positivism, and indeed provides practical norms for evaluating scientific research. Yet Johnson's claim that Popper was a positivist obscures the
fact that Popper devoted a great deal of his philosophical energy to the
investigation of normative claims. Popper would not have said this
work is scientific in itself, but he did believe that science could not proceed apart from a general philosophical framework that articulated
broad views about the nature of reality; the appropriate norms for rig-

orous inquiry; and a general view of ethical goals, duties, and constraints. Mainstream or Popperian philosophy of science accepts that science—including agricultural science—cannot be made wholly independent from ethical norms. There are three important dimensions to this dependence.

First, science is a process of rational inquiry and as such is ethically committed to a procedure of argumentation and consensus formation that presumes rules, common goals, and a conception of fair play. The second of Popper's two key books on the philosophy of science is entitled *Conjectures and Refutations*. He argued that science makes progress by entertaining hypotheses within the larger context of a theoretical approach, and by using mathematics to determine which observable results would be logically incompatible. Experiments can then falsify or refute a hypothesis and scientists can move on to the next conjecture (Popper 1962). The approach demands that scientists be willing to reach both broad areas of agreement on theory and its empirical implications, and specific areas of disagreement that can be subjected to empirical test. The criteria needed to do this can be achieved only when the scientists accept standards for taking a particular result as disconfirming, for using terminology in a particular (and consistent) way, and a host of other values (Rudner 1953). And this amounts to an ethic of scientific inquiry.

Second, there are certain phenomena that can be defined only in light of ethical or normative reasons why they are of interest. This is particularly evident in the social sciences, where, as Max Weber (1949) argued, key value judgments must be made in order to characterize the relevant phenomena. For example, it is a value judgment that distinguishes a specific class of human deaths, suicides, as worthy of distinct investigation and theoretical explanation. Once characterized as a distinct class, suicide can become the object of a sociological or psychological study. But much the same is true for the applied sciences as well. From nature's point of view, being alive and being dead are just two states of affairs. Inquiry into the basis for health, for growth or productivity, requires value judgments that will allow us to characterize these as distinct processes or states of the world of particular significance. Having done so, it is possible to apply the methods of science to their study. Although there are some topics in applied life science that may not be framed by values in this way, the agricultural sciences in general are committed to the scientific study of processes and techniques that contribute to productivity. Productivity is a value-laden concept; hence those domains of the agricultural sciences that understand biological processes—such as reproduction, growth, or nutrition—as components of productivity have incorporated a value judgment into the very definition of the phenomena they study.

Finally, there is the additional matter of getting together the wherewithal to actually undertake a scientific study. The rationale for conducting many scientific investigations is contingent upon the pursuit of personal or social goals. These goals are frequently evaluated in light of ethical conceptions of public need and benefit. Few grant proposals omit justification in light of social benefits, and even fewer request funding for a given bit of research on grounds that it will be positively detrimental to the public good. To suggest that the exercise of justifying the funding of research is somehow not part and parcel of the scientific enterprise is disingenuous. Scientists routinely do make ethical arguments when requesting funding, and these arguments do influence the selection of which research gets done.

Now there is admittedly a sense in which the epistemic or rational values needed to guide inquiry must dominate over those needed to select a topic or secure support. Yet all of these value judgments are needed to support the scientific enterprise, and there is nothing in mainstream Popperian philosophy of science that provides any reason for believing otherwise. Positivism defined narrowly and typified by Ayer's influential book did indeed inveigh against ethics and the possibility of having a meaningful debate over value judgments, but mainstream Popperian philosophy of science need not do so. If Popper is taken to be a positivist, positivist philosophy of science is capable of accommodating debates over the values necessary to inaugurate and continue scientific research. It is, perhaps, a conflation of the way that rigor, quantification, logic, and experiment distinguish scientific from ordinary knowledge on the one hand, and the now abandoned implications of the verification principle on the other, that have created positivism's legacy. If so, this is a logical error and perhaps a historical accident, but it is not part of a coherent philosophy of science.

POSITIVISM IN THE AGRICULTURAL SCIENCES

Both positivism and Popperian falsificationism provide strong rationales for thinking that the careful collection of empirical data is central to science. Both provide rationales for thinking that the characterization of quantitative relationships among data is critical. Both speak to the question of how experiment and further collection of data can be used to test hypothetical characterizations of quantitative relationships, and how successive application of such tests could lead to a notion of progress in science. However, Popper's view provides a better account of these last two features. Both also provide a rationale for understanding scientific hypotheses as predicting certain observable values; though again Popper's view is stronger in this respect. In sum, then, 20th century philosophy of science provides a strong rationale for thinking that if the agricultural sciences are to be truly scientific, they must stress the quantification of data and the use of experiment to test whether hypothesized quantitative relationships between observable variables are borne out.

This suggests that to the extent that professionals in plant and animal research centers and teaching institutions desired that their activities be understood as scientific, they would find both positivism and Popperian falsification to provide a template for action. It is undeniable that the changes wrought in the collective practice of such professionals over the last 50 years have brought their activities closer to the positivist/Popperian ideal of science. Both of these philosophies could have an unintended impact on agricultural research such that topics in which data are easily collected might be favored over topics where data are relatively unavailable. Furthermore, these philosophies might inadvertently favor research on topics where relationships among variables are relatively straightforward over topics where those relationships are plagued by confounding factors. Neither positivism nor Popperian falsification involves doctrines to the effect that gravitating toward available data and easy quantification makes for better science. Yet clearly scientists who choose to work on problems having these features are likely to come closer to the criteria of the positivist/Popperian ideal than scientists who tackle very complex problems, and they will do so more quickly and with fewer resources. If these scientists are rewarded with promotions and more resources, crop and livestock research may be skewed by an availability bias.

Arguably this kind of unintended influence of the positivist/Popperian ideal could have skewed applied research and teaching in agricultural colleges away from problems such as livestock welfare and agroecology. Scientists working on easier problems will publish more, and research administrators anxious to have productive staff might find this attractive. They might, in other words, hire people to work in areas where results would be expected to come relatively quickly. This, however, is an extra-scientific criterion that is not supported by positivist/Popperian philosophy of science. One can imagine how someone might assert that the knowledge base on topics such as nutrition or reproduction was "more scientific" in the sense of being supported by greater amounts of more extensively quantified data. But neither a positivist nor a Popperian solution to the demarcation problem actually supports this claim. On both a positivist and a Popperian view, knowledge is scientific or it isn't. Demarcation does not come in degrees. What does come in degrees is corroboration, and it would indeed be accurate to say that more extensive data and analysis makes for better-corroborated results. Yet research on animal welfare or agroecology cannot be deemed unscientific simply because data collection and experimental tests are complicated by confounding factors.

However, there are differences between positivist and Popperian philosophy of science that could plausibly lead to differences in the professional norms and practices of agricultural sciences. As noted above, 20th century positivism came equipped with a rationale for denigrating talk about norms as values as being mere expressions of emotion and as being unscientific. Taken on face value, this rationale could support the making of statements that science should be value-free, and that those who make evaluative or normative statements are emotional and unscientific. As the preceding section argues, making such a statement involves a performative contradiction. A deeper understanding of positivism might lead scientists to be reticent about making even these kinds of statements, since they were, according to positivist doctrine, purely emotional and unscientific statements about how science should be done. Thus, it would be more consistent with thoroughgoing positivism to simply remain silent about such issues, and to confine one's professional conduct to the collection of data and the testing of quantitative hypotheses.

Popperian philosophy of science, in contrast, recognized that objectivity and truth seeking *are* norms that stand in need of articulate philosophical defense if science is to be logically coherent and intellectually honest. Furthermore, Popper and especially Hempel came to recognize that the selection and characterization of research topics themselves reflected certain collective value judgments about the nature of what is seen to be problematic in a given domain of science. Had Popper and Hempel paid more attention to applied sciences, this important feature of scientific research might have been even more obvious. Research in biomedical sciences is guided by normative conceptions of health and disease. Research in the agricultural sciences is guided by normative conceptions of productivity and efficiency. The selection of where to invest time and money for science is guided by these leading ideas, and a logically coherent and intellectually honest pursuit of applied science would insist upon a public articulation and defense of them.

Here is where the positivist and Popperian conceptions of science would part company. A Popperian agricultural scientist would see that articulation, criticism, and defense of the guiding notions of productivity and efficiency is an integral element of applied production research. These are admittedly philosophical activities, but they are nonetheless essential to rigorous and objective Popperian science. Thus a Popperian would see agricultural science as needing to conduct an ongoing discussion about such topics as the impact of production practices on environment and on livestock welfare. This discussion would be open, public, and unabashedly philosophical. Its goal would be to develop and articulate a conception of the goals for crop or livestock production that could guide empirical research. A positivist would regard all statements about the guiding notions of productivity and efficiency as emotive ejaculations, incapable of being articulated or defended in a manner befitting the scientific enterprise. A shallow positivist would thus respond to the Popperian's call to discuss and debate these philosophical ideas by saying that doing so is unscientific, and might say that responses to such issues should be based on science, rather than emotion. A deep positivist would greet them with dead silence.

This sketch of some plausible ways that ideas about the nature of science could influence conduct and practice in the animal sciences makes no claim about the history of livestock research or the actual beliefs of crop or livestock researchers. Given my own experience with agricultural scientists, the suggestion that a philosophically deep positivism has had a thoroughgoing and longstanding influence is somewhat implausible. A true positivist will have no opinion about whether entities such as soil, water, soybeans, cows, pigs, or laying hens do or do not exist. The existence or non-existence of ordinary livestock is, for the positivist, a metaphysical question lying beyond the data that are given through sensory experience, and I have yet to meet an animal scientist who did not think that plants and animals exist independently from the data they collect about them. Nevertheless, I have personally observed behavior of both the shallow and deep positivist sort when it comes to the question of discussing the underlying notions of productivity that should guide research. Anecdotally, an invited paper on ethics in animal science research was once returned from an external reviewer with the following comment: "This paper reports no data, hence there is nothing for me to say."

NON-POSITIVIST INFLUENCES ON LIVESTOCK RESEARCH

Neither a positivist nor a Popperian philosophy of science can explain all the changes that have taken place within agricultural research and teaching institutions over the 20th century. Not the least of the unexplained variables concerns why such institutions would gravitate toward scientific ideals in the first place. In a recent paper, Bernard Rollin argues that many of the current problems associated with livestock production can be associated with the shift from a mentality of animal husbandry to a mentality of animal science. According to Rollin, husbandry was traditionally conceptualized as a practical art incorporating norms of respect and regard for animals into its basic practices. Industrialization of animal production, he argues, precipitated the abandonment of husbandry ethics and replaced them with a conceptualization of animals reduced entirely to the manipulation of variables relevant to producers' pecuniary interests (Rollin 2001). Though Rollin himself has been one of the most persuasive critics of positivism in the animal sciences, this recent work places far more emphasis on socioeconomic forces.

Over the last 20 years, historians and sociologists have published a number of studies debunking the suggestion that either positivist or Popperian conceptions of science have much influence over what actually happens in the sciences, or on the eventual fate of scientific knowledge in the form of technology and social change. In place of the notion that ideas about what scientific knowledge is or should be influence the direction of research, historians and sociologists have substituted the view that scientific knowledge differs from ordinary knowledge primarily in its capacity to sustain extended social networks (Fuller 1997). This capacity can derive from a number of sources. On the one hand, sciences can produce technologies that are useful to a particular group; hence there is a natural alliance between practitioners of certain disciplines and the corresponding economic interests. It should not be particularly controversial to say that agricultural science in general has been built on its capacity to develop techniques useful to farmers, agricultural input suppliers and the food and fiber industry in general.

On the other hand, some historians and social scientists have stressed the way that technical complexity can give certain groups the power to dominate a given area of commerce or political life. People with certain pecuniary or political interests form networks with working scientists and provide support for the continuation of scientific work. It is often in the interests of all participants of these networks that this knowledge be shuttered up in "black boxes" of expertise. These black boxes exist because they can produce reliable results that are useful to allied actors and because the combination of technical complexity and controlled access creates a domain of power (Latour 1987). So, for example, medical doctors' political authority and commercial niche depends on a certain black-boxing of biomedical knowledge that prevents challenges from competitors and keeps patients coming back to the medical establishment. Extending this approach to the agricultural sciences would produce analogous claims.

This is not the place to offer a sociological analysis of the factors influencing applied crop and livestock production research over the last 50 years. It is, however, worth reminding ourselves that an analysis of why livestock research has tended to emphasize topics such as nutrition or reproduction, while placing fewer resources in areas such as environment or animal welfare, might readily be given in purely social and political terms. Well-financed actors have been able to make money in livestock production, and attending to environmental or welfare externalities would have served only as a drain on profits. Indeed, such analyses have been given both for agricultural sciences in general and for livestock research, in particular (Schillo 1998; Cheeke 1999) Someone offering such an analysis might find the preceding discussion of positivist and Popperian philosophy of science to be naïve and irrelevant. Indeed, within the social sciences, *positivism* is sometimes taken to be any research that is not specifically committed to an agenda of opposing capitalism, opposing dominant power interests, and supporting the liberation of marginalized groups (Comstock 1982).

Someone working this line of argument might suggest more nefarious motives for the conduct that was associated with a positivist view above. Insisting that science has no place for a debate over values, or simply maintaining a stony silence, could easily be the result of a desire to maintain a domain of power. On this view, it is simply the exclusion of troublemaking parties that motivates the claim that a given issue should be based on science, rather than emotion. Science, on this view, is just a synonym for the particular collection of interests that happen to be in power. One advantage that this sociological argument has over the philosophy of science is that it provides an account of why livestock research would move from a paradigm of husbandry to a paradigm of science in the first place. Kloppenburg (1988) provides a detailed sociological study of similar forces operating in the plant sciences during the first half of the 20th century.

People who have participated in applied plant and animal production research and the institutions that support it will certainly have their own opinions about the relative importance of positivism and economic or political forces in motivating the direction of change. The point that bears emphasis in the present context is that from the perspective of someone outside these circles, it is virtually impossible to tell the difference between a pattern of conduct motivated by positivism and the alternative hypothesis that a more suspect pursuit of power is the cause. From my own perspective, this point provides additional support for adopting a Popperian view within the network of actors who perform and support agricultural science. The positivist view is not only plagued by performative contradictions; it is also self-defeating when it comes to providing a science in which outsiders can have confidence. Far from being the kind of neutral and rational stance that Comte envisioned nearly 200 years ago, positivism has become indistinguishable from an entirely cynical view of the possibilities for achieving truth. How can a viewpoint that refuses to dissociate itself from the raw and unfettered exercise of arbitrary power call itself a philosophy of science?

CONCLUSION

Reasoning something like that I have given here led Popper to embrace a sociopolitical philosophy in support of liberal democratic institutions (Popper 1945). It is also what has led me to argue that animal science needs to develop a professional ethic committed to the open and ongoing debate of the philosophical propositions central to its leading ideas. I have argued that key leaders in the animal sciences must be willing to articulate what they find to be problematic about livestock production, and this general argument applies to all aspects of scientific research on the food system. Agricultural and food scientists must be willing to defend the philosophical values that are implicit in seeing a given situation as problematic, and then they must pursue the research that they believe is most likely to address the problems that they have identified. "Defend" can be defined as a sustained exchange of views that follows six key rules:

1. No one who contributes to the discussion may contract him or herself, and different discussants may not intentionally use the same expression with different meanings.
2. Participants may only assert what they actually believe.
3. Participants must be willing to accept the narrow logical implications of their statements, or be willing to revise them.
4. Participants may only assert value judgments or imperatives that they would equally assert in all situations that are the same in relevant respects.
5. Every participant must justify what he or she asserts upon request.
6. The forum for dialog must be open to anyone who will follow these rules. (Thompson 1999).

Organizations such as research institutes, universities, government agencies, producer organizations, and professional societies must provide the forum in which these discussions and debates can occur. Conferences, seminars, and workshops that are currently conducted on scientific or practical topics could become vehicles for ethical discussion, debate, and learning. There should also be opportunities for written dissemination of views and continuing debates in journals and other publications of the animal sciences. This is not complicated; nor are the rules that must be followed entirely foreign to those followed in most scientific debates. It will be more difficult to reward the individuals who contribute to an articulation of the key values underlying food system research, because criteria for employment and promotion in most agricultural science departments do not presently recognize such activity as demonstrating professional competence. Clearly there will be different approaches taken in different institutions. Research groups that consider themselves to be multidisciplinary may have an easier time welcoming people with philosophical, interpretive, and scholarly expertise into their membership, and may find that criteria now applied to social scientists, philosophers, and legal scholars can also be applied in evaluating the work of those who undertake an articulation and defense of the values implicit in agricultural science.

Silence on such issues must not be tolerated on the ground that it is good manners, much less good science. This is not to say that there should not be tolerance of many different points of view, including points of view that are inconclusive and indecisive. There is no sin in finding it difficult to make up one's mind on difficult values questions. Often the philosophy that considers multiple arguments and ends with the conclusion that it is difficult to side with any one argument rather than another represents true wisdom. Yet individuals and research groups that continually fail to articulate any underlying philosophical vision or rationale for their research practice should come to be thought less scientifically competent for that failure. The days when such people could look down on those who heroically struggle with the philosophical values implicit in their disciplinary practice must come to an end. Positivism is dead. Every day its legacy becomes less and less distinguishable from anti-scientific cynicism. It is time to move on.

REFERENCES

Ayre, A.J. 1952. *Language, Truth and Logic*. New York: Dover Publications, Inc.

Burkhardt, J. 1997. The inevitability of animal biotechnology? Ethics and the scientific attitude. In *Animal Biotechnology and Ethics*. A. Holland and A. Johnson, eds. London: Chapman and Hall, pp. 114–132.

Burkhardt, J. 1999. Scientific values and moral education in the teaching of science. *Perspectives on Science* 7:87–110.

Cheeke, P.R. 1999. *Contemporary Issues in Animal Agriculture*. 2nd Ed. Danville, IL: Interstate Publishers.

Comstock, D. 1982. A method of critical research. In *Knowledge and Values in Social and Educational Research*. E. Bredo and W. Feinberg, eds. Philadelphia: Temple University Press, pp. 370–390.

Fuller, S. 1997. *Science*. Minneapolis: University of Minnesota Press.

Hempel, C. 1965. *Aspects of Scientific Explanation*. New York: Harper and Row.

Johnson, G.L. 1976. Philosophic foundations: Problems, knowledge and solutions. *European Review of Agricultural Economics* 3(2/3):207–234.

Johnson, G.L. 1982. Agro-ethics: Extension, research and teaching. *Southern Journal of Agricultural Economics* (July):1–10.

Kloppneberg, Jack Jr. 1988. *First the Seed: The Political Economy of Plant Technology: 1492–2000*. Cambridge: Cambridge University Press.

Kunkel. H.O., and G.R. Hagevoort. 1994. Construction of science for animal agriculture. *Journal of Animal Science* 72:247–253.

Latour, Bruno. 1987. *Science in Action*. Cambridge, MA: Harvard University Press.

Popper, Karl. 1945. *The Open Society and Its Enemies*. London: G. Routledge & Sons.

Popper, Karl. 1959. *The Logic of Scientific Discovery*. London: Hutchinson.

Popper, Karl. 1962. *Conjectures and Refutations: The Growth of Scientific Knowledge*. London: Routledge & Paul.

Rollin, Bernard. 1990. *The Unheeded Cry: Animal Consciousness, Animal Pain and Science.* Oxford: Oxford University Press.

Rollin, Bernard. 2001. *Livestock Production and the Emerging Social Ethics for Animals.* Paper Presented at the 3rd EURSAFE Conference, Florence, Italy, October, 2001.

Rudner, Richard. 1953. The scientist *qua* scientist makes value judgments. *Philosophy of Science* 20:1–6.

Schillo, K. 1998. Toward a pluralistic animal science: Postliberal feminist perspectives. *Journal of Animal Science* 76:2763–2770.

Thompson, P.B. 1999. From a philosopher's perspective, how should animal scientists meet the challenge of contentious issues? *Journal of Animal Science* 77:372–377.

Weber, Max. 1949. *The Methodology of the Social Sciences.* New York: The Free Press.

NOTES

1. Department of Philosophy, Purdue University, West Lafayette, IN 47906–1360.

23

AGROTERRORISM: NEED FOR AWARENESS
Radford G. Davis

The fear that a weapon of mass destruction (WMD), particularly a biological weapon, might be employed against the citizens of the United States has caused considerable concern of late among public health officials and politicians, and a palpable degree of panic throughout the public, which has been inflamed by zealous media coverage. Taking into account the anthrax mailings of late 2001 that claimed the lives of five people and sickened 17,[1] this anxiousness is neither unfounded nor out of place. Indeed, the saying, "It's not if, but when" has been repeated countless times by countless experts who were pressed to talk about the risks of a biological attack on U.S. soil prior to the anthrax letters. But this has always been in reference to humans. Our fears over insecurity and terrorism propelled us to spend a record $315 million on security surrounding the 2002 Winter Olympics held in Utah.[2] Volumes have been written on the biological programs of various countries, the threat to humans, and the lack of preparation within the U.S. Very little, comparatively, has been written concerning the implications for, and the vulnerabilities of, American agriculture to a biological attack. This is changing.

Agricultural bioterrorism—or agroterrorism—has seen more media coverage and garnered more political attention than ever before,[3-8] thrusting this important topic into the public's eye. The purpose of this chapter is to explore agroterrorism and all of its various facets so as to better understand why we must concern ourselves with protecting livestock, crops, food, and water within the U.S. against biological attacks.

TARGETS AND GOALS

The targets of agroterrorism can be anything or anyone, from the farm to the dinner table. This includes, but is not limited to, livestock (including poultry), crops, timber, farmers and producers, grain holding facilities, food processors and processing facilities (including canneries and bottlers), slaughter houses, grocery stores, county and state fairs, transportation modalities (trains, ships, trucks), restaurants and their staff, one or more main sector of agriculture—such as the pork industry —wildlife, municipal water supplies, or the U.S. economy as a whole. In the end, the definitive target is the American citizen. Terrorists will count on Americans to panic, to become alarmed at the discovery and announcement of a serious foreign animal or plant disease or tainted food product or beverage. Terrorists will count on the public to lose confidence in a particular food item or in the ability of the U.S. government to control the situation and return us to normal. The public may see or hear things that this country has never experienced before, just as it did in 2001. Agriculture is a sleeping target, not yet fully recognized for its vulnerability and not yet fully protected.

When large, well-known federal agencies such as the Centers for Disease Control and Prevention (CDC) write about U.S. vulnerability to biological and chemical terrorism, agriculture is never directly mentioned, despite the fact that many of the most effective microbial agents against humans are zoonotic and can be isolated from, and may cause disease in, animals.[9] In a biological attack against people with the smallpox virus, we can expect an incubation period of roughly 7–19 days before cases of rash are seen by physicians.[10] By then the perpetrators will be long gone. So, too, with agriculture there is an incubation period that can extend to many days or even weeks before anyone realizes that something terrible has been carried out. While more people and more agencies are looking at agriculture and assessing the potential for disaster from a biological attack, the attention heretofore has primarily focused on humans and human agents. Direct attacks against people have taken precedent. This leaves the door wide open for terrorists who may hit us where we least expect it, where we are least prepared.

Why would any one person, or any group or nation for that matter, direct biological agents against agriculture? The goals of bioterrorism can be numerous: to induce fear, to paralyze an industry (medical, political, transportation, food), to create economic hardship. Bioterrorism in the traditional sense targets people directly. The overarching goal of agroterrorism, it should be remembered, is not the animals or crops themselves, but the people who rely on them for food and economy— agroterrorism is truly economic terrorism. An agroterrorist attack would focus on terrorizing American citizens indirectly. This terror would come about through the economic targeting of our agriculture, the generation of uncertainty and doubt, and loss of consumer confidence. "The most dangerous weapon could be a single animal," says Dr. David Franz, a noted expert in bioterrorism, a veterinarian, and the former head of medical research at the United States Army Medical Research Institute for Infectious Diseases (USAMRIID).[11] This comment makes reference to the economic devastation that has been wrought by such notorious diseases as foot-and-mouth disease (FMD) and bovine spongiform encephalopathy (BSE, otherwise known as mad cow disease) in other parts of the world.

TERRORISM AND TERRORISTS

To understand the risk that is poised against the U.S. agricultural system and its products, it is necessary to understand, to the degree possible, the person or persons who might resort to such actions. Examining the terrorists, their motivations, and historical terrorist trends may yield information useful in preventing future attacks—or at the least aid us in better preparing ourselves and industry.

TERRORISM

What is terrorism? There is some debate as to how terrorism should be defined. What acts or actions should be classified as terrorism? Should terrorism be defined based upon the goals (political, social, economic) behind the act or the perpetrator's actions, or upon the outcome of the act, or simply limited to the act itself? Should this definition encompass the cause served? We might ask, too, might not some crimes, because of the motive, which is often concealed, be classified really as terrorism? What is the difference between a biocriminal and a bioterrorist?

The definition of terrorism varies among organizations and individuals. The Code of Federal Regulations (28 C.F.R. § 0.85) defines terrorism as " . . . the unlawful use of force and violence against persons or property to intimidate or coerce a government, the civilian population, or any segment thereof, in the furtherance of political or social objectives." Title 22 of the United States Code (U.S.C.), §2656f(d) defines terrorism as " . . . premeditated, politically motivated violence perpetrated against noncombatant targets by subnational groups or clandestine agents, usually intended to influence an audience." These definitions, however, while fit for general categorization, fail to focus sharply enough and do not take into consideration nihilistic cults—those religious organizations who feel they must bring about the destruction of modern day governments or political systems so that a "new world order" may emerge from the apocalypse, one in which only the devoted will become the principal leaders.[12,13] This was the motivation behind the Aum Shinrikyo's release of sarin gas in the subway system of Tokyo on March 20, 1995.[14,15] These cults, along with other religion-based terrorist groups, differ from other terrorist groups in that they tend to be more violent and unconcerned with the consequences of large-scale killing. Terrorism is not a goal—mass murder is. Therefore, more mainstream definitions do not pertain to these types of groups. One thing that seems to remain constant is that noncombatants are usually the targets of terrorism.

When we compare terrorists to criminals, the objectives criminals have for using biological agents tend to be things such as murder, extortion, to induce terror, to cause disruption, revenge, and even for anti-animal or anti-crop uses.[13] About 9 percent of the confirmed threatened uses of biological agents by criminals were for extortion, either from a food or grocery company.[13] Terrorists, on the other hand, tend to have terror, murder, mass murder, political statements, incapacitation, and use against animals and crops as their primary objectives. Note how the objectives of the terrorist and the criminal can overlap. Indeed, the distinction between what is a crime and what is terrorism can be murky. Biocriminals seek to use biological weapons as a means to carry out a criminal act, usually for selfish reasons and likely for profit; there is no higher purpose involved, no elevated social, political, or religious ideal.

BIOTERRORISM

So what is bioterrorism? That is not so easy to delimit because its definition varies depending upon the source. It can be agreed that bioterrorism makes use of biological agents (bacteria, viruses, fungi, rickettsia, protozoa) in some manner. What remains contentious is the categorization of their use as either a bioterrorist attack or a biocriminal event. Criminals may also seek out biological solutions for their felonious deeds. Organized crime syndicates might also become interested in brokering biological agents, information, or possibly even scientists for profit. It's very likely that the majority of cases in which biological agents have been used in the past have been associated with pure criminal activity and intent.[13] Just as with terrorism, bioterrorism has been defined in many ways, too, depending upon the goals, motivations, outcomes sought, and the person or persons perpetrating the incident. The best definition of bioterrorism comes from W. Seth Carus[13] :

> ... bioterrorism is assumed to involve the threat or use of biological agents by individuals or groups motivated by political, religious, ecological, or other ideological objectives.

Notice how this definition takes into account doomsday cults, where terrorism is not the main objective but bringing about the end of the world and mass murder is. This is the definition that will be carried forth throughout this chapter. When the recent anthrax mailings of 2001 are taken into consideration, there appears to be only two known instances within the U.S. in which a lone terrorist (suspected but not yet established for the anthrax cases), or terrorist group (the Rajneeshees in Oregon in 1984) successfully acquired and used biological agents to induce morbidity or mortality in people. Why is this? It is likely due to the fact that obtaining and safely growing an agent to sufficient quantities, and then effectively deploying the agent, is not as simple and straightforward as many believe. It would appear, based on the historical record and upon the demanding exactness and skill requirements, that the successful use of biological agents is limited to well-funded, state-sponsored terrorists, or to well-outfitted, well-financed independent groups. The expertise needed and the ultimate success in using biological agents is intrinsically tied to which agent is chosen and how it is to be disseminated.

AGRICULTURE

The importance of agriculture in our everyday lives cannot be emphasized enough. Americans are quite used to abundant, safe, cheap, wholesome food, even to the point of taking it for granted. We stop at the market almost daily to partake in the endless abundance of produce, baked goods, and quality meats. We visit roadside stands for farm-fresh sundry fares, and take our children to pick berries. We celebrate every significant holiday with feasts, and make a point of going out to eat to the tune of just over $2,000 per household each year.[16] We involve our children in 4-H clubs and county fairs, take the soccer team out for ice cream, and swap garden-grown vegetables with our neighbors.

ECONOMIC IMPORTANCE

Agriculture touches our lives every day. But what does this industry mean economically? To understand the threat we must understand the importance. The U.S. Gross Domestic Product (GDP) for 1999 was $9.3 trillion, for 2000 was $9.9 trillion, and for 2001 was $10.2 trillion.[17] According to Edmondson, in 1999 the U.S. food and fiber system accounted for 16.4 percent GDP, up from 14.7 percent in 1991, and it added $1.5 trillion to the National GDP.[18] The farm sector alone accounted for $69.8 billion. Foreign trade in agricultural products generated $49 billion in exports in 1999 and rose to $51 billion in 2000.[19]

In July of 2000, there were approximately 274 million Americans eating from the civilian food supplies.[19] In 1999, the typical household in America had an average income of $40,761 after taxes and spent $5,031 annually on food, or 12.3 percent of their take-home pay.[16] Of that, $5,031, 42 percent, was spent on dining out. The average person will eat approximately 4.2 meals away from home each week.[20] In 1991, approximately 18.8 percent of the total U.S. workforce was involved in the food and fiber system in some manner; by 1999, this was down to 17.4 percent (24.3 million people).[18]

Our agriculture industry is heavily dependent upon exportation. If we cannot export, we lose money, farms, jobs, and international stature. Export revenues accounted for 20–30 percent of farm revenues over the last 30 years and are expected to continue in this fashion until 2010.[21] In 2000, the U.S. exported meat to the tune of $3.1 billion for beef and veal, $1.3 billion for pork, and $1.8 billion for poultry.[19] All told, we exported nearly $51 billion in agricultural commodities; $39.3 billion came from non-animal products. The leading U.S. export crop for 2000 was corn ($4.6 billion), followed by wheat ($3.6 billion). Japan, Canada, and Mexico, in order, make up our top three leading export countries.[22] In 2000, the U.S. had nearly 60 million pigs worth $4.6 billion, 98 million cattle worth $67 billion, 435 million non-broiler chickens worth $1.1 billion, and 7 million sheep worth $669 million.[19] In 2000, the U.S. was second only to China for egg production, and produced more eggs than the whole of the European Union combined.[19]

The value of agriculture to the U.S. is interwoven into our society, into our very existence. A careful, well-conceived attack carried out with subterfuge against this diverse, almost priceless element could be more devastating than many care to imagine.

Agroterrorism

Agroterrorism has never been officially defined; however, for our purposes it will be defined as the use of biological (to include toxins), chemical, or radiological agents against some component of agriculture in such a way as to adversely impact the agriculture industry—or any segment thereof—the economy, or the consuming public.

Leroy Fothergill, in his 1961 publication, was one of the early proclaimers of the threat our agriculture faces: "We must face the threat of military delivery of truly exotic agents, including zoonotic agents and mixtures, against our human and animal populations and against our agricultural crops."[23]

HISTORY: BIOLOGICAL PROGRAMS
AND AGENTS AGAINST AGRICULTURE

Biological agents have been used in warfare for over 2,000 years. While not meant to be an exhaustive description, this section does cover some of the best described and notorious uses of biological agents against animal and crops yet known. Descriptions of historical biological warfare, biowarfare programs, and potential human agents for use in bioterrorism are numerous and are covered in other publications.[15,24-30]

It has been pointed out that the first use of biological weapons in the 20th century was not against humans, but against animals.[29] One of the first known offensive biological weapons programs was begun by Germany around the time of WWI, as early as 1915, and included such agents as anthrax, glanders, and the wheat fungus *Pucinia graminis*.[28] Through the German Secret Service, German agents, already inside the U.S. by 1915, planned and carried out the infection of horses, mules, and other draft animals that the Allies had purchased from the U.S with glanders (*Burkholderia mallei*) and anthrax.[13,31] There were, apparently, German plans to use plague in St. Petersburg, Russia, in 1915, and anthrax and glanders were delivered to Romania in 1916 to infect sheep scheduled to leave for Russia. In 1917, the Germans managed to successfully use glanders in Mesopotamia, where they infected 4,500 mules[31–33] under the control of British troops, and they attempted to infect reindeer in Norway as well. From 1917 to 1918, Germany infected sheep, cattle, and horses in Argentina slated for shipment to England and the Indian Army with glanders and anthrax, resulting in the death of around 200 mules.[13,28,33] Horses were also infected with glanders in Argentina that were shipped to France and Italy, and it seems that because of this the shipment of further animals was stopped.[13,31] Germany also had plans to make use of a wheat fungus, but this idea appears to have been scrapped before being carried out.[13,28] Germany's enthusiasm for biowarfare continued into the years of WWII, during which time they experimented with airborne foot-and-mouth disease virus.[28]

France, whose biological program began in 1921, had researched the use of rinderpest against cattle,[34] and late blight (the cause of the Irish potato famine that killed one million in 1845–46) and the Colorado beetle as a means to devastate potato crops.[35] Not only did Germany also research these two pathogens, but they also investigated leaf-infecting yellow and black wheat rusts, rapeseed beetle, and the corn beetle.[35] According to what is known, the Germans were very close to using the Colorado beetle against British potato crops by June 1944, but it was too late in the season and the war ended before they could make use of it. While the German offensive BW program was never really successful or robust, it was nevertheless a path down which several nations were later to follow with significantly better, and more terrifying, results.

One of the world's most active and frightening offensive programs was that of Japan in the 1930s and 1940s. The infamous Unit 731 focused on human biological agents and conducted large-scale human testing, led by military physician General Ishii Shiro, Japan's near equivalent to Joseph Mengele. Unit 100, a smaller, lesser–well-known biological outpost located primarily in the Japanese-controlled small village of Mokotan and the larger city of Changchun in northern Manchuria, China, was led by veterinarian Major Wakamatsu Yujiro.[27] The focus of Unit 100 was research on anti-plant and anti-animal biological agents, but it experimented on humans, too, although not to the extent that Unit 731 did. Research by Unit 100 was carried out under the establishment in 1936 of the Kwantung Army Anti-Epizootic Protection of Horses Unit. It is known that Unit 100 carried out research on plague, anthrax, and glanders, and very likely many other agents. Unit 100 aided Unit 731 in 1939 in carrying out attacks on Soviet troops, their animals, and pastureland in what turned out to be a very bloody battle near the Inner Mongolian, Soviet, The People's Republic of Mongolia, and the Manchurian borders. This battle would later come to be called the Nomonhan Incident.[27,33] Japan's anti-crop efforts are less detailed, but it seems they concentrated on wheat smut and nematodes of wheat. They also had the ability to produce 90 kilograms of cereal rust spores per year.[35] In a macabre twist on ethics and global politics, Ishii's audacity and ruthlessness in bioweapons engineering and research would be rewarded. The U.S., in a rush to get its offensive program moving ahead, gave Ishii Shiro amnesty for his crimes against humanity in exchange for his records and expertise. This, indeed, was poor judgment on the part of the Americans because, although they were highly interested in the human trials, they learned very little from Ishii due to his poor records and unscientific methods. Ishii's daring would also go on to infuse new growth into the Soviet program as well. In the end, a true monster was allowed to get away with mass murder because of greed.

The Canadians had their offensive biological program underway and ahead of the American program by 1940, thanks to the efforts of Fredrick Banting, a physician and the recipient of the 1923 Nobel Prize for the discovery of insulin.[26] The U.S. began, albeit in simply an investigatory fashion, its offensive biological warfare research in the fall of 1941, with an official report from the formal committee, the WBC, looking into the potential of using biological agents for warfare being submitted in early 1942.[26] With all the agents one typically associates with a biological program targeting humans included, the committee also suggested there should be serious consideration into the research of Newcastle disease virus, fowl plague, foot-and-mouth virus, hog cholera, rice blast, cereal stem rust, wheat scab, and late blight of potatoes. It was in May 1942 that Franklin D. Roosevelt approved a biological warfare research program, and George W. Merck, president of the Merck & Co. pharmaceutical firm, was appointed chairman of the War Research Service and head of the U.S. biological program.

While Japan and Germany carried on with their biological programs, England, too, began its BW research program before the U.S. and launched the first successful biological weapon test known in the Western world on July 15, 1942, with the release of anthrax on Gruinard Island. Tethered sheep were the target and a crude bombshell filled with anthrax spores was exploded upwind, proving successful. But England, although ahead in research, did not have the facilities nor the equipment to bring about full-scale biological weapons that they immediately needed to fend off Germany, so they turned to the U.S. for help. In late 1942, England put in a request for large quantities of anthrax and three kilos of dried botulinum toxin, to be delivered as soon as possible.[26] With the U.S. BW program officially housed at Camp Detrick in Fredrick, Maryland, as of early 1943, and the military and their recruited civilian scientists still settling in, the Americans set to work in filling the U.K. order. The British, however, were not sitting idly by. In late 1942 and into 1943, they were putting the finishing touches on their first completed bioweapon—cattle cakes filled with anthrax spores.[26] It is interesting to note that Britain's first choice of a biological weapon was one geared toward livestock and not humans. The reason, it is thought, is that the British simply couldn't generate the number of functional bombs needed in time for the war effort, so they went with the next best thing.[15] The goal of Operation Vegetarian, as it was named, was to drop the linseed meal cakes from the air onto German pastures where cattle and other herbivores would consume them and die.[26] Perhaps even a German might die if they were to consume the meat from an animal that had died from this disease or that was incubating the illness.

Early on in its offensive program, the United States developed anti-crop agents, which had the potential to inflict some measure of economic disruption and depression on the country against which they were used.[36] Some anti-crop agents developed by the U.S. included the fungal pathogens rice blast, rye stem rust, late blight of potatoes, and wheat stem rust.[26,30,35] The U.S. produced and stockpiled 30,000 kilograms of *Puccinia graminis tritici* spores, which causes stem rust of wheat, between 1951 and 1969. Additionally, the U.S. chose *Piricularia oryzae* as its primary pathogen against rice and had roughly one ton stored up by 1966.[35] Other crop targets of the U.S. program included soybeans, sugar beets, sweet potatoes, and cotton.[35] Wheat in the Soviet Union and rice in Asia were identified hard targets. Other agents the U.S. worked on included some very notable zoonotic microbes, such as anthrax, brucellosis, tularemia, and psittacosis (*Chlamydia psittaci*); botulinum toxin also figured prominently.[15] Through the 1950s new agents were added to the list: Q fever, cholera, plague, Venezuelan equine encephalitis (VEE), dengue, shigella, and glanders. In addition to the main site at Camp Detrick, there were other facilities built; a very sizable facility was built at Pine Bluff, Arkansas, in 1942 (at a cost of

$90 million) and initially focused on brucellosis and tularemia. The U.S. even flirted with using insects to spread biological agents and had the capacity in 1953 of producing half a million *Aedes aegypti* mosquitoes, a common vector for yellow fever, dengue, and possibly VEE, in a month's time.[15] By the end of WWII, the U.S. was spending $60 million on BW, and from 1943 until the time the offensive program shut down, a grand total of approximately $726 million had been spent by the U.S. on biological warfare.[15]

The earnestness with which the U.S. undertook anti-animal biological warfare can be seen in the military's testing of two agents through unique, if not completely obvious and useless, means in 1951 at Eglin Air Force Base, Florida. Turkey feathers were coated with hog cholera virus, sealed inside E-73 bombs (a type of cluster bomb) and detonated 1,500 above ground level. Of 115 pigs penned in and exposed during the trial, 93 became ill.[29] A similar trial was undertaken using Newcastle disease virus on poultry. Turkey feathers have the potential to carry roughly 10 percent of the total feather weight in agent;[35] however, it is not likely that the use of hydrogen filled balloons to deliver these bombs, as was done in these experiments, would be of any practical application today.

Poisonous gases used during WWI killed over 100,000, and injured almost ten times that. The Geneva Protocol of 1925 prohibited the use of poisonous gases and bacteriological methods during warfare. The 1972 Biological Weapons Convention (BWC) on the Prohibition of the Development, Production and Stockpiling of Bacteriological (Biological) and Toxin Weapons and on Their Destruction was ratified by 141 countries and signed by another 18; the U.S. was not one of them.[37] The end of the U.S. offensive biological program (National Security Decision 35) was announced on November 25, 1969, by then-President Richard Nixon, and finalized on February 14, 1970, with the amendment of including toxins in the final ban (National Security Decision 44).[15,30,37] The BWC treaty was ratified by President Gerald Ford on January 22, 1975, the same day he also signed papers ratifying the 1925 Geneva Protocol.[15,37] The reason behind the U.S. eliminating their BW program was from a practical, military standpoint; there were other weapons that could be used where the results were virtually guaranteed. Biological weapons were generally unpredictable compared to conventional weapons: They required great skill, familiarity, good timing, and were essentially unproven; and they could be lethal if winds shifted. The top brass within the U.S. military did not think bioweapons were very effective. Add to this the extreme cost of continuing to operate the program and the aging of U.S. BW facilities, halting the program only made good sense—plus it carried with it the added bonus of appearing as a peace gesture to the rest of the world. By banning biological and chemical weapons, that left only nuclear weapons as the last remaining WMD. Given the terrific expense associated with developing a nuclear program, the U.S. could guarantee that very few countries could compete, leaving it an elite member of a small number of countries that still retained the threat posed by weapons of mass destruction.

In 1952, Mau Mau freedom fighters in a region of Kenya poisoned 33 steers at a mission station with what is believed to be the toxic African milk bush plant, *Synadenium grantii* Hook. f., by inserting the milky latex into cuts made into the skin of the cattle; eight animals died.[13,38]

South Africa's National Party and apartheid regime of the late 1970s sought to control influences and revolutionary factions that advocated the overthrow of the government and the elimination of racial discrimination. White South Africa was fighting a war against what was felt to be a communist-backed African National Congress (ANC), which had been banned. The government was involved in spying in countries nearby as well as in some as far away as Europe. They used guerilla warfare and specially trained, secret groups that used biological and chemical weapons to fulfill their goals.[15] These efforts included covert attacks throughout neighboring nations, to include Rhodesia (later renamed to Zimbabwe), using Rhodesian and South African forces. It is believed that, within Rhodesia, from 1979–80 a thousand cattle died from intentionally introduced anthrax and that 10,738 people contracted this disease, resulting in 182 deaths.[13,15,39,40] The evidence that this epidemic was manmade comes from examining the entire epidemiological picture. From 1950–78, there had been a total of only 334 human cases of anthrax documented in Zimbabwe.[40] Add to this the fact that the 1979–80 outbreaks were almost wholly limited to Tribal Trust Lands, swaths of land that had been set aside for blacks when the country was divided into black and white districts by the Land Apportionment Act of 1930.[40] The goal, it appears, was to kill the cattle (the primary source of wealth) of the blacks, the tribesmen, thus eliminating food supplies for them and for the Rhodesian black guerrillas (such as the ZANLA, the

Zimbabwe African National Liberation Army) who operated inside Rhodesia against white rule.[40] There were, interestingly, no reports of anthrax in white Zimbabweans during this same time period. As a result of this concealed attack, anthrax has remained an endemic problem since the end of the war in February 1980. A more direct method of targeting guerillas was the injection of thallium (a heavy metal) into tins of corned beef slated to be eaten by them.[15]

Under the leadership of P.W. Botha's Africa's National Party, South Africa in the mid-to-late 1980s had, by all accounts, the second largest offensive program in the world. In addition to the many agents that one might expect in an offensive arsenal (plague, cholera, tularemia, 45 strains of anthrax, ricin, botulinum toxin, brucellosis, and others), South Africa's illegal program also researched Ebola, Marburg, and Rift Valley fever viruses.[15] South Africa signed the BWC on November 3, 1975.[15]

The Soviet offensive biological warfare program is perhaps the best described and most highly publicized of all programs due to the defection of two of its top ranking officials (Vladimir Pasechnick in 1989 to England, and Kanatjan Alibekov in 1992 to the U.S.), due to its sheer vastness and size, and due to its production of enormously large quantities of many different biological agents, most of which were weaponized. The modern Soviet program was considered to have been the largest ever and began in 1972 and ended in 1992; however, it was as early as 1928, under Bolshevik rule when typhus was ordered to be developed as a weapon, that the fledgling Soviet interest in biological warfare took form.[15] By WWII, the Soviets had not only typhus, but also tularemia, Q fever, plague, anthrax, and cholera. And they made use of them on occasion. There is strong suspicion that the Soviets used tularemia on advancing German troops near Stalingrad, evidently infecting and causing illness on both sides.[15,25] Q fever was also used, according to sources.[25] It was from Ishii Shiro that the Soviets learned how to design and construct their BW plants. They borrowed Japanese maleficent know-how and, under Stalin's strict guidance, built the Sverdlovsk plant, which later was responsible for killing at least 66 in the 1979 accidental release of anthrax.[15,25] In 1973, Biopreparat, a civilian pharmaceutical company operating initially under the Ministry of Defense, was founded and tasked to make bioweapons. Under Biopreparat, the Soviet program included dozens of installations masquerading as pharmaceutical or medical research companies spread out all over the country. There were other facilities built and existing ones that were made to detour their path of research to come into line with Ministry of Health's BW needs.

In addition to the Ministry of Health, the Ministry of Defense, the Ministry of Medical and Microbial Industries (under which Biopreparat was housed from 1980 on), and the Ministry of Agriculture were also involved in overseeing BW facilities. Anti-agricultural biological weapons, overseen by the Ministry of Agriculture, were developed by the Soviet Union in the late 1940s to early 1950s under the project name "Ecology."[25,41] The Ministry of Agriculture's Main Directorate of Scientific Research and Production Facilities oversaw several facilities that produced agents such as anthrax, psittacosis, wheat rust, rye blast, an unspecified anti-corn agent, rice blast, and anti-livestock agents such as African swine fever, foot-and-mouth disease, rinderpest, and others.[25,41] Ken Alibek (Kanatjan Alibekov) notes that glanders was likely used against Afghanistan mujahedin forces sometime between 1982–84.[15,25,41] Anthrax and psittacosis had been designed to be employed as dual-target agents, having both anti-personal and anti-animal capabilities. Biopreparate employed approximately 30,000 people, but the total employed in the Soviet BW program might have actually been closer to 60,000 when all other military activities were taken into account.[15] Testing was also conducted, and over a 20-year time span BW simulants were placed into ballistic missiles and tested over the Pacific Ocean. As for funding, the program seemed to want for nothing: In 1990, the Soviets spent close to $1 billion on BW development.[25] Finally, Ken Alibek, the one time First Deputy Chief of Biopreparat, speculates that the only reason that the Soviets ever became interested in anti-agricultural bioagents was because the Americans were developing them.[41]

Cuba has recently been brought to the attention once again of the American public for its human rights violations and lack of free elections,[42] and for speculation that it has a small, limited offensive biological weapons research program,[43] though the Bush administration admits it has no evidence to back its claims. Former President Jimmy Carter's visit to Cuba shortly after Bush's comments prompted Castro to denounce the claim and offer Carter access to any facility he wished to verify this. Since Fidel Castro took power in 1959, Cuba has claimed that the U.S. has carried out at least 24 biological attacks against it, but only one complaint has ever been lodged formally to the U.N. The official complaint by Cuba, brought before the U.N. in 1997, alleges that in 1996 the U.S. dispersed *Thrips palmi*, an insect that feeds on various agricultural products and subsequently causes significant damage.[44] This incident was brought before a meeting of the BWC State Parties and was subsequently dismissed. No determination was officially made. Of the 24 attacks alleged to have occurred, all but three targeted agriculture. A partial list includes the introduction of Newcastle virus, African Swine Fever, tobacco blue mold, sugar cane rust, Dengue Hemorrhagic Fever (DEN-2), and many others.[44]

Some reports and opinions in the literature quite obviously point a finger at the U.S. for one or more biological attacks on Cuba, but with very little to sustain their argument other than coincidence and conjecture.[45] While Castro has claimed that the U.S. has repeatedly targeted Cuba using biological agents, some in the U.S. are looking hard at the recent outbreak of citrus canker in Florida and wondering if it might not be a product of Cuba's suspected offensive program.[5]

Agricultural products have come under direct attack before. In 1974, a Palestinian terrorist group called the Revolutionary Command said they had contaminated grapefruit from Israel destined for Italy.[34] In 1978, the Arab Revolutionary Council, another Palestinian terrorist group, injected liquid mercury into lemons, grapefruit, and oranges; this contaminated fruit made it all the way to the Netherlands, Germany, Belgium, the U.K., and Sweden, resulting in a 40 percent drop in the citrus export market for Israel.[29,34] Once again, in 1988, Israeli grapefruit exports were threatened with contamination.

In November 1989, a group who called itself the "Breeders" claimed it had released thousands of Mediterranean fruit flies (Medflies) around Southern California in response to aerial pesticide spraying, which the group wanted stopped. While Medflies were found in some areas, it was never concluded that this group was real or that its actions were substantial.[13,34] The Mediterranean (*Ceratitis capitata*) and Mexican fruit flies are a very serious threat to U.S. agriculture, attacking many types of fruit as well as some vegetables, and could cost the citrus industry over $33 million per year.[46] The total costs to the U.S. could top $821 million if we were unable to eradicate this pest and were forced to live with it.[40]

A chemical attack against cattle occurred in 1981 when an unknown individual climbed to the top of a silo and dumped in a bag of organophosphate insecticide, killing 135 animals in 24 hours. The person who did this was never identified, but it was felt that he/she might have been acting on feelings of animosity against the farmer.[47]

Another attack on agriculture took place in late 1996 with the purposeful contamination of liquid fat at a rendering plant in Wisconsin with chlordane, an organochlorine pesticide that was manufactured in the U.S. from 1948–88, when it was finally banned.[47] This fat found its way to feed manufacturers (one being National By-Products, a supplier to a Purina Mills plant in Fond du Lac, Wisconsin)[48] and then eventually on to nearly 4,000 farms in Wisconsin, Minnesota, Michigan, and Illinois. Chlordane-contaminated milk from dairy cows was collected and shipped for the making of items such as butter, ice cream, and cheese. This situation had the potential to ruin the dairy industry of more than one state and may have generated significant health concerns in people had higher levels of chlordane been achieved in the milk. Luckily, samples taken from some of the herds that had eaten the affected feed were negative or contained levels well below those that pose a health hazard to humans. The individual responsible, Brian W. Lea, was finally caught in September 1999 and indicted on two counts of agricultural tampering.[29] He was sentenced to three years in prison and made to pay $2.2 million in restitution.[49] Total cost for disposing of the contaminated feed and fat was almost $4 million; however, as numerous state and federal agencies became involved in dealing with this issue, the final price tag was likely considerably much higher. Was this terrorism or a criminal act? One might argue that the rendering company, the feed companies, the dairies, the farmers, and even the consumers were terrorized into thinking that tainted and potentially deadly dairy products were on grocery store shelves. Perhaps the goal was to strike terror into the dairy industry by killing cows, which might have happened had the levels of chlordane been high enough. Perhaps the goal was economic terror pointed at the farmers, who would lose milk and potentially cows as a result. While this was not an instance where biological agents were used, it does point out how easily agriculture can be targeted.

In an anonymous letter in January 1984, Peter Vivian Wardrop threatened to release foot-and-mouth disease in Queensland, Australia, if prison reforms were not carried out within 12 weeks.[13,48] Two letters in total were sent and told how wild pigs, infected with the virus, would be used to introduce the agent into cattle and pigs. Wardrop was eventually found out, but it turned out he was already in prison serving a life murder sentence.

VULNERABILITIES OF U.S. AGRICULTURE

American agriculture has changed dramatically over the last 100 years. In the first half of the 20th century, the size of the typical farm increased, but in the second half of the century, owing to improved farming techniques and better overall productivity, the acreage of farms actually shrank some. The average size farm grew from 147 acres in 1900 to 487 acres in 1997 (the average farm shrank to 434 acres by 2000[19]), with the result that we have approximately one-third as many farms today as back then. There has been movement away from farms, too. It is estimated that in 1900, roughly 39 percent of the U.S. population lived on farms; today that figure is less than 2 percent.[50] Agriculture has become big business. Gone, for the most part, are small family farms. Profits belong to those who can work large tracks of land efficiently. Machinery has taken the place of horses and mules and human labor. The U.S. has seen the demise of oat farming (45.5 million acres in 1921 to less than 3 million acres in 1998) and the rapid growth of soybeans (less than half a million acres in 1920s to nearly 60 million in the 1990s). Indeed, diversity on farms today is scarce. Where 100 years ago a farm might raise a few head of cattle, a few dozen chickens, a few pigs, and grow corn, hay, potatoes, and oats, most farms today specialize in growing one or two crops or raising only one species of animal. Farms of long ago were more self-sufficient than today's counterparts. The food of America is now produced by fewer farmers located in more geographically defined parts of the country: Pigs are raised in Iowa, North Carolina, and Illinois; beef cattle are fairly widespread, but Texas, Kansas, and Nebraska are the country's biggest producers; dairy operations are widespread, too, but this industry is heavily tied to the states of Wisconsin, California, Pennsylvania, and Minnesota; egg layers are farmed in Iowa and Ohio, and broiler chickens in Georgia, Alabama, and Arkansas; corn and soybeans are raised in Iowa and Illinois; and wheat is grown in Kansas and North Dakota.[51] This consolidation means that use of anti-animal or anti-crop agents could be used in a relatively focused region of the U.S. but have widespread national repercussions.

Consolidation in agriculture has also taken place in related off-the-farm businesses. In the U.S., we slaughter, on average, 130,000 head of cattle per week[52] and, due to the consolidation that has taken place over the last 25 years, this is now a business that is dominated by three large companies.[53] In 1997, 63 percent of all cattle slaughter took place in companies that slaughter a minimum of 1 million head per year.

Decreased genetic diversity is not limited to plants that have been genetically enhanced to resist pests and disease, but includes animals as well.[54] Animals that are genetically similar offer the producer predictable operating costs and end results, and they offer the consumer a predictable-quality slice of meat or a consistent dozen of eggs. However, these finely tuned plants and animals are also more at risk from a designer pathogen engineered to exploit this lack of diversity.

To date there has never been an intentional introduction of a foreign animal or plant disease in the U.S. that has been documented. That is not to say it has not happened, or that it has not been tried. The FBI claims to have stopped the intentional release of a fungal agent against wheat in 1999, although it gives no further details.[55] The emergence of West Nile virus in the United States has been speculated by some to have been an intentional event, although never proven. In March of 2000, rabbit hemorrhagic disease appeared from nowhere on a farm in central Iowa. This is the same virus that accidentally escaped Australian research facilities in 1995,[56,57] was smuggled into New Zealand in 1997, and broke out in Cuba in 1997. This calicivirus is foreign to the U.S., and while all rabbits on this farm either died or were euthanized, the epidemiological investigation conducted by U.S. Department of Agriculture (USDA)-Animal and Plant Health Inspection Service (APHIS) failed to discover the source of infection.[58,59] Then in August of 2001, RHD emerged in Utah. Tracebacks in this outbreak led to the quarantine or slaughtering of rabbits in four states. Once again the origin of the infection was never discovered.[60,61] A captive exotic animal facility in Flushing, New York, experienced an outbreak of this disease in their display rabbits, and once again no source was ever found.[62] The U.S. rabbit industry is valued at $25 million per year, a paltry amount compared to other animal-based sectors of agriculture, but significant still.[59] What does this example imply? It explicitly shows that a foreign animal disease can occur within the U.S., and that we may never know the source, the route of introduction, or whether its appearance was of natural/

accidental means, or borne of manmade malevolence. This demonstrates that we are vulnerable.

We have seen how important agriculture is to our country, both financially speaking and in how it unites us a society. We have already seen that anti-agriculture biological agents are attractive enough that several countries went to great effort and cost to develop offensive biological programs geared toward their production, testing, and weaponization—that in itself says a good deal about their usefulness, however baleful. We have seen, too, that terrorists are ever evolving, looking for new ways to strike at what is perceived by them to be injustice. The fact that very few attacks against agriculture have ever been carried out, and the fact that humans are rarely afflicted secondarily in these attacks, should not lead us into complacency, into thinking anti-agricultural agents are of no importance or concern.

In 2000, the U.S. was home to 2.17 million farms that encompassed 942 million acres of land.[19] Can we protect them all? Should armed sentries be posted on every farm? Obviously not, but logical, useful steps must be taken to ensure better security, in one form or another, to not just farms but to our entire agriculture industry. U.S. agriculture is vulnerable in a number of ways, including the following:[63]

- Little immunity in our livestock to foreign animal diseases
- High-density husbandry methods
- Rapid transportation over long distances of animals, feed, agriculture products, and food for human consumption
- Centralized feed suppliers/manufacturers
- Lack of foreign animal disease training and recognition by U.S. veterinarians/veterinary students
- Lack of education of farmers on foreign animal diseases
- Livestock sale and transportation practices
- Poor-to-nonexistent on-farm biosecurity
- Porous national borders
- Limited traceability of animal movements
- Lack of a federal mandatory animal identification system

Since almost all severe, highly contagious diseases of livestock have been eradicated from the U.S, and the use of any vaccine for them is either nonexistent or rarely employed, livestock in America are at high risk for foreign animal diseases. To this add the rapid transportation systems we have developed; the centralized feed suppliers (the implications of which were seen in the discussion about Wisconsin earlier); the crowding of animals in feedlots (large feedlots may hold up to 100,000 head of cattle), swine confinement units, or egg laying facilities, and the susceptibility of our livestock increases greatly. Unlike developing countries that raise livestock on pastureland or in less-confined facilities, the U.S. has become the archetype for the large-scale raising of animals in confined quarters. Feed, as has been noted, can be an ideal vehicle in which to distribute an agent or compound, and it is made all the more ideal when its manufacture has little or no security or checks and balances. One feed manufacturer can supply hundreds or thousands of farms. Of even more concern is the lack of biosecurity that our animals and plants face and the inefficiency with which we can trace animal movement.

MOVEMENT, BORDER THREATS, AND BIOSECURITY

Biosecurity entails the efforts and measures employed in keeping out unwanted microbial agents and in minimizing the spread and risk of these disease agents to our livestock, crops, and food supply. This is not an easy task. Biosecurity can be applied to a farm, a laboratory, a county, a state, a country—anything really. This is also a concept that is brought up in discussions centering around human infectious diseases. One key component and concern in agricultural biosecurity is movement. Moving animals, plants, people, food, or anything else can spread disease. Restricting movement limits the spread of disease, limits the damage done by the disease, and limits the scope of the problem of combating it.

The U.S. is a very mobile society, both in terms of business and pleasure. The U.S. Department of Transportation offers these statistics:[64]

- In 1999, there were 220 million vehicles on America's highways, up 22 million from a decade previously. Of these, 7.8 million were large commercial trucks.
- From 1989–99, there was a 30 percent increase in the number of commercial aircraft.
- Prior to Sept 11, 2001, the unlawful interference with civil aviation had been on the decline since the 1970s. In 2000, attacks against civilian aircraft accounted for just 2 deaths and 27 injured.

- As of 2000, there were 5,317 public-use airports (29 large hubs) and 13,964 private ones.
- In 1999, there were 81 passenger and freight air carriers that carried 13.9 billion ton-miles of freight.
- In 1999, there were 511,000 interstate freight carriers traveling our highways that were responsible for 1.1 trillion ton-miles of freight carried.

Over 290 million people entered the U.S. from Mexico at the borders in 1999, and 90 million entered from Canada. The busiest gateways were El Paso, Texas, and San Ysidro, California, for crossing from Mexico, and Detroit, Michigan and Buffalo-Niagara Falls, New York for Canadian crossings.

Trade utilizing containers is especially important, because as much as 90 percent of the world's cargo moves in this manner. Indeed, over 200 million containers move between the world's seaports each year.[65] In a five-year period (1996–2001), U.S. Customs saw a 48 percent increase in trade entries into the U.S. to where it now stands at 23.7 million.[66] In 2001, U.S. Customs processed 25 million container entries that entered the U.S. via sea, truck, or rail, and the number of containers moving through the nation's 102 seaports has nearly doubled since 1995.[65] The U.S. had 301 ports of entry as of 2001. And if trade was not enough to keep U.S. Customs officials busy, in FY2001, they processed 472 million pedestrians and passengers—65.9 million of whom arrived via commercial airlines, 11 million by ship, 307 million by automobile, and 53 million crossing our borders on foot. The remainder (142.5 million) came by trains, buses, or private or corporate planes.[66]

But this mobility is not limited to just humans. We also imported 2.4 million head of cattle in 2001, and it is expected that we will import nearly 6.1 million pigs in 2003.[67] And with the introduction of these new animals comes the risk of introducing new diseases. It is said that a pound of meat travels almost 1,000 miles on the hoof before finally finding its way to your dinner plate.[29] Sale barns, where livestock are bought and sold, can serve as a sort of "meeting place" for animals from all over a region. Adult cattle and calves, for example, presented at sale barns mix with potentially hundreds of other animals that have come from around the state or from neighboring states, each bringing with them their own diseases and exposing others in a crowded setting. Once sold, animals are loaded onto trucks, bathing each other with respiratory and salivary secretions as well as urine and feces. The trucks, the handlers, their clothing, and anything else not properly disinfected that the animals contact all become contaminated and serve as reservoirs for further dissemination of microorganisms to other animals or to other fomites (inanimate objects). It is also feasible to imagine that any animal secretions or excretions that emanate from the truck while in transport with its animal cargo can serve to infect domestic livestock or wildlife that come into contact with them along its travels. Animals are shipped hundreds, sometimes thousands of miles, from the point of purchase and sale, allowing for the potential spread of a foreign animal disease, or some other disease of concern, across a very large region of the U.S. Sale barns are an excellent *milieu* for the exchange of infectious diseases —such as FMD, hog cholera, or Nipah—and might serve as a staging area for terrorists to launch an attack against our livestock industry. Over 5 million cattle pass through sale barns annually, presenting many opportunities for the terrorist.[29] Wilson et al. point out that the U.S. military carried out several test runs at sale barns back in the 1950s and 1960s, using water as a biological simulant and using small hand-held spray cans as their means of dispersal.[29] These men carried out their "attacks" at sale barns without ever being discovered.

On-farm biosecurity may be quite good—as is more likely the case in large, well-run livestock facilities—or it can be poor to nonexistent—as one might more commonly see in medium-to-small operations. Husbandry and operating methods may contribute to the spread of infectious agents brought onto the farm. Some large swine producers may have 10,000 animals in a single barn, with several barns located not far from each other; and a dairy may have up to 10,000 animals. Quarantine is crucial to keeping out disease in livestock operations. Wilson et al. note that 70 percent of producers do not quarantine incoming calves or heifers at dairies, and one-third of dairies allow calves access to animals of various ages.[29] Nationally, milk trucks, often part of a local co-op, will go farm to farm picking up milk until the truck is full, taking with it the potential to spread an infectious agent to multiple farms and thousands of animals. The dairy industry is taking steps to educate producers about the importance of biosecurity.

Biosecurity means limiting access of visitors and wildlife, making visitors wear clean rubber/plastic boots and leaving them on the premises when done, disinfecting hard boots and supplies brought onto the farm, wearing of clean coveralls and removing them before leaving, setting a rule that says workers or visitors must be free from contact with livestock outside of the farm for a given length of time (e.g., 1–2 days) before being allowed on, limiting employee movements to only certain areas of the farm or production facility, and having workers wash their hands before leaving for the day. Biosecurity means restricting the movement of vehicles such as feed, rendering, and milk trucks, as well as personal vehicles, to defined areas on the premises so as to limit spread.

The need for biosecurity is so important that veterinarians in the U.K. who were sent to farms were required to "lay off" for at least two days if they walked onto a foot-and-mouth–positive premise before being allowed to visit another farm. Part of this reason was the concern that humans may be able to harbor this virus in their respiratory tract for a short period of time, and officials did not want veterinarians spreading this agent any farther through means such as coughing or sneezing. To try and prevent this, mouthwash was used to clear their throats. No known transmission from veterinarians to livestock occurred.

To prevent the transmission of microbial agents, such as FMD, federal law requires the cooking of all garbage before being fed to swine (9CFR Chap 1, part 166, Sec 166.2, 2002 edition). Many states, however, prohibit the feeding of garbage outright. It is this route that is speculated to have led to the FMD outbreak in the U.K. in 2001.[68,69] For swine production facilities, Amass and Clark, who give a thorough study of swine biosecurity, point out that, while traditionally feed, rodents, insects, vehicles, and even people are thought to pose biosecurity risks, little research has been done on the actual risk of transmission of microbial agents and how best to combat them.[70]

Beef cow-calf operations tend to compensate for poor biosecurity measures by increasing the use of vaccines.[71] In 1996, 38.7 percent of producers imported cattle into their herd; yet these same operations will put these new animals into quarantine only 32–53 percent of the time. It is also not uncommon for different herds of cattle to mix on pasture, and 12.4 percent of producers used manure handling equipment to also feed heifers at least once a week.

Biosecurity at sale barns and on farms is paramount to preventing the spread of diseases of livestock and poultry. While we are continually taking more steps in the right direction to think about and enhance biosecurity in America's livestock operations, there is still need for improvement. Biosecurity measures will likely need to be tailored to each farm or production facility, and cost-benefit analysis will need to be conducted to ensure that such security does not bankrupt the producer. Individual farm analysis of biosecurity measures, however, does not readily take into consideration national or state concerns over foreign animal disease incursion or emerging diseases, and some universal precautionary measures will need to be adopted by all farms for the greater good of agriculture.

We must also envisage biosecurity beyond the boundaries of individual farms, on a broader, more national scale and how we might prevent a foreign animal disease crossing our borders. The USDA has the difficult job of keeping foreign animal and plant pests out of our country, whether they come by air, car, mail, sea, train, or walking carriers. About 3,000 USDA inspectors search baggage at airports and cargo at major ports of entry to ensure compliance with animal and plant import restrictions.[72] USDA-APHIS-PPQ (Plant Protection and Quarantine) has 1,800 inspectors stationed at over 100 ports, who carry out roughly 2 million interceptions of illegal agricultural products each year.[73] Passengers arriving in the U.S. with fruit or meat can expect to have this food confiscated. In the event someone attempts to smuggle in banned items, the Beagle Brigade, a group of highly trained beagles trained to sniff out produce and meat, will alert officials. There are approximately 130 dog teams at 21 airports in the U.S., as well as beagles stationed at one Mexico border crossing.

The illicit trade of exotic animals or animal parts is a multibillion dollar industry, just behind the illegal trade of arms and drugs. Obviously then, the smuggling of these lucrative animals into the U.S. poses the risk of bringing in foreign animal diseases. The U.S. Fish and Wildlife Service in its 2000 Annual Report notes an increase in the smuggling of pet birds, usually concealed in luggage or in a person's clothing.[74] Exotic birds are big business and have the potential to bring in velogenic viscerotropic Newcastle disease, which could infect our poultry and cost large sums to control.[75]

While the topic of agroterrorism is highly concerned with the intentional introduction of foreign agents, there is the possibility that international travelers might bring one or more of these microbial agents into the U.S. accidentally. When the outbreak of FMD occurred in the U.K. in 2001, many thought that the U.S. would surely be next to get it. While undoubtedly the risk of acquiring FMD did increase for our nation, many failed to realize that over two dozen countries across the globe at that time also had FMD to contend with, and that we were already welcoming visitors from those countries. In addition, the USDA took further steps to reduce the chances of FMD making its way into the U.S.: Inspection personnel at ports of entry were increased by 40 percent and the number of inspection dogs doubled over what it had been two years previously. A 1998 USDA assessment entitled "The Potential for International Travelers to Transmit Foreign Animal Diseases to U.S. Livestock or Poultry" concluded that while it is possible for humans to transmit List A diseases to animals, the risk, after taking into account various factors and circumstances, is negligible.[76] High on their negligible list for mechanical transmission by humans to animals was Newcastle disease and swine vesicular disease. Only two diseases, avian influenza and foot-and-mouth disease, had any potential for biological transmission and this was considered very low.

TRACEABILITY OF U.S. LIVESTOCK

The ability to trace back an animal's movements and identify its contacts is vital in controlling or eradicating a disease of concern, especially if we are talking about an exotic disease. Identification of individual animals is critical, beyond disease control and eradication, for the following:

- Monitoring zoonotic disease throughout the country or in a given region/state
- Development of emergency response programs
- Food safety and public health welfare
- Meeting consumer demands/concerns at the market and maintaining their trust
- Meeting trade requirements internationally
- Improving production efficiency
- Regulating quality at the retail and consumer level
- Addressing growing concerns over drug residues in meat

Currently in the U.S. there is no federal animal identification and tracking system in place for livestock, as is seen in some other countries; however, identification of animals may occur under federal disease control programs such as have been established for brucellosis, tuberculosis, scrapie, or pseudorabies. A big concern is being able to track individual animals when an emergency arises. Livestock, especially cattle, may be traded several times before slaughter, and tracing individual movement could prove difficult. In response to the BSE epizootic, the U.K. implemented the "passport" system for cattle, an identity document that follows an animal for life and allows officials to know exactly where it has been at what age.[77,78] In fact, it is estimated that at least 25 countries are in the process of implementing individual identification systems on a national level.[79] The U.S. is not one of them.

In the U.S., current identification systems are geared toward animals going to market, their sale, and the interstate shipment of such animals. Many, if not all, states require livestock to have a health certificate and/or state permit to enter their state from another state. Each animal in a herd being shipped may be required to be identified by an ear tag, tattoo, ear notching, backtag, paint marks, tail tags, neck chains, freeze brands, leg band, or some other method. Exceptions may be allowed, such as in the federal allowance for steers, spayed heifers, or cattle of any age to be moved from one state to another by the same owner, so long as the destination premise is owned, rented, or leased by that same person, without the need for individual identification markers (9CFR 71.18, 2002 ed.). State laws may fill federal loopholes and allow for better traceability of animals and animal movement. Under the new federal Scrapie Eradication Program, many, but not all, sheep and goats are required to be individually tagged. The sale of livestock, for the most part, is also to be accompanied by papers that identify the animals. However, traceback of animals with the current U.S. process can be tedious because there is no central database that keeps track of all relevant animal data and movements. A paper trail, sometimes incomplete, must be followed and this could take time—time that in the face of an animal health crisis we might not have. This delay might easily result in excess numbers of herds being quarantined, adding to the costs of fighting an outbreak—or worse still, result in missing infected animals and spreading the disease. Intrastate movements can go completely unrecorded. Owners may not be required to keep a record of where the animal has been previously, or records that do exist may not be kept current.

Sometimes a tag falls off an animal and it may be impossible to identify the farm or producer from which it came. In some instances, animals are simply lost. A case in point: In October 2001, ProMED Listserve sent out a plea from the Texas Animal Health Commission, which was working to find all animals that had been commingled with a TB positive herd. One Charolais bull was unaccounted for and could not be located. Its last appearance was at a livestock market.[80] This incident is of concern since Texas has TB-free status and this one animal could have potentially lost the state its standing, not to mention millions of dollars in revenue.

Ideally it would be nice to see the U.S. adopt an identification system that could keep track of individual animals throughout their entire lifespan and locate that data in readily accessible computer databases. A permanent identification method, possibly using either an electronic ear tag or ruminal bolus transponder, could be used to identify the animal's herd of origin, dates of movements, sales, etc. This type of system is being looked at carefully by the European Commission in a system called IDEA (Identification Electronique des Animaux).[81] This system consists of an electronic identifier—a database containing identification and other vital data on all animals—the ability to have this information automatically downloaded at the slaughterhouse, and the printout of a tag by the retailer to be placed on the meat package for sale in the grocery store for consumer benefit.[81] The European Commission is looking to replace its current system, ANIMO, which has flaws in data standards for most animal identifiers and has no multilingual capacity.[82] Even Canada has created a national identification program for its cattle and small ruminants out of concern it could not adequately trace and respond well to animal health or food safety crises.[79] As with IDEA, the Canadian system allows information and history about the animal to be carried through life and through slaughter.

With BSE and FMD in the public's eye, consumers no longer want to buy anonymous beef, pork, or chicken; they are demanding to know where the meat in the grocery store display case came from. Farmers in the U.S., apparently fearful of litigation should a meat product involved in a foodborne illness or death be traced back to them, are blocking efforts to develop a nationwide animal identification system.[83] However, one group, the Holstein Association, is at the leading edge for developing a national ID system. This organization established the National Farm Animal Identification Records (FAIR) program as far back as 1996 to improve dairy operations and continues to expand its use in conjunction with state partners such as New York's Cattle Health Assurance Program, Ohio's Johne's eradication program, and Michigan's TB Eradication Project.[84] Florida is already moving toward better animal tracking by converting health certificate information into an electronic form that can be quickly searched.

The U.S. may find that in order to export to Europe or to other countries in the future, they must adopt a national individual animal identification and tracking system. Such a system will serve many purposes, from tackling issues of animal health; to addressing concerns of public health associated with foodborne diseases, zoonotic diseases, and food recalls; to dealing with animal welfare, tax, or government subsidy issues.[78]

ANIMALS, AGENTS, OUTBREAKS, AND THE ECONOMIC OUTLOOK

Some of the earliest literature pointing toward the risks our agriculture faces in bioterrorism goes back 50 years.[85] The livestock of the U.S. faces not only new, emerging diseases that emanate from outside its borders, but these industries also face the emergence of new disease pathogens from within.

Characteristics of an Anti-Animal Agent

An extremely large number of microorganisms are listed in the literature and asserted to have significant anti-animal potential, but they are certainly not all equally useful or effective. In fact, some are rather poor choices. The effect desired by a criminal or terrorist would depend upon the agent selected, of course. Causing vast economic disruption of the beef and pork industry would require the use of very specific agents distributed in a very well-crafted manner. Ideally, to cause significant economic harm to animal agriculture on any meaningful scale, an anti-animal agent would have to have many of the characteristics in the following list.[23,29,85] However, the targeting of a single feedlot operation or egg laying facility by a disgruntled employee could, since the objectives are not so grandiose, be easily achieved with markedly less effort and skill by selecting an agent that meets very few of the below criteria. The ideal anti-animal agent would have the following characteristics:

- Be easily obtainable by terrorists
- Be easily grown
- Be highly contagious to the animal target
- Be highly pathogenic and virulent
- Have a predictable response by the target species
- Survive reasonably well in the mode of dissemination chosen, and preferably also in the environment
- Not cause disease in the terrorists working with it

- Cause morbidity or mortality in more than one animal species, impacting more than one industry
- Have an incubation period of several days, allowing the terrorists time to escape the area
- Result in an illness with concomitant shedding of the microorganism in high numbers
- Result in undetected latent carriers who may shed the organism
- Result in marked economic losses
- Have the potential for deniability as a natural outbreak by perpetrators

Obviously there is no one microorganism that fits all of these criteria, but there are a few that do meet several of them. And it is possible for skilled scientists today, given recent advances in biotechnology, to genetically modify an organism to more fully express certain traits or desired qualities, or to give it new characteristics it never had before, such as antibiotic resistance, making it all the more dangerous. It is not so preposterous to think of someone designing a new zoonotic agent to infect human and animal targets simultaneously, or of someone creating a chimera, marrying the best characteristics of two already very hazardous agents to produce an almost unstoppable one.

AGENTS OF CHOICE

Which agents might be used against animals? The following is a list of agents thought to be useful against livestock:[29,86-89]

> African horse sickness
> African swine fever
> Anthrax
> Bluetongue
> *Chlamydia psittaci* (psittacosis)
> *Cowdria ruminantium* (Heartwater)
> Foot and mouth disease (FMD)
> Highly pathogenic avian influenza
> Hog cholera (Classical swine fever)
> Lumpy skin disease
> Lyssaviruses
> Newcastle disease
> Porcine reproductive and respiratory syndrome virus

Pseudorabies
Rabies
Rift valley fever
Rinderpest
Screwworm
Sheep and goat pox
Teschen disease (porcine enterovirus)
Venezuelan equine encephalomyelitis (VEE)

To this list we should add BSE and Nipah virus, both of which will be discussed in further detail shortly. Only two of the agents in the list above (anthrax and VEE) are on the CDC's Critical Biological Agents List (Categories A, B, or C) for concern to humans.[9] This CDC list, however, does include multiple zoonotic agents, such as plague, tularemia, Lassa, Q fever, brucellosis, glanders, Nipah virus, hantaviruses, and others that may also infect animals. A recent survey of chief livestock officials in all 50 states noted that 44 required reporting of anthrax, 46 reporting of brucellosis, 46 reporting of "all" foreign animal diseases, 25 the reporting of plague, and 23 reporting of Q fever.[90] Animals may be the first to show signs if an attack against people were carried out with a zoonotic agent. It is disheartening to see that more states do not require the reporting of zoonotic diseases in animals. Hopefully this will change.

Founded in 1924, the Office International des Epizooties (OIE), with its headquarters in Paris, France, is recognized by 162 member countries as being the world's premier agency in dealing with animal health issues. As such, OIE has set out to track internationally many diseases of concern in animals, to collect, analyze, and disseminate valid scientific information on animal health around the world, to provide expertise in controlling animal disease, to promulgate standards for diagnostic testing and vaccines, and to harmonize trade standards. OIE's efforts are focused primarily on livestock, including poultry. OIE has created two lists of diseases of concern to animal health: List A diseases, of which there are 15, are those diseases that are usually highly contagious, spread quickly, have severe economic or public health consequences, and are of great importance in international trade:

African horse sickness
African swine fever

Bluetongue
Classical swine fever
Contagious bovine pleuropneumonia
Foot-and-mouth disease
Highly pathogenic avian influenza
Lumpy skin disease
Newcastle disease
Peste des petits ruminants
Rift Valley fever
Rinderpest
Sheep pox and goat pox
Swine vesicular disease
Vesicular stomatitis

All member countries are required to report the occurrence of a List A disease to OIE within 24 hours of its detection. The result is often the swift implementation of a trade embargo by many countries against the afflicted animal species and affiliated meat derived therefrom. The U.S. has two List A diseases, vesicular stomatitis and bluetongue, which pose little economic hardship. It should be noted that, of the List A agents, foot-and-mouth disease, swine vesicular disease, avian influenza, vesicular stomatitis, rift valley fever, and Newcastle disease are all zoonotic. However, with the exception of rift valley fever (and possibly avian influenza), human cases attributed to these diseases are rare and of no threat to human health. This is of importance to the terrorist who might make use of these agents. Terrorists could obtain, grow, and handle almost all List A agents with relative impunity and disregard for their health and not fear taking ill or dying in the process. This would not be the case if the terrorist were to chose from the CDC's Category A, B, or C agents. Vaccines have been made for many of the List A diseases, though they may not be employed within a country or region.

List B diseases, of which there are approximately 90, are diseases that are generally considered to be less contagious, less severe, and of less economic importance than List A diseases; however, this is not entirely true. BSE is a List B disease but has created severe economic hardship for the U.K. since its discovery in 1986. Other diseases of List B include anthrax, brucellosis, bovine tuberculosis, porcine reproductive and respiratory syndrome of swine, glanders, and acariosis of bees. The widespread reemergence of brucellosis or tuberculosis among our cattle would create heavy monetary losses in trade and cost vast sums to eradicate fully. Wilson et al. give a very in-depth review of some of the most feared foreign animal diseases, complete with transmission, signs, pathology, diagnosis, and zoonotic potential.[88] In addition to these, agents that target seafood (fish, shellfish, etc.) or freshwater aquatic life of commercial importance should also be considered. Private aquaculture in the U.S. generated $63.4 million in sales for trout and $469 million for catfish in 2000.[16] Diseases such as channel catfish virus, enteric septicemia of catfish, herpesvirus salmonis of trout and salmon, infectious hematopoietic necrosis and viral hemorrhagic septicemia of salmon (both are found in the U.S. and are notifiable to OIE) are just some diseases that threaten U.S. aquaculture and fisheries.

It should be realized that the selection of an anti-animal agent will likely be based, in large part, upon its ability to inflict economic harm and create panic or fear, and not for its sheer morbidity or mortality numbers, as is done with human bioagents. Agents that are not contagious through respiratory secretions, saliva, or urine or feces are not likely to spread and are not going to make a severe impact on the livestock industry. Anthrax, while effective in inhalational form, requires great know-how and technical means to make it an effective biological weapon. Historically, imported bonemeal has been a source for anthrax infection in pigs and, to a lesser extent cattle;[42] therefore, it is possible that feed could serve as a source by which anthrax, as well as other agents such as foot-and-mouth, might be intentionally or unintentionally disbursed. Remember, too, that inhalational anthrax is not contagious person-to-person, or animal-to-animal. While a terrorist or criminal might effectively use this pathogen against one or two livestock facilities, and might even kill thousands of animals, it would not impact the agriculture industry on a large scale and would not adversely affect the economy of the nation or our society in large. The generated terror would be limited in scope and not full scale. Also, there is a vaccine for anthrax that is currently used in livestock. Anthrax, therefore, would not be the agent of first choice of an agoterrorist.[63]

Nipah virus first emerged in late 1998 in Ipoh, Malaysia, as fatalities in pig farmers. Nipah is the name of the village where this virus was first isolated, and is of the same family of viruses that causes measles, canine distemper, and Hendra—a fatal disease of horses and people that was first seen in Australia in 1994.[91] In the end, 1.1 million pigs (roughly 40 percent of the entire swine population in Malaysia) were slaughtered to control this outbreak. In pigs, this disease causes severe respiratory symptoms, nicknamed the "one-mile cough," and the rare case of neurological disease. Adult pigs were found to have died suddenly without any signs of illness.[91] Pigs within three miles of an infected farm were slaughtered by the military, although some impatient farmers carried out massive slaughter on their own by either clubbing the animals with planks or burying them alive.[92] What were some of the factors that helped in the emergence and spread of this disease? The routes of transmission have not been conclusively established, but it is felt that respiratory aerosol and direct contact between pigs were the primary, although not the only, means. The sharing of boars or boar semen, adding newly acquired pigs to a herd without the use of a quarantine period, the illegal movement of pigs by desperate, panicked farmers in the face of the outbreak, poor to no disinfection of feed trucks going onto farms, and the movement of dogs, cats, and people between farms all contributed to the spread of this virus and the very near demise of the pork industry.[91,92] Humans have been found to harbor this virus in their throats and urine,[92] and dogs, cats, horses, and goats were also found infected in this outbreak.[93]

The reason for the appearance of Nipah is unknown, but its effects were long lasting. The demand for pork, at one point, had fallen over 90 percent, and the $395 million a year pork industry was expected to take a full five years to recover.[92] Slightly disturbing is the realization that this virus was brought into the U.S. inside carry-on luggage on a commercial flight from Malaysia, destined for the CDC for further characterization.[92] All told, there were 265 cases of human viral encephalitis with 105 deaths due to Nipah.[93]

Many consider foot-and-mouth disease virus to be the ultimate biological agent to be used against livestock. FMD is a highly contagious vesicular disease of cloven-hoofed animals (cattle, pigs, sheep, goats, deer, elk, and many zoo or exotic species) that causes fever and the formation of vesicles in the mouth, on the tongue, muzzle, feet, teats, and vulva.[88] The result of FMD infection is that animals do not eat well, they lose weight, and milk production drops markedly. Deaths are few, but are more common in calves and piglets. Animals may never return to their previous weight or production level. Sheep and goats often have very mild signs and cases may be missed if not examined closely. FMD can be transmitted by saliva, respiratory aerosol, direct contact (for example, through unwashed hands), and vehicles (contaminated feed, and fomites such as coveralls, shoes, instruments, etc). It has also been shown that humans can harbor FMD virus in their respiratory tracts for up to two days, posing a theoretical risk for transmitting this agent to uninfected animals. The importance of vehicles in the introduction of FMD cannot be overstated. In addition to the U.K. outbreak of 2001 being attributed to the feeding of raw garbage to pigs, the introduction of FMD into Canada in 1951 was thought to have occurred as a result of a West German immigrant passing on the virus from his clothes and a discarded sausage.[94] This outbreak cost Canada $722 million (Canadian).

FMD is the most feared of all livestock diseases because of the losses incurred by the farmer/producer. Being a List A disease, any case of FMD discovered in the U.S. would need to be reported to OIE within 24 hours. The consequences would be immediate and harsh. Countries around the world would close their doors to our exports of beef, pork, mutton, cattle, pigs, sheep, and dairy products. This means that the $3.1 billion in beef exports and the $1.3 billion in pork exports each year would vanish unless we could control this disease very quickly. There have been many estimates as to the impact of a FMD outbreak in the U.S. Old estimates claimed that the U.S. would lose $27 billion. Paarlberg et al., in their recent analysis of a FMD outbreak in the U.S., estimated that $14 billion would be lost in farm income.[95] Taken into account was public fear and that 10 percent of the U.S. public would stop buying beef, pork, and dairy products. Had there been no losses to public fear, lost farm income was predicted to be only $6.8 billion. However, had 20 percent of the public stopped buying these products, the costs could have reached $20.8 billion. It is interesting to note in this paper that crop prices were also adversely affected, but rebounded with the increase of the poultry industry and government support. Livestock exports dropped $6.6 billion. What is not taken into consideration in this scenario is the impact to other industries that rely on those commodities hit by FMD. An example would be tourism. When the U.K. experienced its FMD outbreak, people were restricted in where they could go and what they could do. Some footpaths, parks, and zoos were closed out of fear of spreading the virus. Events were canceled and people stayed home. A degree of the resulting impact was borne out of ignorance: Some people thought they would become infected with FMD, or confused it with the human illness hand, foot, and mouth disease. Fundraising events were canceled and some charities lost millions of pounds. National elections were postponed. Estimates of the FMD impact on the U.K. put overall economic losses between £2.4–£4.1 billion for 2001.[101] Tourism in 2001 lost £2–£3 billion,[96] and is expected to lose a grand total of between £4.5 to £5.3 billion by 2005.[97] It was predicted that FMD alone would result in the loss of £750 million in sports-related sales and activities.[98] The total cost of FMD compensation to farmers for slaughter livestock was around £1.1 billion,[96] and the total economic effect of FMD on agriculture and the food chain in the U.K. is expected to be approximately £3.1 billion by 2005.[97] Indeed, while it is known that 6 million animals were slaughtered in the U.K. to control this disease, resulting in their reaching FMD free status in less than one year, the true costs will likely never be known. The U.K. is not alone in dealing with this dreadful disease. Taiwan's 1997 FMD outbreak cost the country roughly $15 billion[29], and Italy's 1993 outbreak cost over $130 million.[75] There is a good chance that a significant FMD outbreak in the U.S. could go well beyond estimated figures.

The impact of a FMD outbreak could mean that milk on dairies is thrown out because milk trucks aren't allowed to go farm-to-farm. Tracing movement, animal and human, to and from infected farms could take dozens of people, or more, just for one busy farm. It has been estimated that it took approximately 25 veterinarians to handle all matters concerning FMD on just one farm in England, including tracebacks, diagnosis, slaughter oversight, and so forth. In the end, the U.K. did not have enough veterinarians, so veterinarians came from other countries to help out. The question arises: Does the U.S. have enough veterinarians to handle a large-scale FMD assault?

Highly pathogenic avian influenza is also of great concern. In 1983–84, an outbreak in Pennsylvania cost $63 million to eradicate, the costliest ever in U.S. history,[75] leaving 2 million birds dead.[99] Estimates put the cost of living with this disease at nearly $5.6 billion. During the first six months of this outbreak, poultry prices rose $349 million. The ongoing 2002 outbreak of low pathogenic H7N2 avian influenza in Virginia, where poultry sales at the farm gate contributed $745 million to the state's economy in 2000[100], has resulted in the depopulation of roughly 4.7 million birds.[101] The cost for this control operation is not yet known.

An outbreak of classical swine fever that occurred in the Netherlands in 1997–98 forced the destruction of 11 million pigs and resulted in short-term economic costs of $2 billion.[102]

BSE is considered by some to be an agent that might be used by agroterrorists. BSE is transmitted to cattle mainly through ingestion of contaminated feed made from the poorly rendered tissues, primarily nervous tissue, of cattle harboring the BSE prion. BSE is not directly contagious animal-to-animal and has an incubation period of several years. While BSE is a List B disease, it carries with it the impact of FMD. Few, if any, countries would want to import beef or cattle from the U.S. if we should ever discover a case. The extent of damage to our export market would in large part be decided by the number of BSE cases that were discovered, where they were located in the U.S., and how quickly we were able to assure the world we had eliminated this agent from our country, but overall the finding of one or more BSE cases would not bode well for U.S. trade. How might a terrorist introduce BSE into the U.S.? Introduction of the agent into feed would need to be done on a large scale from a centralized feed manufacturer/distributor so as to best ensure infection of animals and development of disease. Even if successful at this, it would take years for cattle to develop disease. If this type of attack were carried out, there would be no guarantees that cases would develop or, if animals were subclinical, that cases would be diagnosed as BSE. The other scenario involves terrorists bringing in dairy cattle already harboring BSE. This would be a more direct, faster approach. No matter the scenario, the establishment of BSE in cattle throughout the U.S. as a direct result of a terrorist introduction would be very remote. The economic consequences of BSE appearing in the U.S. could be crippling. Discovered in 1986 in England, the BSE epidemic there reached its peak in 1992–93. By 1996, consumption of beef in the U.K. had plummeted 40 percent. Exports to other countries were shut off completely. Even today, now that the EU has officially lifted the export ban on the U.K., many countries refuse to buy their beef. To date over 5.6 million cattle in the U.K. have been slaughtered and kept from entering the human food chain. The cost of BSE to the U.K. economy has been billions of pounds, and some attribute the fall of the Conservative government of the mid-1990s to be, in part, due to BSE.[29]

There are many agents that may potentially be used against animals, but only a few that would effectively target the economic foundation of agriculture. It can be seen that when the properties of an agent are combined with the consequences of that same agent, we begin to understand that, from a terrorist's point of view, the best agent with which to wage agricultural bioterrorism against the U.S. would be a foreign animal disease, (most likely a List A agent).

CROPS, FIBER, AND FORESTS

Today, it is estimated that the U.S. currently loses $33 billion a year to plant diseases, $21 billion (65 percent) of which is attributed to nonindigenous plant pathogens.[103] We also lose approximately $7 billion in forest products due to pathogens of forest plants, $2.1 billion of this is due to nonindigenous pathogens.[103] In addition to providing food, we depend upon various crops for other reasons: cotton to make many of our clothes; canola, soybean, and sunflower for oil; and tobacco for pleasure.

The major crops of the U.S. in 2000 are listed in Table 23.1.[19]

One of the earliest recorded attempts of bioterrorism against crops took place during the Civil War when the Union was accused of having introduced the harlequin bug, *Murgantia histrionica*, into the southern United States in hopes of destroying Confederate crops.[104] In other historical anti-crop events, the U.S. was accused of dropping the Colorado potato beetle (*Leptinotarsa decemlineata*) on Germany during WWII, and again on East Germany in 1950.[104] The U.S. even considered, briefly, destroying Cuba's sugarcane crop with Fiji disease, a pathogen transmitted by leafhoppers.[104]

Many pathogens have been listed by various individuals and organizations that are of concern to U.S. crops and forests. The American Phytopathological Society (APS) has created two lists of threatening plant pathogens—one in which the agent is already partly established in the

TABLE 23.1. Major crops of 2000

Crop	Production	Value
Corn for grain	9.9 billion bushels	$18.6 billion
Soybeans	2.7 billion bushels	$13.1 billion
Wheat	2.2 billion bushels	$6 billion
Cotton	13.1 million acres harvested	$4.8 billion
Tobacco	485,000 acres harvested	$2 billion
Rice	3 million acres harvested	$1 billion

U.S., and another of foreign agents.[105] It is difficult to have every expert agree upon which pathogen or pest is of most concern to the U.S. It is also hard to determine which criteria to use in basing a decision as to an agent's threat potential. Of the dozens of agents listed by APS, several were highlighted as being especially threatening to U.S. plants/crops (Table 23.2).

The U.S. does have some significant plant pathogens within its territory already, but not widely distributed. Further spread, however, of these could spell disaster. A less-aggressive, non-Asiatic strain of soybean rust has been in Puerto Rico for some time. Accounts of losing 10–30 percent of soybeans in a region to the Asiatic strain[105] are frightening when one considers that the U.S. produced 46 percent of the world's soybeans from 1999–2000.[19] MacKenzie et al. speculate that the release of soybean rust spores one mile above ground 200 miles off the coast of Louisiana into 35 mile-per-hour winds blowing up the Mississippi Valley would infect St. Paul-Minneapolis, Minnesota.[106] It is estimated that the burden of soybean rust could run as high as $8 billion[106,107] and result in crop losses of as much as 50 percent.[107]

The spread of plant pathogens (and also animal and human pathogens to some degree) is highly dependent upon meteorological conditions: Season, temperature, rain, sun, and wind all play a role in how well a new agent will survive, take hold, and spread. And while the agent may be effective against a particular plant, there may be limitations in the stage of growth of the plant and its susceptibility. A good example is Karnal bunt of wheat. Karnal bunt is caused by *Tilletia indica* and found in Pakistan, India, Mexico, the U.S, and other countries and is spread mainly by the planting of infected seeds. Infection occurs during flowering. The organism survives and infects plants best in cool climates that have good rainfall and high humidity during the time of heading of wheat.[108] APS did not designate this disease as being of significance, but it is a pathogen often in the news and of enough concern to merit closer attention here. While Karnal bunt does not drastically reduce crop yield, it does convey a fishy taste to the final wheat product making it undesirable. This pathogen also carries international trade restrictions, thus making it a great concern to the U.S. $3.6 billion dollar wheat export industry. Karnal bunt was first documented in the U.S. in 1996 in the wheat fields of Texas, California, and Arizona. In 2001, this disease had spread outside regulated areas and was discovered in Young County, Texas. This pathogen can remain in the soil for up to five years,

TABLE 23.2. Potential anti-crop and anti-forest agents

Common Name	Scientific Name	Plants Affected	Geographic Distribution	Threat
Brown stripe downy mildew	*Sclerophthora rayssiae* var. *zeae*	Corn, sorghum	India, Nepal, Pakistan, Thailand	Significant
Rice bacterial leaf streak	*Xanthomonas oryzae* pv. *Oryzicola*	Rice	S.E. Asia, India, China, N. Australia, W. Africa	Significant
Rice bacterial leaf blight	*Xanthomonas oryzae* pv. *Oryzae*	Rice	S.E. Asia, India, China, N. Australia, W. Africa, **Louisiana**	Significant
Philippine downy mildew	*Peronosclerospora philippinensis*	Corn	Philippines	Significant
Potato war	*Synchytnum endobioticum*	Potato	Canada, other countries	Significant
Huanglongbing (citrus greening)	*Liberobacter africanus, L. asiaticus*	Citrus	S.E. Asia, S. Africa	Significant
Sweet orange scab	*Elsinoe australis*	Citrus	Southern S. America	Significant
Citrus variegated chlorosis	*Xylella fastidiosa*	Citrus	Southern S. America	Significant
Black spot	*Guignardia citricarpa*	Citrus	Asia, S. Africa, southern S. America	Significant
Apple proliferation	*Phytoplasma*	Apple	Southern Europe, possibly India and S. America	Significant
Soybean rust	*Phakopsora pachyrhizi*	Soybean	India, Australia, **Hawaii, Puerto Rico**	Significant
Citrus canker	*Xanthomonas axonopodis* pv. *Citri*	Citrus	Asia, S. America, **Florida**	Significant
Sharka (plum pox)	*Plum pox virus*	Stone fruits	Middle East, Europe, Chile, Canada, **Pennsylvania**	Significant

continues

391

TABLE 23.2. Continued.

Common Name	Scientific Name	Plants Affected	Geographic Distribution	Threat
Potato golden cyst nematode	*Globodera rostochiensis*	Potato	Europe, India, N. Africa, Asia, S. America, **New York**	Significant
Tomato yellow leaf curl	*Tomato yellow leaf curl virus*	Tomato	Dominican Republic, E. Mediterranean, Jamaica, **Florida**	Significant
Larch Canker	*Lachnellula willcommii*	Larch	Europe	Significant
Brown rust of larch	*Triphragmiopsis laricinum*	Larch	China	Significant
"Oak disease"	*Phytophthora quercina* sp	Oak	Europe	Significant
Needle blight	*Cercospora pini-densiflorae*	Pine	Japan	Significant
Resin top disease	*Cronartium flacidum*	2-needled pines	Europe	Significant
Pine twist rust	*Melampsora pinitorqua*	2-needled pines	Europe	Significant
Blue stain of beech	*Ceratocystis nothofagi*	Nothofagus	S. America	Significant
Armillaria root disease	*Armillaria novae-zelandiea*	Hardwoods and pine	Australia, New Zealand, PNG	Significant
Blue stain of beech	*Ophiostoma valdivianum*	Nothofagus	S. America	Significant
Phytophthora disease	*Phytophthora alni*	Alder	Europe	Significant
Root disease and wilt	*Leptographium truncatum*	All pines	New Zealand, S. America, Canada	Significant
Watermark disease	*Erwinia salicis*	Willow	Europe, Japan	Significant
Pink disease	*Erythricium salmonicolor*	Fruit trees, woody perennials	Tropics, **Florida, Louisiana, Mississippi**	Significant
Sudden oak death	*Phytophthora ramorum*	Oaks, Ericaceae	Germany, **California, Oregon**	Significant
Scleroderris canker (European)	*Gremmeniella abietina* (European strain)	Pine, conifers	Europe, Canada	Significant

and spores can be carried on various plant parts, farm equipment, tools, vehicles, and can be found on the surfaces of buildings.[108]

Another agent, *Puccinia graminis*, a fungus that causes stem rust of wheat (also called black rust or black stem rust), might have excellent weapon potential because it retains viability in cool storage for >2 years and spreads rapidly after being released. A single infected wheat kernel can contain 12 million spores, each of which can infect one more plant.[35]

Other pathogens and pests of concern are citrus canker and the Asian longhorned beetle. Citrus canker, introduced in 1993, was discovered in Miami in 1995. This disease agent has spread to well over 1300 square miles in Florida and has cost over $200 million in eradication efforts.[105] To date, 2.1 million trees have been cut down in order to stop the spread of this pathogen. Florida has a citrus industry valued at $8.5 billion, so controlling this disease is very important to them and to consumers. The Asian longhorned beetle (*Anoplophora glabripennis*) is listed by USDA-APHIS-PPQ as an insect that is " . . . a serious threat to hardwood trees and has no known natural predator in the United States. If the Asian longhorned beetle becomes established here, it has the potential to cause more damage than Dutch elm disease, chestnut blight, and gypsy moths combined, destroying millions of acres of America's treasured hardwoods, including national forests and backyard trees. The beetle has the potential to damage such industries as lumber, maple syrup, nursery, commercial fruit, and tourism accumulating over $41 billion in losses."[109]

Although this insect has already entered the U.S. and established itself temporarily in New York City and Chicago, intentional spread of this pest could wreak havoc on many fronts. Eradication efforts are underway, including the ban of wooden packing material from China and Hong Kong, which is suspected to be the source through which the insect arrived. Two other nonnative wood-boring pests of concern are the Spruce bark beetle (*Ips typographus*) and the Mediterranean pine engraver beetle (*Orthotomicuserosus*).

Timber is used for a great many things in our society, from framing our houses to supplying us with disposable diapers, toilet tissue, newsprint, mailing envelopes, and many other items. It is just as crucial that our forests be examined and considered in the discussion of agroterrorism, because any significantly detrimental attack against this commodity would be felt by us all just as readily as an attack on a food crop might—perhaps even more so. Forest products in 1999 accounted for $2.9 billion in farm cash receipts.[16] Lumber consumption in the U.S. has steadily increased over the entire country as well as on a per capita basis, and paper products are also in more demand.[16] Timber-related manufacturing, lumber and wood products, and paper products all contributed nearly $194 billion to the GDP in 1999.[16] Biological agents or pests that target timber and their intentional release should demand our attention.

FOOD TAMPERING

Based upon history, ease of access, availability, and impact, there are many who feel that direct contamination of food itself would be the easiest approach for waging bioterrorism, that it would be much easier than growing and aerosolizing an agent.[21,110] And they may be right. Although the overarching goal of agroterrorism is weakening the economy, a by-product of this terrorism would be fear and potential hysteria. Nowhere would this be more apparent than in the indiscriminate tampering of food. While the intentional tampering of a food by a terrorist or criminal might be conducted only after significant thought and planning, the public may perceive illnesses or deaths to be random and may feel they have no control over their exposure—they may feel helpless. Everything eatable may be considered deadly by them. It is technically less demanding to utilize direct food contamination than it is to process an agent and distribute it via aerosolization.

There are several vulnerabilities in our food supply: the low rate of inspection of imported foods; the processing of large quantities in a central facility; the rapid transport of these foods; and the vast distribution network to tens of thousands, or perhaps even millions, of consumers. Contamination at a source point could cause massive numbers of illnesses.[111] The U.S. processes enormous quantities of food and beverages, including milk and soft drinks, that are vulnerable to intentional contamination. This concern was voiced over 40 years ago.[23] In today's realm of processing, packaged foods such as hot dogs might be cooked, packaged, and shipped to ten states as well as overseas all in a matter of just a few days. By the time a foodborne illness would be connected with this product and a recall initiated, a large portion of the total product may already have been consumed. Food processing has become more consolidated and more global in scope, with several extremely large companies dominating world distribution and sales. Hazard Analysis and Critical Control Points (HACCP) identifies points in food processing where control measures can be instituted to reduce hazards such as microbial contamination or overgrowth. Terrorists attempting to contaminate a food product at the beginning of the process may find the introduced agent is killed by one or more measures implemented at a critical control point, such as cooking. Another consideration to take into account when thinking about contaminating food is that only foods not destined to be cooked make the best targets, as cooking would naturally kill the applied biological agent(s). An alternative for determined terrorists might be to use a toxin that is heat-stable, like a staphylococcus enterotoxin, which can survive cooking, or contaminate cooked or baked foods after the heating process, after the critical control point.

The list of bacteria and viruses that could be grown and used to intentionally contaminate food to make people ill is extremely long. Indeed, over 200 diseases are thought to be transmitted through food.[112] Discussion of these agents is beyond the scope of this chapter; however, Khan et al. give several lists of agents that might be of consideration.[113]

The safety of imported foods has also been called into question. In 1999, 68 percent of consumed fish and shellfish, 40 percent of fresh fruits, 32 percent of fruit juice, and 11 percent of fresh vegetables were imported into the U.S.[16] In addition to numerous episodes of foreign fruits or vegetables being responsible for foodborne outbreaks within the U.S., other foods should also be considered as means of introducing a biological or chemical agent. Lead poisoning in the U.S., for example, has been associated with imported spices and Mexican candy.[114]

Another concern and vulnerability of our food system is the widespread use of illegal immigrants in agriculture. Might not terrorists pose as immigrants, make their way into California as thousands of others do to work for the season, and then contaminate a crop at harvest as they work? This is plausible. Some have suggested that terrorists can contaminate produce abroad destined for U.S. tables and cause mass illness this way.[110] The problem here is that causing 500 cases of salmonellosis by contaminating a food or beverage might not generate a lot of panic when one takes into consideration that over 1 million cases occur already each year in the U.S. Many of these 500 cases might even go unnoticed. It is not enough to just contaminate a food and cause illness anonymously. Finding the culprit (a food item or person shedding) responsible for a foodborne outbreak does not always occur; in fact it is quite common to never know the source. The human target must be

made afraid—terrorized into no longer trusting his/her food supply, the
government inspectors and/or regulators, or the grocery store. The impact of contaminating food with a biological agent must be severe above
and beyond what we live with daily; otherwise, the risk is of no more
concern than what we now face in our average lives.

Consumers can do a great deal to minimize their risk from contaminated foods. Adequate cooking, proper handling and storage, and
proper hygiene in dealing with foods, especially of meats, would go a
long way to reducing the number of foodborne illnesses and deaths each
year. Other measures consumers might take include thorough washing
of produce; examination of cans, jars, bottles, or other containers and
discarding damaged or suspicious ones; and maintaining alertness to
strange odors, tastes, or the abnormal appearance of foods. The USDA
Food Safety and Inspection Service (FSIS), under the Federal Meat Inspection Act, the Poultry Products Inspection Act, and the Egg Products
Inspection Act, inspects all meat, poultry, and egg products sold in interstate commerce and reinspects imported products to ensure that they
meet U.S. standards. FSIS employs more than 7,600 inspectors in
roughly 6,500 meat, poultry, and egg processing plants.[115] FSIS has also
recently published *FSIS Security Guidelines for Food Processors* in an
effort to assist federal and state inspected plants processing meat, poultry, or egg products to fortify their biosecurity.[116] Until now the FDA has
managed to physically inspect <1 percent of import entries.[117] To help
ease this burden, the FDA is adding 420 inspectors to be stationed at
border points or working on imports.[118] By September 2003, there will
be roughly 4,000 port and border agriculture inspectors, 55 percent
more than in 1998.[8] Some feel this is still not enough. The FDA has also
published detailed recommendations for food producers, processors,
transporters, retailers, importers, and filers on food security preventive
measures.[119,120] These guidelines were also featured in a *Wall Street
Journal* piece.[6]

The food system in the U.S. is extremely vast and complex. We cannot monitor every worker and every step along the chain. We cannot
station police in the kitchens of restaurants. We will remain vulnerable
to some extent. Inspecting foods before they are shipped from another
country would add tremendously to prevention efforts. Detecting foodborne illnesses early, establishing good reporting and communication
systems among public health workers and agencies, and rapid agent
identification are critical to identifying an outbreak, limiting illnesses

and deaths, stimulating an investigation, and combating the terrorist who might contaminate our food.

To better address foodborne illness in the U.S., the CDC has developed PulseNet and FoodNet, both of which would be of tremendous use in the event of an intentional foodborne contamination. PulseNet is a network of public health laboratories that uses pulsed-field gel electrophoresis to genetically fingerprint foodborne pathogens found in clinical specimens or food samples. Finding the same organism with the same genetic makeup in both the food and the patient allows for making a definitive link between the two. The genetic pattern of an organism can also be compared through an electronic database, allowing for the identification of cases in other states due to the same food item. Currently, all 50 states and Canada participate in PulseNet. FoodNet is a joint CDC, FDA, and USDA program to conduct active surveillance on foodborne diseases in the U.S. FoodNet conducts surveillance for laboratory-confirmed cases of foodborne disease due to nine pathogens (seven bacterial and two parasitic) in all or part of nine states. In all, roughly 10 percent of the U.S. population is included in this surveillance network.[111,121]

Food Poisoning Incidents

There are many examples of the intentional contamination of food by criminals for various reasons:

- One prime example is Diane Thompson, who worked at the St. Paul Medical Center hospital in Dallas, Texas, and who, on October 29, 1996, intentionally made 12 people ill by anonymously offering them doughnuts and muffins contaminated with *Shigella dysenteriae* type 2.[13,122] Eleven people ate the pastries at work and became ill. The twelfth person took a muffin home and shared it with a family member, who reportedly also became ill. While there were no deaths from this malicious act, four people were hospitalized. The *Shigella* was apparently obtained from the laboratory's freezer deposits and grown up by Ms. Thompson, who is now serving a 20-year prison sentence.
- One large-scale episode of food contamination was carried out by a group called Nakam, a group of European Jews who, after the end of WWII, carried out attacks on Germans for revenge. In April 1946, members of this group infiltrated a bakery at a U.S.-run POW camp that housed hundreds of German prisoners. Nakam members spread

an arsenic poison over loaves of bread that were later eaten by the prisoners. It is estimated that hundreds died as a result.[13,123]

- The first of only two confirmed instances (the second being the anthrax letters of 2001) of successful biological terrorism was conducted by the religious cult, the Rajneeshees, against the townsfolk of The Dalles, Oregon, in 1984.[13] This attack has been well described in several other places. The cult had built a town, called Rajneeshpuram, and eventually overtook control of an adjacent smaller town, named Antelope. Hostilities were growing between the cult and the citizens of The Dalles, which numbered only 10,000. Lawsuits were frequently being filed by the cult to contest unfavorable rulings by the Wasco County Court on various matters. When two of the three commissioner's positions came up for re-election, the Rajneeshees decided to make townsfolk ill so that one of their members would win the seat. To win the election, the Rajneeshees ordered *Salmonella. typhimurium* from VWR Scientific. Once the organism had been grown up sufficiently, the group visited a town grocery store and sprinkled salmonella solution over lettuce. They stopped just short of injecting milk cartons. In September 1984, cult members visited ten restaurants in The Dalles and poured Salmonella over food held on salad bars. Coffee creamers at two restaurants were also contaminated. In the end, 751 people were made ill, at least 45 were hospitalized, and there were no deaths.[124] The plans of the Rajneeshees failed and they did not win the election.

- In 1971 Eric Kranz, a postgraduate student in parasitology at MacDonald College in Montreal, Canada, contaminated the food of his roommates, making four of them seriously ill. The agent selected was *Acaris suum*, a parasite of swine. All four victims required hospitalization and eventually recovered. Kranz was arrested.[13,125]

- In 1964, Dr. Mitsuru Suzuki tainted a sponge cake with *Salmonella typhi* and offered it to 4 co-workers, all became ill. In 1965, he laced bottles of a milk-based beverage again with the same organism and gave them to 16 colleagues where he worked, again making them all ill. Suzuki contaminated another cake in late 1965, and that same day made many nurses and physicians ill with spiked bananas. Bananas were used on other occasions by Suzuki, too, including against his own family. He even laced medication and mandarin oranges. It is speculated he did these acts to further his research and allow him to complete his dissertation. Dr. Mitsuru Suzuki was believed responsible for 120–200 illnesses and 4 deaths. His arrest in April 1966 charged him for just 66 illnesses, but no deaths.[13,113]

- A man was killed on Labor Day in 1986 when he ate a cyanide-tainted package of Lipton Cup-a-Soup chicken-noodle soup.[126]
- An epidemic among schoolchildren of various complaints—including nausea, vomiting, abdominal pain, dizziness, and headache—in June 1999 was associated with Coca-Cola in Belgium, with cases showing up later in France. This outbreak lead to the mandatory withdrawal of 15 million crates (65 million cans) of the soft drink from Belgium, France, and Luxembourg. Hydrogen sulfide and carbonyl sulfide contamination of carbon dioxide used in the manufacturing process was confirmed by Coca-Cola and affected only some bottles. There was also contamination of the outside of some cans delivered on pallets that had been sprayed with a fungicide. A good portion of the cases were also attributed to mass sociogenic illness—illness that resulted from friends and classmates becoming ill and propagating sickness and anxiety.[127]
- Not all tampering involves biological or chemical agents. A man in Sioux City, Iowa, admitted in May 2000 to sticking needles in meat, fruits, and bread at a local grocery store. The man worked in one grocery store and visited a competing store on several occasions where he would insert needles. No one was seriously injured and he pleaded guilty to the tainting of consumer products with the intent to cause serious injury to a business.[128]

THE THREAT TO THE U.S.

To a terrorist, the attractiveness of targeting agriculture is manifold, but one attribute that stands out is that a military reprisal is not likely. The attacks on the Pentagon and World Trade Towers brought full-time media coverage, 24 hours a day, 7 days a week. With the illnesses and deaths that resulted from the anthrax letters of 2001, we also saw heavy media coverage. Direct attacks of terrorism against people necessarily bring a will and desire to respond militarily. We want to strike back, to bomb something or someone—and we did. However, what is the likelihood that a few cases of BSE, or a full-scale outbreak of FMD would result in 24-hour coverage by CNN? Very low. What is the probability that our military would launch a massive offensive against a country that purposefully introduced FMD, Karnal bunt, soybean rust, hog cholera, or Nipah virus? Again, almost nil. While terrorists call attention quickly to their cause by attacking and killing people, they also draw attention to themselves. The world will stop and take notice if the attack is shocking enough. An emotive heart-wrenching factor results from seeing fellow humans lying dead from a suicide bomber. We want action, retribution. The use of biological weapons evades this somewhat. The delay between release of an agent and public awareness can be many days or weeks—time enough for the terrorists to be long gone. The use of biological weapons against agriculture also carries with it the ability of a nation or terrorist organization to deny involvement and point toward natural circumstances as the cause—"plausible deniability," as it has been labeled. We have seen that rabbit hemorrhagic disease emerged in the Central U.S. without so much as a clue as to its source. This is the type of bewilderment the terrorists who will use anti-agriculture agents are counting on.

Threats must be taken seriously, as can be seen from the chlordane episode in Wisconsin in 1986 in which the dairy industry narrowly averted wide-scale catastrophe. While it has been said that biological weapons are the poor man's nuke, they do require some education, thought, and technical know-how, albeit much less than the creation of a nuclear weapon.

Response

With the possibility of the intentional release of biological agents against our livestock and crops, we must be able to respond in a decisive manner. We must know what to do before the real event happens, but this is often easier said than done. The lead federal agency in safeguarding American livestock and poultry health, and in responding to a foreign animal or emerging or reemerging disease, is the USDA-APHIS, Veterinary Services division (VS). For plants it is the USDA-APHIS, Plant Protection and Quarantine (PPQ) division. Both entities have an Emergency Programs division (EP). The USDA-APHIS is to animal and plant health as the CDC is to human health—but with less funding.

For animal diseases, the EP division, founded in 1972, prepares and trains for response to outbreaks of foreign or newly emerging diseases. They continually monitor animal health and strive to detect an exotic disease quickly. EP provides training to many veterinarians and support personnel whose work requires knowledge of foreign animal diseases. These may include federal and state veterinarians, diagnosticians, animal health technicians, epidemiologists, port veterinarians, foreign veterinary medical officers, VS staff, or others.[129] In the event of a foreign animal disease outbreak, these trained professionals would be called upon to help in diagnosis, control, monitoring, and eradication. Four USDA-APHIS laboratories comprise the National Veterinary Services Laboratory (NVSL) and provide services for the diagnosis of domestic or foreign animal diseases, import/export testing of animals, training, and testing for eradication or control programs. All NVSL laboratories are located in Ames, Iowa, with the exception of the Foreign Animal Disease Diagnostic Laboratory (FADDL) at Plum Island, New York, where foreign animal diseases are studied and veterinarians are trained to become foreign animal disease diagnosticians. There are roughly 400 trained foreign animal disease diagnosticians—private, state, federal, and military veterinarians—spread out across the nation that could be called upon in the event such a need arose.[130] All suspected foreign animal disease (FAD) outbreaks must be investigated within 24 hours of notification. Built into the EP are two (eastern and western) Regional Emergency Animal Disease Eradication Organizations (READEO) that are each composed of 38 people.[29] These teams are to be immediately called together and dispatched in the event of a significant FAD incursion. An Early Response Team composed of three veterinarians may be put together to aid the USDA Area Veterinarian in Charge and Regional Director when a disease situation might require further evaluation or the potential for READEO activation.[130] READEOs operate in the field; the headquarters for eradication efforts are located in the Emergency Response Center in Riverdale, Maryland.[40,146] The legal unit, composed of the staff of the Office of the Inspector General, FBI, and other law enforcement personnel, is also an important component of READEO and would provide forensic and investigative capabilities should evidence point toward terrorist or criminal activity.[29]

The USDA-APHIS-VS works in conjunction with each state department of agriculture, the state veterinarian, private veterinarians, universities, the Department of Defense, producers and farmers, national industry groups, and other federal agencies, such as the Federal Emergency Management Agency (FEMA) to address animal health emergencies.[29] While the USDA is the lead federal agency in responding to animal health threats, the reality is that they would not be able to do it all by themselves. Since the U.K. outbreak of FMD, many states have been encouraged, and are completing, their own FMD response plans—identifying resources, personnel, and drafting contingency plans in case the disease is ever found in their state. With the discovery of an FAD or other threatening disease, the USDA would provide oversight and interact with state officials on the matter, but would remain in mostly an advisory type of role until such time as the disease became unmanageable for the state. USDA would then step in and provide federal-level assistance and control. States are using their FMD response plans to draft plans addressing other exotic diseases now. The USDA is recognizing, as are most in the animal industry, that response to an animal emergency should be considered at the local, state, and federal levels to be most effective.

There are approximately 60,000 accredited veterinarians who sign the health papers of animals for sale or for movement nationally or internationally and are key to early detection of exotic diseases and in preventing their spread. There is much concern among veterinary profession leaders and the USDA that many of these veterinarians would not be able to recognize a foreign animal disease and are in serious need of FAD education. As it stands, FAD education makes up very little of the curriculum in most veterinary medical schools.

The simultaneous release of three to four highly contagious, foreign animal pathogens in several locations around the country at key points would, most likely, be enough to overwhelm response capabilities and present a severe hardship to the U.S. agriculture system and to the U.S. economy. Establishing an enzootic wildlife foci of at least one foreign animal disease (the most detrimental being foot-and-mouth) would ensure that resultant trade disruption would be almost permanent. The incursion of rinderpest, heartwater, or FMD might be devastating to our wildlife or zoo species, and has the potential to adversely affect endangered species.[75] In April 1985, an outbreak of FMD among mountain gazelles (*Gazella gazella*) in a nature reserve in Israel resulted in the deaths of approximately 1500 animals and had a 50 percent case-fatality rate. The virus also spread from the gazelles to unvaccinated livestock on neighboring farms resulting in cases in cattle, sheep, and goats. Spread beyond these farms was prevented due to the prior use of the vaccine in livestock.[131] Control in zoo or endangered species would not be a simple matter of slaughtering, as would be used in our domestic livestock.

One last component of a national effort to address emergency animal health issues, be they of intentional or unintentional means, is the National Animal Health Emergency Management System (NAHEMS). NAHEMS is a partnership between federal, state, industry, and professional associations and focuses on prevention, response, recovery, and preparedness. Its overall objective is: "To have in operation in the United States, a world class National Animal Health Emergency Management System by 2005."[132]

For crops and other plants, the National Plant Board (NPB) is an organization of the plant pest regulatory agencies of each state and Puerto Rico. Formed in 1925, the NPB works with federal and other state agencies to address issues and concerns about plant diseases, regulations, quarantine, enforcement, and compliance issues, and carries out the instructions of regional plant boards. There are four regional boards: Eastern, Western, Central, and Southern. Within USDA-APHIS-PPQ there has recently been a National PPQ Pest Detection Committee and Regional Pest Detection Committees. The New Pest Advisory Group (NPAG) within PPQ is alerted when a new pathogen is found in the U.S. The NPAG convenes experts and makes a recommendation to PPQ as to how it should proceed to deal with this new pest or disease. The NPAG is involved in evaluating the significance of new plant pests, determining the appropriate response, and constructing recommendations on how PPQ should respond to a new plant pest. It is also involved in relaying PPQ's position on plant pests new to the United States.[133] In the end, PPQ may continue to monitor a plant pathogen, do nothing, or begin eradication efforts and declare quarantine and other containment measures.[134]

Concerning plants, the USDA-APHIS does conduct risk assessments for selected pathogens and commodities, but cannot complete such a study on every plant, foreign and domestic.[105] The ability to identify a deliberate release of an anti-plant agent is hampered by the absence of a rapid reporting system.[135] It may not be possible to predict the total outcome of a foreign plant pathogen upon a vital U.S. crop.

Impact

The impact and consequences from a foreign animal disease such as FMD in the U.S. has the potential to be worse than most can imagine and a very dark picture can be painted. The Draconian measures needed to control and eradicate such a disease would be shocking to a good portion of the public. Harsh restrictions on movement would be enacted, inflaming Americans who are used to going wherever they want, when they want. We would see road closures and the quarantining of farms and farm homes. Access to campsites, state parks, wilderness areas, lakes, city parks, and zoos may be denied. Fishermen and hunters may be barred from taking part in their favorite hobby. School buses traveling rural roads could be diverted, and children forced to live with relatives in the city to attend school. Small towns might be cordoned off completely with police or military checkpoints. Animal movement would stop. Sporting events—fun runs, horse racing, golfing, soccer tournaments, bike races—might all be postponed. Depending on the extent of the outbreak, meat prices could skyrocket in some areas and drop in others. Exportation of livestock could be halted, and so could the import. Tourism would drop, just as was seen in the U.K. Jobs would be lost, unemployment would increase, the GDP might fall, a recession might ensue. In other words, a substantial incursion of a foreign animal disease—and possibly, although less likely, even a foreign crop disease—would change the way we live and eat almost overnight. In 1999, the National Park System counted 287 million visits, and in 2000, State Parks and Recreation Areas counted 787 million overnight visitors. Total expenditures for hunting, fishing, and wildlife watching activities in the U.S. in 1996 amounted to $87.6 billion,[16] although some estimates put that figure at closer to $100 billion when camping and other outdoor activities are included.[29] In 1999, the national forests and grasslands, which host 43 percent of outdoor recreation on public lands, including 60 percent of the nation's skiing, contributed $134 billion to the GDP.[136] A good portion of all of this could be lost, at least temporarily, by the incursion of a List A disease.

There are direct and indirect ties to agriculture by allied and reliant industries, and a significant event would have a domino effect, each event impacting the next. Crop farmers buy equipment to plant, harvest, and move their crops, and structures to store their commodities; but if their crops are eradicated to control disease, they will have little reason to buy or use any of this. In the face of a large-scale disease outbreak, either brought about by terrorists or by the hand of nature, companies such as John Deere would face a downturn in sales. Livestock producers would no longer need to purchase feed, nor would they need the services of a veterinarian. Pharmaceutical companies that sell the drugs to treat livestock diseases would see their businesses fall. Restaurants, ice cream shops, food processors and distributors would be affected, as would transporters. Small communities that thrive on farming could see soaring bankruptcy rates for surrounding farms, foreclosures, and the movement of residents to larger cities for employment. Small farming towns in some areas could turn into ghost towns. Seed companies may actually give away seed to help stimulate the farming sector and to stay in business.

Then there are the direct costs of eradication and control: disease surveillance, diagnostic testing, tracebacks on animal movement, implementing and maintaining quarantines, depopulation costs, indemnity paid to the farmer for his/her crop or the producer for his/her animals, overtime for law enforcement, hiring additional veterinarians, overtime for USDA and other federal or state employees . . . the list goes on, and so do the costs. Losses due to a foreign animal disease may take three to five years to fully realize.[29] The outcome of it all is almost too vast to comprehend.

The positive side of growing crops and raising animals in such focused regions of the country as we do means that they are theoretically easier to protect.[54] Should a biological agent make its way into one segment of our agriculture, there is the real likelihood that the entire industry will not be shut down. An outbreak of soybean rust in Iowa would very probably leave soybean farmers in Ohio reaping record profits. The soybean industry would still remain standing, however bruised, and the impact to our agriculture and to our society would be light. Should soybean rust turn up in four or five states at one time, then we might expect to see some alarm.

CONCLUSION

U.S. agriculture, so robust and inestimable in value, is at risk for the deliberate use of biological or chemical agents. What's more, attacks against our livestock, crops, food, and water would not need to be overly elegant or technically sophisticated to be effective. The type of weaponization required for aerosol delivery of a biological agent against a large human audience would not be necessary in many instances when targeting agriculture. Anti-animal and anti-crop biological agents are easy to find and pose little health risk to the terrorist, and production can be accomplished by crude, rudimentary means.[41] A state-of-the-art laboratory and millions of dollars are not needed to carry out a successful attack. This is not to say that anyone can be a successful terrorist. The fashioning of a biological agent into a successful weapon of terror, while easier than constructing a nuclear device, is not so simple that the average person could do it; it would require some amount of skill and training. The Aum Shinrikyo (now renamed Aleph), despite almost unlimited funding, dedicated personnel, and laboratory space, failed miserably on multiple occasions in their efforts to kill with biological agents. In early 1992, they traveled to Zaire, under the guise of wanting to assist in the medical care of Ebola patients, to acquire the Ebola virus, but failed. Four times in 1993, they sprayed anthrax (the vaccine strain) from roof tops or from a moving truck in an attempt to kill people, but were unsuccessful. And once in 1993, then again in 1994, they tried to kill using botulinum toxin, again without success. While the Aum was successful at killing with VX and sarin (a total of 33 fatalities), they could not master biological agents.[137]

Everyone wants to know what the chances are that a biological attack targeting our agriculture will occur. The fact is no one really knows. It is probably safe to assume that the direct targeting of humans is still the highest threat, but as terrorists learn and evolve, the likelihood that agriculture will fall under their nefarious plans to inflict harm and destruction will draw ever closer. The world has seen how nation-sponsored BW programs placed considerable importance on anti-agriculture agents. But we must also be vigilant of other, less obvious people or groups who might wish to use these agents against us. During the FMD outbreak of 2001 in the U.K., the president of People for the Ethical Treatment of Animals (PETA) was quoted as saying, "I openly hope it comes here"—meaning the U.S.[138] Might it not be possible for an organization such as this to "help" the introduction of certain foreign animal diseases along?

Chances are high that if an attack against our livestock or crops were ever carried out, we would never know it. We would simply respond as usual to a disease crisis and give our best effort at control or eradication, assuming it was due to happenstance or a natural phenomenon unless evidence pointed to the contrary. It is most probable that any attack on agriculture will be covert, as any overt attempts will draw attention and result in quick action by agriculture and law enforcement officials to contain the outbreak and catch the perpetrators. There's a saying, "The only criminals you catch are the stupid criminals." That might be true of terrorists, too. No known biological attacks have taken place against American agriculture.[29] There are some who feel that since agriculture has not been targeted by terrorists to date that it is either not an attractive target, or that we are doing something right and need not concern ourselves with additional protective measures. Ignoring the vulnerability of our agriculture and downplaying the threat will only lead to greater losses and devastation later. At one time, it was believed no foreign terrorists would dare strike on U.S. soil, but as 1993 and 2001 have shown, things change. Terrorists have the upper hand. Preventing a terrorist attack from happening in the first place is ideal and is what we strive for, but it is unrealistic to think that we as a nation are inviolable. Terrorism has been around for thousands of years and will continue to remain well into our future.

Most, if not all, local, state, and federal agencies are drawing up response plans to biological and chemical attacks. But what of agriculture? States and counties are now recognizing that they can't, and shouldn't, leave it up to the federal government to take care of everything. Plans must be in place that involve resources and personnel at all levels. An alert and well-informed lay public on the issues of agroterrorism, biosecurity, and zoonoses is needed. This will help in reducing the chances of introducing foreign diseases into the U.S., it will help in early identification, and it will reduce the overall impact on our society. The farmer, the veterinarian, the state veterinarian, and the state agricultural department will all be the first responders when it comes to a biological attack directed against animals. It is imperative that farmers and producers be suspicious and alert to new signs or symptoms in their animals. They must remove complacency and call their veterinarian to help identify the causative agent and address the disease accordingly. Farmers and producers must also become educated on foreign animal and plant diseases, their impact, who to call, implications of breaking quarantine, and their potential for causing human illness. And they must also take greater care to ensure strict biosecurity on their farms. Industry should play a role in this awareness effort, and it is. Even more important is the training of veterinarians to recognize foreign animal diseases and to know what do to, what not to do, and who to call. Law enforcement and the FBI will be working closely with agriculture officials in the event of an agroterrorist attack. It is necessary to educate them now on the disease agents, the workings of agriculture, and the wide scope of impact this event would carry.

The United States is gearing up to better handle exotic animal and plant pathogens by building upon an already excellent system and adding in much-needed components. The overall impact of such diseases, or from a direct attack on our food supply, is not fully known and depends upon many, many variables, but could severely impact the economy, our society, and even public health. Terrorism targeting agriculture has the potential to weaken a workforce and destabilize government. The reality is that terrorism does not need to infect all of our soybean crops or all of our beef cattle to create economic hardship, loss of consumer confidence, or panic; it would take only a small attack using the right agent at the right time in the right way. We just need look back and see how a few grams of anthrax nearly paralyzed our country to realize that an agent used correctly can cause catastrophe. That said, it is unlikely that an agroterrorist attack would create mass food shortages; we have enough food to feed ourselves and most of the planet. We are better prepared to handle the deliberate introduction of a single known agent in a single location than if that agent were introduced in multiple locations simultaneously. And we face the unknown when it comes to battling agents that have never been seen before.

The U.S. is the leader in providing food for the world, but in order to maintain this position we must be prepared and ever vigilant. Agroterrorism, now being looked upon in a more serious light, could have a grave humanitarian impact. Plans have been laid and are continually being updated and modified for prevention and response. But certainly there are some things, some issues, that we have not fully thought of or addressed; there are some repercussions not fully conceived of and some scenarios not yet considered. We must hope these are minor. U.S. agriculture is a sleeping target for terrorism and must be protected to our fullest potential lest we lose what we have taken for granted for so long.

NOTES

Radford G. Davis is a member of the Department of Veterinary Microbiology and Preventive Medicine College of Veterinary Medicine, Iowa State University; Ames, Iowa 50011.

1. CDC. 2001. Update: Investigation of bioterrorism-related anthrax—Connecticut, 2001. *Morbidity & Mortality Weekly Report* 50:1077–1079.
2. CNN. 2002. Ridge: More needed to make air travel safe. *CNN.com*. February 8. http://www.cnn.com/2002/US/02/08/ridge.security/index.html.

3. Science Daily. 1999. Bioterrorism may be a threat to U.S. agriculture, expert says. *Science Daily.* August 10. www.sciencdaily.com/releases/1999/08/ 990810065844.htm.

4. Kilman, S. 2001. Special report: Aftermath of terror. U.S. defense against agroterrorism still suffers from plenty of weak spots. *Wall Street Journal.* December 26.

5. Windrem, R. 2001. U.S. to launch war on "agroterror." *MSNBC.* October 22. www.msnbc.com/news/314627.asp?cp1=1.

6. Carroll, J. 2002. FDA issues guidelines to protect food. *Wall Street Journal.* January 8.

7. Kilman, S. 2001. Farms present a target susceptible to terror. Agricultural resources are remote, appear easy to attack. *Wall Street Journal.* December 26.

8. O'Driscoll, P. 2002. Farm country confronts "undeniable" threat. *USA Today.* May 8. http://www.usatoday.com/news/nation/2002/05/08/footandmouth.htm.

9. CDC. 2000. Biological and chemical terrorism: Strategic plan for preparedness and response. *Morbidity & Mortality Weekly Report* 49:Rf-4.

10. APHA. 2000. Smallpox. In *Control of Communicable Diseases Manual.* J. Chin, ed. Washington D.C.: American Public Health Association.

11. Picard, A. 2002. Bioterrorists prey on public's anxiety. *The Globe and Mail.* February 18. http://www.theglobeandmail.com/servlet/articlenews/printarctile/gam/20020218/uterrm.

12. Lesser I., B. Hoffman, J. Arquilla et al. 1999. *Countering the New Terrorism.* Washington D.C.: Rand.

13. Carus, S. 2001. *Working Paper: Bioterrorism and Biocrimes. The Illicit Use of Biological Agents Since 1900.* Center for Counterproliferation Research, National Defense University. 2001.http://www.ndu.edu/centercounter/ Full_Doc.pdf.

14. Olson, K. 1999. Aum Shinrikyo: Once and future threat? *Emerging Infectious Diseases* 5:513–516.

15. Mangold T., and J. Goldberg. 1999. *Plague Wars: The Terrifying Reality of Biological Warfare.* New York: St. Martin's Press.

16. U.S. Census Bureau. 2001. *Statistical Abstract of the United States: 2001.* Washington, D.C.: U.S. Government Printing Office.

17. U.S. Bureau of Economic Analysis. 2001. *National Accounts Data 1929–2001.*

18. Edmondson, W. 2001. Food and fiber system share of GDP remains robust. *Rural America* 16:56–57.

19. U.S. Department of Agriculture, National Agricultural Statistics Service. 2001. Agricultural Statistics 2001. Washington, D.C.: U.S. Government Printing Office.

20. National Restaurant Association. 2002. *Industry at a Glance.* http://www.restaurant.org/research/ind_glance.cfm.

21. U.S. Department of Agriculture. 2002. *U.S. Agricultural Trade.* Economic Research Service, Briefing Room. http://www.ers.usda.gov/briefing/agtrade/ usagriculturaltrade.htm.

22. U.S. Department of Agriculture. 2001. *Electronic Outlook Report.* U.S. Department of Agriculture, Economic Research Service. November 30.

23. Fothergill, L. 1961. Biologic warfare and its effects on foods. *Journal of the American Dietetic Association* 38:249–252.

24. Zilinakas, R. 1997. Iraq's biological weapons: The past as future? *Journal of the American Medical Association* 278:418–424.

25. Alibek K. 1999. *Biohazard.* New York: Random House, Inc.

26. Regis E. 1999. *The Biology of Doom: The History of America's Secret Germ Warfare Project.* New York: Henry Holt and Company, LLC.

27. Harris S. 1994. *Factories of Death.* London, England: Routledge.

28. Hugh-Jones, M. 1992. Wickham Steet and German biological warfare research. *Intelligence and National Security* 7:379–402.

29. Wilson T., L. Logan-Henfrey, R. Weller et al. 2000. Agroterrorism, biological crimes, and biological warfare targeting animal agriculture. In *Emerging Diseases of Animals.* C. Brown and C. Bolin, eds. Washington, D.C.: ASM Press.

30. Christopher, G., T. Cieslak, J. Pavlin et al. 1997. Biological warfare: A historical perspective. *Journal of the American Medical Association* 278: 412–417.

31. Hall W., and A. Peaslee. 1944. *Three Wars with Germany.* New York: Putnam's Sons.

32. Stockholm International Peace Research Institute. 1971. *The Problem of Chemical and Biological Warfare, vol. 1.* New York: Humanities Press.

33. Robertson, A., and L. Robertson. 1995. From asps to allegations: Biological warfare in history. *Military Medicine* 160:369–373.

34. Cameron, G., J. Pate, and K. Vogel. 2001. Planting Fear: How Real Is the Threat of Agricultural Terrorism? *Bulletin of the Atomic Scientists* 57:38–44.

35. Rogers, P., S. Whitby, and M. Dando. 1999. Biological warfare against crops. *Scientific American* 280:70–75.

36. Carus, S. 1998. Biological warfare threats in perspective. *Critical Reviews in Microbiology* 24:149–155.

37. Biological Weapons Convention. 1972.

38. Thorold, P. 1953. Suspected malicious poisoning. *Journal of the South African Veterinary Medical Association* 24:215–217.

39. Lawrence, J., C. Foggin, and R. Norval. 1980. The effects of war on the control of diseases of livestock in Rhodesia (Zimbabwe). *Veterinary Record* 107:82–85.

40. Nass, M. 1992. Anthrax epizootic in Zimbabwe, 1978–1980: Due to deliberate spread? *Physicians for Social Responsibility Quarterly* 2:198–209.

41. Alibek, K. 1999. The Soviet Union's anti-agricultural biological weapons. *Annals of the New York Academy of Sciences* 894:18–19.

42. CNN. 2002. Transcript of Bush remarks on U.S.-Cuba relations. May 20.

43. CNN. 2002. Cuban biotech boom: Risk or rescue? U.S. bioweapon concerns highlight Havana's life sciences. May 13.

44. Zilinakas, R. 1999. Cuban allegations of biological warfare by the United States: Assessing the evidence. *Critical Reviews in Microbiology* 25:173–227.

45. Schapp, B.U.S. 1982. Biological Warfare: The 1981 Cuba dengue epidemic. *Covert Action* 17:28–31.

46. Fruit Fly, Program Information. 2002. USDA, Animal and Plant Health Inspection Service, Plant Protection and Quarantine. http://www.aphis .usda.gov/ppq/ispm/ff/ndex.html.

47. Neher, N. 1999. The need for a coordinated response to food terrorism. The Wisconsin Experience. *Annals of the New York Academy of Sciences* 894:181–183.

48. Agroterrorism: Chronology of CBW attacks targeting crops and livestock 1915–2000. 2000. Center for Nonproliferation Studies, Monterey Institute of International Studies. http://cns.miis.edu/research/cbw/ agchron.htm.

49. 2000 year in review. 2000. Wisconsin Ag Connection. http://www.wis-consinconnection.com/news-3rd-2000.html.

50. U.S. Department of Agriculture. 2002. *Trends in U.S. Agriculture.* http:// www.usda.gov/nass/pubs/trends/index.htm.

51. U.S. Department of Agriculture. 2001. *Statistical Highlights 2000/2001 of U.S. Agriculture.* U.S. Department of Agriculture—National Agricultural Statistics Service. http://www.usda.gov/nass/pubs/stathigh/content.htm.

52. U.S. Department of Agriculture. 2002. *USDA Market News.* May 31.

53. MacDonald, J., M. Ollinger, K. Nelson et al. 2000. *Consolidation in U.S. Meatpacking.* U.S. Department of Agriculture, Food and Rural Economics Division, Economic Research Service.

54. Deen, W. 1999. Trends in American agriculture: Their implications for biological warfare against crop and animal resources. *Annals of the New York Academy of Sciences* 894:164–167.

55. Federal Bureau of Investigation, Counterterrorism Threat Assessment and Warning Unit Counterterrorism Division. 1999. *Terrorism in the United States, 1999.*

56. Commonwealth Scientific and Industrial Research Organization (CSIRO). 1995. Australia and New Zealand Rabbit Calcivirus Disease Program: One mainland rabbit dies of rabbit calcivirus.October 16. http://www .csiro.au/communication/rabbits/pr16oct.htm.

57. Commonwealth Scientific and Industrial Research Organization (CSIRO). 1995. Australia and New Zealand Rabbit Calcivirus Disease Program: Rabbit calcivirus on Wardang Island. October 10. www.//www.csiro.au/ communication/rabbits/pr10oct.htm.

58. ProMED postings. 2000. Rabbit hemorrhagic disease virus—USA, Iowa. April 13 and 19.

59. U.S. Department of Agriculture. 2000. *Viral Hemorrhagic Disease of Rabbits, Iowa. Impact Worksheet.* U.S. Department of Agriculture, Animal and Plant Health Inspection Service.
60. U.S. Department of Agriculture. 2001. *Viral Hemorrhagic Disease of Rabbits, Utah. Impact Worksheet.* U.S. Department of Agriculture, Animal and Plant Health Inspection Service.
61. ProMED postings. 2001. *Rabbit Hemorrhagic Disease Virus—USA, Utah.* October 24.
62. ProMED postings. 2002. *Rabbit Hemorrhagic Disease Virus—USA, New York.* December 13, 2001, and January 31, 2002.
63. Franz, D. 1999. Foreign animal disease agents as weapons in biological warfare. *Annals of the New York Academy of Sciences* 894:100–104.
64. U.S. Department of Transportation. 2000. *Transportation Statistics Annual Report 2000.* U.S. Department of Transportation, Bureau of Transportation Statistics.
65. U.S. Department of State. 2002. *Fact Sheet: U.S. Customs Makes Progress on Container Security.* U.S. Department of State, International Information Programs. www.usinfo.state.gov/topical/pol/terror/02080802.htm.
66. U.S. Customs Service. 2001. *Annual Report Fiscal Year 2001.* U.S. Customs Service.
67. U.S. Department of Agriculture. 2002. *Livestock, Dairy, and Poultry Outlook.* U.S. Department of Agriculture, Economic Research Service.
68. Guardian. 2001. Farmer 'not to blame' for foot and mouth. *Guardian.* May 7.
69. Gibbens, J., C. Shapre, J. Wilesmith et al. 2001. Descriptive epidemiology of the 2001 foot-and-mouth disease epidemic in Great Britain: The first five months. *Veterinary Record* 149:729–743.
70. Amass, S., and K. Clark. 1999. Biosecurity considerations for pork production units. *Swine Health and Production* 7:217–228.
71. Sanderson, M., D. Dargatz, and F. Garry. 2000. Biosecurity practices of beef cow-calf producers. *JAVMA* 217:189.
72. U.S. Budget for 2003. 2002.
73. U.S. Department of Agriculture. 2002. *USDA's Detector Dogs: Protecting American Agriculture.* 2002. U.S. Department of Agriculture, Animal and Plant Health Inspection.
74. U.S. Fish and Wildlife Service. 2000. *Annual Report, FY 2000.* U.S. Fish and Wildlife Service, Division of Law Enforcement.
75. Brown, C., and B. Slenning. 1996. Impact and risk of foreign animal diseases. *JAVMA* 208:1038–1040.
76. U.S. Department of Agriculture. 1998. *The Potential for International Travelers to Transmit Foreign Animal Diseases to U.S. Livestock or Poultry.* U.S. Department of Agriculture, Animal and Plant Health Inspection Service, Veterinary Services, Centers for Epidemiology and Animal Health.
77. Mckean, J. 2001. The importance of traceability for public health and consumer protection. *Revue Scientifique et Technique (Office International des Epizootics).* 20:363–371.

78. Pettitt, R. 2001. Traceability in the food animal industry and supermarket chains. *Revue Scientifique et Technique (Office International des Epizootics)* 20:584–597.

79. Stanford, K., J. Stitt, J. Kellar et al. 2001. Traceability in cattle and small ruminants in Canada. *Revue Scientifique et Technique (Office International des Epizootics)* 20:510–522.

80. ProMED posting. 2001. *Tuberculosis, Bovine—USA* (Texas). October 7.

81. Caporale, V., A. Giovannini, C. Francesco et al. 2001. Importance of the traceability of animals and animal products in epidemiology. *Revue Scientifique et Technique (Office International des Epizootics).* 20:372–378.

82. McGrann, J., and H. Wiseman. 2001. Animal traceability across national frontiers in the EU. *Revue Scientifique et Technique (Office International des Epizootics)* 20:406–412.

83. Brasher P. 2002. Farms fight food tracing. *Des Moines Register.* September 22.

84. Holstein Association USA. 2002. *National F.A.I.R.—A Proven National Identification.* Holstein Association USA. http://www.holsteinusa.com/html/newsrel_0102.html.

85. Huxsoll, D., W. Patrick, and C. Parrott. 1987. Veterinary services in biological disasters. *JAVMA* 190:714–722.

86. Todd, F. 1952. Biological warfare against our livestock. *North American Veterinarian* 33:689–693.

87. Gordon, J., and S. Bech-Nielsen. 1986. Biological terrorism: A direct threat to our livestock industry. *Military Medicine* 151:357–363.

88. Wilson, T., D. Gregg, D. King et al. 2001. Agroterrorism, biological crimes, and biowarfare targeting animal agriculture. *Laboratory Medicine* 21:549–591.

89. Morbidity & Mortality Weekly Report. 2000. Biological and Chemical Terrorism: Strategic Plan for Preparedness and Response.

90. Fitzpatrick, A., and J. Bender. 2000. Survey of chief livestock officials regarding bioterrorism preparedness in the United States. *JAVMA* 2217:1315–1317.

91. Bunning, M. 2001. Nipah virus outbreak in Malaysia, 1998–1999. *Journal of Swine Health and Production* 9:295–299.

92. Gibbs, W. Trailing a virus. 1999. *Scientific American* 281:81–87.

93. Nor, M., C. Gan, and B. Ong. 2000. Nipah virus infection of pigs in peninsular Malaysia. *Revue Scientifique et Technique (Office International des Epizootics)* 19:260–265.

94. Sellers, R., and S. Daggupaty. 1990. The epidemic of foot-and-mouth disease in Saskatchewan, Canada, 1951–1952. *Canadian Journal of Veterinary Research* 51:457–464.

95. Paarlberg, P., J. Lee, and A. Seitzinger. 2002. Potential revenue impact of an outbreak of foot-and-mouth disease in the United States. *JAVMA* 220:988–992.

96. BBC News. 2001. FMD report: Outbreak's economic impact. *BBC News.* August 29.

97. Department for Environment. 2002. *Economic Cost of Foot and Mouth Disease in the U.K. A Joint Working Paper.* Department for Environment, Food and Rural Affairs, Department for Culture, Media, and Sport. http://www.defra.gov.uk/corporate/inquiries/lessons/fmdeconcost.pdf.

98. Anonymous. 2002. Sport market forecasts show consumer spending on sport tops £15 billion. *Sport Market Forecasts.* http://www.thesportslife.com.

99. American Veterinary Medical Association. 2002. *Avian Influenza Strikes Virginia Poultry Farms.* July 1. http://www.avma.org/onlnews/javma/jul02/020701a.asp.

100. Virginia Agricultural Statistics Service. 2002. *Livestock Statistics.* Virginia Agricultural Statistics Service. October 13. http://www.nass.usda.gov/va/lvstkplty2.htm.

101. Lidholm, E. 2002. *Update on Avian Influenza.* Virginia Department of Agriculture and Consumer Services. September 10. http://www.vdacs.state.va.us/news/releases-a/avianupdate.html.

102. Horst, H., C. De Vos, F. Tomassen et al. 1999. The economic evaluation of control and eradication of epidemic livestock diseases. *Revue Scientifique et Technique (Office International des Epizootics)* 18:367–379.

103. Pimentel, D., L. Lack, R. Zuniga et al. 2000. Environmental and economic costs of nonindigenous species in the United States. *BioScience* 50:53–65.

104. Lockwood, J. 1987. Entomological Warfare: History of the use of insects as weapons of war. *Bulletin of the Entomological Society of America* 33: 76–82.

105. Madden, L. 2001. What are the nonindigenous plant pathogens that threaten U.S. crops and forests? *American Phytopathological Society.* http://www.apsnet.org/online/feature/exotic/.

106. MacKenzie D., M. Marchetti, and C. Kingsolver. 1984. The potential for the willful introduction of biotic agents as an act of anticrop warfare. The movement and dispersal of agriculturally important biotic agents: An International Conference on the Movement and Dispersal of Biotic Agents. pp. 601–608.

107. Dunn, M. 1999. The threat of bioterrorism to U.S. agriculture. *Annals of the New York Academy of Sciences* 894:184–188.

108. U.S. Department of Agriculture. 2001. *Karnal Bunt: A Fungal Disease of Wheat.* U.S. Department of Agriculture, Animal and Plant Health Inspection Service, Plant Protection and Quarantine, Industry Alert.

109. U.S. Department of Agriculture. 2001. *Asian Longhorned Beetle (Anoplophora glabripennis).* U.S. Department of Agriculture, Animal and Plant Health Inspection Service, Plant Protection and Quarantine.

110. Cunnion, S. 2002. The meat-and-potatoes approach to bioterrorism. *Journal of Homeland Security.* May 2002.

111. Sobel, G., A. Khan, and D. Swerdlow. 2002. Threat of a biological terrorist attack on the U.S. food supply: The CDC perspective. *Lancet* 359: 874–880.

112. Mead, P., L. Slutzker, V. Dietz et al. 1999. Food-related illness and death in the United States. *Emerging Infectious Diseases* 5:607–625.

113. Khan, A., D. Swerdlow, and D. Juranek. 2001. Precautions against biological and chemical terrorism directed at food and water supplies. *Public Health Reports* 116:3–14.

114. CDC. 1998. Lead poisoning associated with imported candy and powdered food coloring—California and Michigan. *Morbidity & Mortality Weekly Report* 47:1041–1043.

115. U.S. Department of Agriculture. 2001. *Protecting the Public From Foodborne Illness: The Food Safety and Inspection Service.* U.S. Department of Agriculture, Food Safety and Inspection Service. http://www.fsis.usda.gov/oa/background/fsisgeneral.htm.

116. U.S. Department of Agriculture. 2002. *FSIS Security Guidelines for Food Processors.* 2002. U.S. Department of Agriculture, Food Safety and Inspection Service.

117. Food and Drug Administration. 2000. *FDA Talk Paper: 2001 Budget Request for FDA.* Food and Drug Administration.

118. Food and Drug Administration. 2002. *FDA's Sentinel of Public Health: Field Staff Safeguards High Standards.* Food and Drug Administration.

119. Food and Drug Administration. 2002. *Guidance for Industry—Importers and Filers: Food Security Preventive Measures Guidance.* Food and Drug Administration, Center for Food Safety and Applied Nutrition.

120. Food and Drug Administration. 2002. *Guidance for Industry—Food Producers, Processors, Transporters, and Retailers: Food Security Preventive Measures Guidance.* Food and Drug Administration, Center for Food Safety and Applied Nutrition.

121. CDC. 2001. Preliminary FoodNet data on the incidence of foodborne illnesses—Selected sites, United States 2000. *Morbidity & Mortality Weekly Report* 50:241–246.

122. Kolavic, S., A. Kimura, S. Simons et al. 1997. An outbreak of *Shigella dysenteriae* type 2 among laboratory workers due to intentional food contamination. *Journal of the American Medical Association* 278:396–398.

123. Falkenrath R., R. Newman, and B. Thayer. 1998. *America's Achilles' Heel. Nuclear, Biological, and Chemical Terrorism and Covert Attack.* Cambridge, MA: MIT Press.

124. Török, T., R. Tauxe, R. Wise et al. 1997. A large community outbreak of salmonellosis caused by intentional contamination of restaurant salad bars. *Journal of the American Medical Association* 278:389–395.

125. Phills, J., A. Harold, G. Whiteman et al. 1972. Pulmonary infiltrates, asthma and eosinophilia due to *Ascaris suum* infestation in a man. *New England Journal of Medicine* 286:965–970.

126. L.A. Times. 1986. 2nd soup packet in box poisoned; Random act seen. *Los Angeles Times.* September 4.

127. Gallay, A., F. Van Loock, S. Demarest et al. 2002. Belgian Coca-Cola–related outbreak: Intoxication, mass sociogenic illness, or both? *American Journal of Epidemiology* 155:140–147.

128. Anonymous. 2002. Man admits he stuck needles in food. *Des Moines Register*. May 10.
129. U.S. Department of Agriculture. 2002. *U.S. Department of Agriculture, Animal and Plant Health Inspection Service, Veterinary Services, Emergency Programs*. http://www.aphis.usda.gov/vs/ep/.
130. Bowman, Q., and J. Arnoldi. 1999. Management of animal health emergencies in North America: Prevention, preparedness, response, and recovery. *Revue Scientifique et Technique (Office International des Epizootics)*. 18:76–103.
131. Shimshony A., U. Orgad, D. Baharav et al. 1986. Malignant foot-and-mouth disease in mountain gazelles. *Veterinary Record* 119:175–176.
132. United States National Animal Health Emergency Management System. 2002. *2001 Annual Report*.
133. U.S. Department of Agriculture. 2002. *U.S. Department of Agriculture, Animal and Plant Health Inspection Service, Plant Protection and Quarantine, Emergency Programs Manual*.
134. Sequeira, R. 1999. Safeguarding production agriculture and natural ecosystems against biological terrorism: A U.S. Department of Agriculture Emergency Response Framework. *Annals of the New York Academy of Sciences* 894:48–67.
135. Schaad, N., J. Shaw, A. Vidaver et al. 1999. Crop biosecurity. *American Phytopathological Society*.
136. U.S. Department of Agriculture. 2000. *Agriculture Fact Book 2000*. U.S. Department of Agriculture, Office of Communications.
137. Center for Nonproliferation Studies. 2001. *Chronology of Aum Shinrikyo's CBW Activities*. Center for Nonproliferation Studies, Monterey Institute of International Studies.
138. Anthan, G. 2001. PETA statements spark agroterrorism concerns. *Des Moines Register*. April 12.

ACKNOWLEDGMENTS

I would like to thank Dr. Kristina D. August and Dr. Terry Wilson for their great effort in reviewing this manuscript and for their suggestions.

24

AGRICULTURAL BIOTECHNOLOGY IN DEVELOPING COUNTRIES

Greg Graff,[1] *Matin Qaim,*[2] *Cherisa Yarkin,*[1]
David Zilberman[1]

The harnessing of new scientific knowledge to enhance agricultural productivity has enabled feeding a sixfold increase in human population (from 1 billion to 6 billion) between 1800 and 2000, a momentous increase in yield per capita and a large increase in yield per acre. Discovery of the internal combustion engine and a better understanding of the basic principles of mechanics have led to a wide array of mechanical innovations and substitution of farm labor with capital. Basic knowledge of organic chemistry was crucial in the development of synthetic chemicals that increased soil fertility and reduced pest damages. The discovery of the basic principles of genetics was a major force behind the green revolution, and it led to a systematic breeding process that increased crop productivity. Knowledge of molecular and cell biology and, in particular, discovery of the principles of the genetic code, formed the foundation for agricultural biotechnologies that have been commercially introduced and widely adopted in the late 1990s. These latest technologies have shown significant promise, but they are also subject to serious concerns and debate.

While farmers in the United States, Canada, Argentina, and several other countries have embraced the new agricultural biotechnology innovations and adopted them extensively, concerns about these innovations have impeded their introduction and constrained their use in Europe. In spite of the different perspectives in both Europe and the United States about the potential benefits and risks of agricultural biotechnology, both Europe and the United States have excess capacity of production in their agricultural sectors. Their agricultural policies of recent decades have been largely aimed at controlling supply, not enhancing it. The actual value, and to a larger extent the future potential, of modern agricultural biotechnologies may instead perhaps rest in their application and utilization in the developing world. The high rates of population growth in many developing countries, problems of malnourishment and deforestation, and major constraints on their agricultural and natural resources suggest that developing countries are likely to benefit substantially from technologies that enhance land productivity, that are easily accessible, and that are relatively environmentally benign. Moreover, realization of the value of biotechnology in the South will affect attitudes toward it in the North.

It has been suggested that the role and potential of current agricultural biotechnology innovations in developing countries are limited, and that developing country agricultural sectors should instead consider alternative strategies to enhance their production system, such as modernizing their agricultural economies. This paper draws on new empirical evidence, the findings of theoretical economic research, and the emergence of new institutional innovations to explore at a new level some of the major objections to the introduction of agricultural biotechnology in developing countries. We show that agricultural biotechnology innovations have the following characteristics:

- Significant potential exists to enhance the yield productivity of agriculture in developing countries.
- Some of the worrying constraints in terms of access and ownership of technologies can be overcome.
- With appropriate management environmental concerns and ecological risks associated with agricultural biotechnology can be minimized and biodiversity maximized.

The objections to development and adoption of agricultural biotechnology in developing countries stem from understandable concerns about its lack of impact on productivity, access to technology protected by intellectual property, and the environmental and biodiversity risks that the technology presents. Each section of this paper addresses one of these issues. The concerns about the productivity stem from the fact that evidence of the yield effects of agricultural biotechnology in many developed countries has not been very high. We will use both economic theory and empirical evidence from India to explain why we may expect a much higher yield effect in developing countries.

The concern about access to technology stems from the high concentration of intellectual property rights (IPR) over agricultural biotechnology in a small number of private hands. We will explain the research-generation process that led to the evolution of agricultural biotechnology innovations and the current pattern of IPR ownership. We will also suggest that technology market mediating or "clearinghouse" mechanisms can and are being introduced to provide access to IPR for crops that serve the poor in developing countries.

The third objection to agricultural biotechnology stems from the concern that it will lead to the contamination of genetic materials and the depletion of biodiversity. We will show that appropriately managed biotechnology can actually be a mechanism to enhance crop biodiversity rather than reduce it. The final section of this chapter will examine the long-run effects of agricultural biotechnology and assess the policies being instituted to manage and take advantage of biotechnology. We should recognize that this technology is still in its infancy. The challenge of the developing world is to capture some of the benefits of biotechnology in the near future by developing and disseminating new farming practices, recognizing that agricultural biotechnology is still evolving.

PRODUCTIVITY AND IMPACTS OF AGRICULTURAL BIOTECHNOLOGY

Altieri (2001), who argues against agricultural biotechnology, claims that the technology was developed by the North to meet the needs of the agricultural economies of the developed countries. Therefore, they suggest biotechnology may not be appropriate for the agricultural sectors of the South, which grow different crops and have different technical and economic needs. On the surface it seems that there is some real economic logic to this argument. The theory of induced innovations (Hayami and Ruttan 1985) suggests that new technological solutions emerged in different locations in response to particular economic situations and constraints. Thus, for example, U.S. agriculture in the 19th century developed labor-saving technologies, while at the same time Southeast Asian agriculture developed labor-intensive practices. Induced innovation theory and the recognition that agricultural systems are heterogeneous—that no one solution fits all—may lead to the suggestion that a technology that originated in one environment may not be appropriate for another.

While it may be incorrect to presume that a uniform technological solution applies to all problems, it is equally untrue that technologies cannot be exported and transferred across domains. Many products and technologies have migrated across locations throughout history and, after some modification and adjustment, have become established in new regions. This is obviously the case for plants and crops. Maize was originally domesticated in the Americas, but is now grown throughout the world. Similarly, wheat originated in the Fertile Crescent but has been diffused globally. The automobile, the tractor, and the combine have spread worldwide. We are living in a world where there has, to a large extent, been a flow of technologies from the North to the South and in some cases a flow in the reverse direction. Transported technologies have been adapted successfully many times, but obviously there have been cases of failure. While the current set of agricultural biotechnologies have originated in the North, they may in some cases, after adaptation and modification, be quite appropriate for solving technical and economic problems in the South. In any case, their Northern origin is not a fatal flaw that prevents their use in developing countries.

Furthermore, studies of adoption and impact of new technologies suggest that the impact of the same technology may vary across locations, reflecting heterogeneity of economic conditions and constraints. New applications may lead to unforeseen outcomes and benefits not initially detected when a technology is introduced. Companies that specialized in pest protection products have introduced genetically modified (GM) crops in the United States. Most of the investments in developing these biotech products have been prompted by desires and policy incentives to find substitutes for chemical pest controls or to switch to low-tillage practices (in the case of Roundup Ready seeds) (NRC 2000). Indeed, studies that assess the technology's impact in the United States show that it led to a reduction in pesticide use, but to only modest changes in yield, mostly between 0 and 10 percent (Ameden and Zilberman 2003).

Altieri (2001) suggested that the potential benefits of GM varieties with Bt insect resistance or Roundup Ready herbicide tolerance are limited in the developing world because of their supposed modest yield-increasing effects. Developing countries are in need of new varieties that can substantially increase crop yields; if the experience from the United States is an indicator of a genetic property, the pest-controlling GM varieties may not be very helpful for the developing world. A recent study by Qaim and Zilberman (2003), however, argues that the yield effect of GM varieties in different locations may vary according to the severity of pest problems in those locations as well as the availability and use of alternative pest controls. Lichtenberg and Zilberman (1986) distinguish between actual and potential crop output per acre. Actual output is potential output minus pest damage, and the pest damage is a function of the pest population as well as use of pest control. If, for example, untreated pest damage causes a 50 percent loss in yield, the introduction of pest-control activities that eliminate this pest damage will double the realized yield. Pest damage losses of 50 percent or more are not unheard of in many parts of the world. The introduction of effective pesticides may lead to spectacular yield gains. The magnitude of pest damage is especially high in countries with humid climate, large levels of pest infestation, and minimal use of chemicals or other pest-control strategies caused by lack of knowledge, lack of access, or high cost. Production of GM pest-controlling varieties in such locations may have a drastic yield-increasing effect.

Qaim and Zilberman (2003) found that, based on 157 field trials in India where fields were planted with varieties of Bt cotton and their non-Bt counterparts, Bt had an 80 percent higher yield on average. Their study was conducted in three major cotton-producing states in India where fields were highly vulnerable to the American bollworm. While chemical treatments had been used with traditional varieties, they were not very effective. Bt cotton also led to reduction in pesticide use, and thus farmers benefited both from higher yield as well as lower cost of pesticide use. In another related study, Qaim and Zilberman (2003) found profits per hectare with Bt cotton to be five times higher than profits without Bt cotton, and that farmers will be better off, even if they have to pay a higher seed price for the Bt varieties.

The higher yield effects in India are in contrast to the less than 10 percent yield effect of Bt in China and the United States. Note, however, that crop damages on conventional field crops in India is about 60 percent, while the crop damages without Bt are around 12 percent in the United States and China. Thus, the presumably low intensity of pests and higher intensity of alternative chemicals in the United States and China have caused Bt to have a much lower yield effect than in India. In other studies where the performance of Bt cotton was compared to non-Bt cotton counterparts in South Africa and India, the initial conditions were more similar to those of India than the United States. The yield effect of experimental plots in this country were very substantial (40 percent and above).

The results of Qaim and Zilberman (2003) show that, in spite of the need to pay more for the Bt cotton in India, farmers will substantially gain from their adoption and be consistent with evidence elsewhere. Several studies, including Moschini and colleagues (2000) as well as Marra and Carlson (1990), show that the economic gain from the introduction of GM varieties was distributed among the companies, the farmers, and the consumers. The multinationals that introduce GM varieties in the United States gain about 40 percent of the surplus generated, farmers gain about 40 percent or 50 percent, and the rest goes to consumers (through lower prices). Similar results were obtained for Canada and South Africa (Ameden and Zilberman 2003). Thus, in spite of the monopolistic pricing, there was quite a substantial volume of gains to the individual farmers.

The comparison of the performance of Bt cotton in various locations supports the theory that GM pest-control varieties will have a significant yield effect in locations where pest damage is substantial and agricultural technologies are not being used. That may suggest that introduction of Bt to some other crops—for example, cassava or corn in various locations in a developing country—may lead to substantial increases in yield. While we expect that in some countries the pesticide cost-saving effect that we witnessed in the United States and China may not be fully materialized, there may be significant yield effects that we have not seen in previous applications of GM varieties. Thus, applications of some of the existing technologies in the developing world may provide the answers to food security concerns and enhance food supplies in locations where these are significant problems.

While most of the applications of GM varieties thus far have been in pest control, research labs in the United States and elsewhere are in the advanced stages of developing new varieties that have traits that may be especially desirable in developing countries. These food varieties are drought-tolerant and can effectively utilize low-quality saline water and withstand significant variations in conditions. Scientists are working on developing other varieties that enhance food quality by inserting micronutrients into major food products to enhance their nutritional value. The availability of basic mechanisms for transferring genes and increased knowledge of genetic maps of different crops have reduced the cost of developing a new GM variety. As many researchers in universities and companies around the world are working on improving these technologies, in the next 10 to 20 years we will likely see new trends in product development. These improved technologies will be especially beneficial to developing countries that suffer from acute weather conditions and populations that may be particularly vulnerable to low nutritional intake. However, before these technologies can be utilized in the developing world, crop scientists and Agricultural Experiment Stations and farmers in these countries need to have access to these technologies; this issue is addressed in the next section.

ACCESS TO IPR AND REGULATORY CONSTRAINTS

In the past the public sector has been the predominant provider of seed varieties in developing countries and in many agricultural sectors of the developed countries. Research conducted at universities, international centers, and agricultural experiment stations in different countries, supported mostly by public research monies, developed major strains of seeds and provided the source for the seeds used daily by individual farmers. Public sector institutions established an open access policy for gene banks and sharing of knowledge and technologies, which led to the global development of new varieties, contributed to the green revolution, and overcame several outbreaks of plant diseases throughout the world.

The public sector, however, has not been dominant in the introduction of agricultural biotechnologies. Some of the crucial basic innovations did occur at major universities supported by public monies, but these inventions were frequently patented and transferred exclusively to private companies. The companies funded research to further develop these innovations. They invested in testing and regulatory requirements, processes for industrial production of the GM seeds, and the distribution and marketing of GM varieties. Private companies would not likely have made these large investments in agricultural research without exclusive rights to the core technologies, and the opportunity to earn a return on the investment by capturing the profits made by having a monopoly in the resulting seed technologies. While the privatization of agricultural biotechnology innovations in the United States does have social costs, it is very unlikely that the public sector would have invested even close to the same funds to develop agricultural biotechnology. Left just in the public sector without major investments of several billions of dollars made by companies such as Monsanto, Syngenta, and DuPont, the technology would not have been developed and brought to market as quickly.

The privatization of seeds might cause some loss of comfort in the agricultural community. Farmers who felt in the past that the public sector was "working for them" to produce new varieties now feel that they are dependent on the income-generating activities of small numbers of private companies that dominate the seed business. Growers are also concerned that they have to pay more for seeds, even though in most cases the extra cost will more than pay for itself. We have to question whether the public provision of seeds has been an anomaly. Multinational agribusiness companies have been the dominant providers of chemicals and mechanical equipment for agriculture. A small number of multinationals have been the major providers of veterinary and genetic inputs for animal agriculture. To address equity concerns, in some cases purchases of farm inputs have been subsidized, even though it led to efficiency.

The increased privatization of the technology generation of genetic materials in agriculture, however, leads to changes in the rules of the game. Public sector researchers are now worried about their IPR and may become more proprietary about their genetic materials. This may lead to a reduction in the exchange of knowledge and genetic materials among scientists and economic welfare losses associated with the decline in "open science." Furthermore, Heller and Eisenberg (1998) suggest that the different ownership of IPR among many owners may lead to the emergence of an intellectual "anticommons" slowing the research and development (R&D) of biotechnology products, where high transaction costs and lack of access to essential IPR hamper the development of new technologies. Public sector research efforts in agricultural biotechnology cannot ignore the new realities and should develop mechanisms to cope with them.

While the public sector is likely to have a smaller role in generating seeds with biotechnology, it still must play a major role in their development. Sunding and Zilberman's 2001 survey argues that the private sector has always tended to under-invest in R&D; thus, it is unlikely that the multinationals will sufficiently invest in development of genetic materials, especially for subsistence food for the poor in developing countries.

The work of Traxler and colleagues (Traxler and Falck-Zepeda 1999; Traxler et al. 2001) convincingly argues that the public sector is likely to be the dominant provider of genetic materials for most of the major crops used in developing countries for years to come. One of the major challenges is to design the institutional mechanism that will allow crop breeders in the South to utilize the technologies and the knowledge that is largely accumulated by the private sector in the North. This is especially important as we live in an era of globalization where countries will slowly introduce their own IPR regime, and there is an increasing need to develop a smooth system to transfer access to technologies across nations in a way that will respect the rights of technology owners. The difficulties and high cost to transfer rights to develop golden rice varieties for developing countries underscore the importance of securing easy access to intellectual property for developing agricultural biotechnology in the developing world.

Recent research and institutional developments identified avenues to reduce the challenge and enable developers of biotechnology crops in developing countries to take advantage of the technology base controlled by companies and universities in the North. It is useful to recognize that the private sector has to deal with access to IPR constraints, and companies have developed alternative mechanisms for obtaining the "freedom to operate" so that development of new product innovations can rely on technologies that are legally available. Private companies in various sectors have developed arrangements of patent rights swapping and de facto clearinghouses where companies pay each other for the use of essential property rights. The key for each of these arrangements is a clear documentation of who owns what and the ability to capitalize on intellectual property owned by one organization in order to gain easier access to intellectual property controlled by others.

Graff and colleagues (2003) argue that public sector R&D organization in the agricultural biotechnology area should learn from the experience of the private sector and develop mechanisms that will enable them to better perform in the reality of highly privatized IPR. In particular, they need to have a clear knowledge of the IPR they and others possess. They need to cooperate and establish a critical mass of IPR to obtain access to private IPR at better terms, and they need expertise in IPR and technology transfers to reduce their transaction costs in this area. Such institutional reforms are especially important for public research organizations working on the provision of agricultural biotechnology for crops that are likely to be neglected by the private sector. Namely, subsistence crops in developing countries are specialty crops throughout the world.

Graff et al. (2003) showed that 24 percent of the U.S. patents claimed between 1981 and 2001 were assigned to private sector institutions, and public institutions claimed about one-quarter of the agricultural biotechnology patents throughout the world during this period. Furthermore, the share of patents registered by public sector inventors in major technology classes of agricultural biotechnology varies from 8 percent to 56 percent. Thus, the dependence on public technologies varies by class of technology. The rights to many of the university patents have been transferred to private companies; however, the rights to many others have not been exclusively transferred. Public sector institutions at the same time have also developed many viable technologies that have not been patented but have been published in the research literature and thus are in the public domain. The last two categories of technologies, which are controlled by the public institutions, could be used as a platform of accessible IPR for public sector researchers and seed developers, once mechanisms for documentation and access to these technologies are established.

Collective action by the public sector may provide a reliable, efficient, and cost-effect mechanism to license back (or sublicense) from companies the rights to use technologies for applications that will simply not be pursued by the private sector. For example, public research institutions working on developing new varieties for subsistence crops in the South (like cassava) or minor crops globally (like strawberries) could seek rights to use private sector technologies. Multinationals are likely to be willing to assign rights to use technologies in applications that they do not pursue because of the public relation gain, as long as they are not liable for misusing these technologies or at risk of jeopardizing their existing or future markets. Furthermore, in case of privately controlled technologies that originated in the private sector, the consent of both the private companies and the faculty member who owns the patent may be needed to facilitate assignment of rights to a particular application.

Graff and Zilberman (2001) suggest the establishment of "IPR clearinghouses," which are organizations that aim to reduce the transaction costs to exchange and use IPR in developing agricultural biotechnologies. These organizations could have several additional functions, including establishing a critical mass of publicly controlled IPR and documenting it to reduce search cost for potential users, negotiating access to additional IPR from other sources, and providing assistance to crop breeders in the public sector to achieve "freedom to operate" with agricultural biotechnology techniques. In particular, it may assist in establishing private-public partnerships to allow smooth transfer of rights of privately owned technologies to developers of subsistence crops in developing countries. This clearinghouse may also assist public researchers in developing countries to transfer rights to use their technologies to the public sector.

Several institutional arrangements that have characteristics of the clearinghouse have been established; for example, in the early 1980s ISAAA arranged a public-private partnership between Monsanto and Mexican agricultural research institutes to develop virus-resistant potatoes (Qaim 1998). The Rockefeller Foundation has been the catalyst in establishing the African Agricultural Technology Foundation, a private sector consortium that aims to facilitate a public-private partnership for crop applications in Africa. Rockefeller, together with the McKnight Foundation, is also sponsoring the development of the Public Sector Intellectual Property Resource for Agriculture (PIPRA), which if successful will provide full documentation of the existing and available IPR from participating public sector institutions and may then begin to license access to these technologies for specialty crop and subsistence crop development.

REGULATORY CONSTRAINTS

Farmers in developing countries may not have access to GM-modified varieties because of high costs and complex registration requirements. These registration requirements are important tools for protection against negative environmental side effects and misuse of new technologies. Yet, the benefits of registration may be outweighed by the cost of the constraint that it imposes. Thus, it is crucial for registration requirements to be cost-effective without duplication and redundancy.

Unfortunately, excessive registration requirements may be used for political-economic rent seeking. Regulatory efforts may raise the cost to introduce new technologies and prevent the emergence of new input suppliers who may take advantage of new biotechnology innovations. It may lead to a high concentration of market power in the pest-control supply network (NRC 2000). Thus, the challenge is to develop efficient regulatory regimes for the developing world by avoiding generic and often repetitious testing of new technologies. Also, since borders are arbitrary, countries can take advantage of regulatory clearances granted elsewhere and concentrate on unique local problems and issues.

Because of the high cost of regulatory requirements and lack of expertise, it is important for countries to establish regional alliances for

regulatory purposes. For example, Central African countries that operate under similar conditions should consider establishing integrated regulatory organizations for processing and utilizing regulatory
information from developed countries addressing generic problems, and
to conduct tests that are relevant to the region. Furthermore, procedures
for easy transfer and communication of regulatory information across
countries should be established.

BIOTECHNOLOGY AND BIODIVERSITY

A major concern about introducing agricultural biotechnology is its impact on crop biodiversity. In particular, there may be concerns that a
GM variety will take over a large acreage of land previously planted
with many varieties that each fit a specific agroecological condition. The
introduction of GM varieties is perceived to follow a pattern of modern
agriculture where a large number of traditional varieties is replaced by
synthesized high-yielding varieties that may produce more output but
may be less resilient to pests and changes in weather conditions.

We will suggest here that properties of agricultural biotechnology can
actually be used to sustain and, in some cases, restore crop biodiversity.
The extent to which it can fulfill its potential depends on the way it will
be managed.

One of the main differences between traditional crop breeding and
crop biotechnology is that with traditional crop breeding, existing varieties are selectively combined to develop new varieties. Selective breeding is a long process, and the end product has a unique genetic makeup
that varies from traditional varieties. On the other hand, GM varieties
are based on incremental changes when one or a small number of genes
is inserted or modified. In principle, every local maize variety can be
simply turned into Bt maize. Thus, a region with a large number of traditional varieties can sustain the same number of varieties, but each variety is slightly modified. Furthermore, the diversity may actually
increase if each variety is partially used in its original form and partially
used in a modified form.

Qaim and colleagues (2003) argue that the extent to which GM varieties can be used to preserve existing varieties is an economic and institutional question. Identifying the genetic sequence of a trait that can
modify crop behavior is costly. On the other hand, inserting and modi-
fying genes of an existing traditional variety require a relatively small

fixed cost, and the cost per seed is almost negligible. If the extra gain from introducing a GM trait to an existing variety through increased yield and reduced pesticide use is greater than the cost of the transformation, this variety should be used and the modified seeds should be planted. In cases where the land planted with an existing traditional variety is substantially heterogeneous in terms of severity of pest problems, only those locations where the gain from the new varieties is greater than the cost of modification will adopt the new variety. Of course, the fixed cost associated with the variety is justified if the overall extra profit covers this extra fixed cost.

The economic reality, however, results in outcomes that are sometimes different than what is socially optimal. If private firms control the production of the new varieties, they will charge extra per seed. The price is likely to be greater than the cost to modify the variety, if the firms have some monopolist power. Thus, under a private provision of seeds, there may be under-adoption of the new variety (relative to the socially optimal outcome) because some growers may not afford to pay for the seeds, even though the gains of the technology are greater than the actual cost of modification. Furthermore, in order to cut costs, the private seed supplier may not modify all the varieties that should be modified under the optimal resource allocation. Thus, with private suppliers with some monopolist power, there may be both underutilization of GM technology as well as a tendency to have less diversity. If the relative costs of modification are small compared to the gains, the loss of biodiversity because of the monopolistic production of GM is smaller.

Qaim, Yarkin, and Zilberman (2003) also suggest that other factors will determine the extent to which GM varieties are introduced and biodiversity is preserved. One is *access to traditional varieties.*

Access to traditional varieties may be limited both because of IPR constraints or transaction cost. If the producer of a GM variety modifies only a small subset of varieties, many growers may switch from the traditional varieties to the modified varieties, and there will be a significant loss in biodiversity. For example, when only one variety is modified and it increases yield substantially because of significant pest damage, farmers may switch from many traditional varieties to this modified variety. Increase in access to traditional variety may allow the developer of new technologies to offer a larger choice of modified varieties, and thus the negative effects of biodiversity are likely to be reduced or eliminated.

Another factor is the strength of the breeding sector in a given country. The capacity to modify a large number of varieties is more substantial, and the cost is relatively smaller when the breeding sector is advanced. Thus, developing countries with more advanced breeding sectors (China, Brazil, and India) will be able to modify a large number of varieties (given that it is economically viable), while limited amounts of modification of local varieties may occur in countries with a weak breeding sector.

Based on these factors, Qaim, Yarkin, and Zilberman (2003) suggest an emerging pattern:

- A large number of modified varieties is expected to emerge in countries (e.g., United States and Canada) with a well-enforced IPR, low cost of access to traditional varieties, and a strong breeding sector, resulting in relatively low cost of modifications. Indeed, there are more than 1,000 varieties of Roundup Ready soybeans, and more than 700 GM varieties of Bt cotton in the United States.

- A large number of modified varieties is expected in countries with strong domestic breeding sectors, low transaction cost to obtain access to traditional varieties, and minimal enforcement of IPR of multinational companies. One example is China. In this situation, there is a high spread of GM varieties and also a significant diversity of modified varieties. Twenty-two varieties of Bt cotton are grown in China, more than in the United States, and on a smaller total acreage.

- Minimal introduction of GM varieties is expected when the cost of genetic modifications (because of regulation) is very high, as is the case of Europe, even though the breeding sector is strong and the IPR are respected. If and when the barriers to GM varieties in Europe are reduced, we expect similar patterns that we observe in the United States, where a significant number of varieties is modified.

- A relatively modest diversity of GM varieties is expected when the breeding sector is capable and there is low enforcement of IPR and some access barriers to traditional varieties. This may be the case for some crops in India and Argentina. The multinational may not be inclined to invest in overly diverse GM offerings there because of limited profit potential. In Argentina, 45 varieties of Roundup Ready soybeans are grown on 10 million acres. More than 1,000 varieties are grown on about 20 million acres in the United States.

- There is a risk that only a small number of GM varieties will be introduced, and much biodiversity will be lost in countries with limited crop-breeding sectors and low enforcement of IPR. These countries, which include many of the poor countries in Asia and Africa, are not able to attract private sector investors to develop a large variety of GM varieties. The private sector may import or develop a small number of GM varieties for these countries. Public sector or NGO investment in development of GM varieties for these countries may be needed to introduce these technologies in a diversified manner.

As the cost of genetic modifications is likely to decline over time, the diversity of GM varieties is likely to increase in all countries as long as mechanisms to introduce access to traditional varieties are introduced. Moreover, without biotechnologies, most of the new crops would have been developed with selective breeding that tends to reduce crop biodiversity; thus in almost all circumstances the introduction of biotechnology will present a significant improvement in terms of crop biodiversity in the long run.

ENVIRONMENTAL RISKS OF AGRICULTURAL BIOTECHNOLOGY

Concerns about environmental side effects of agricultural biotechnology have led several critics to suggest "moratoriums" on the application of the technology. Of course, research should continue, but use of the technology should not be disallowed until it is clear that it is safe. This perspective, which may be shared by supporters of the correct EU approach to agricultural biotechnology, has been justified by the "precautionary principles." However, this perspective, which views biotechnology in isolation without comparing the risks and benefits of its deployment to the outcomes that may result from its ban, is likely to lead to suboptimal decisions. As Dake and Wildavski (1990) and others argue, regulation in the pursuit of safety may be excessive, and policies for control and regulation of "risky" technologies should balance their risks and benefits. Furthermore, the works of Arrow and others suggest that technological development is a dynamic process, and thus producers, users, and regulators of the technology will grow from learning and the regulatory process should be adoptive. These ideas guide our perspective on the assessment and control of environmental risks of agricultural biotechnology, as discussed in the following sections.

Flexible Regulation, Not Outright Bans

Environmental concerns can be addressed by flexible regulation, not by outright bans. There is a significant amount of heterogeneity of the environmental and health risks as well as economic impacts of agricultural biotechnology applications. There are likely to be fewer risks when genetic modification may turn a gene on or off than when it inserts a new genetic material into a plant. A genetic transformation of genes from one plant to another may cause fewer constraints than a transformation of a gene from a fish to a plant. Cultural practices may also affect the overall environmental side effects of agricultural biotechnology. Resistant buildup can be controlled by refugia. Planting a modified variety may cause more damage to beneficials in some locations than in others. Similarly, the economic benefits vary and depend on agroeconomic situations. For example, Bt corn may have a much higher yield effect in a location where pest damage is substantial.

Segerson (1999) shows that when there is economic and environmental heterogeneity, uniform policies are suboptimal. Instead, socially optimal outcomes are obtained by flexible policies that adjust to specific situations. It is the role of the regulatory process in agricultural biotechnology to develop flexible policies that vary according to circumstances. Risk considerations can be introduced in the regulatory framework that will be based on quantitative assessments of outcomes. In each circumstance, policymakers will have estimates of risks and benefits associated with agricultural biotechnology applications. Agricultural biotechnology applications with high benefit-risk ratios are more desirable then with low ratios. The policymakers should establish a (shadow) cost of risks, so that only applications with positive benefits minus expected risks will be allowed.

The risks of environmental side effects can be quantified as probability distribution of negative outcomes. Policymakers do not know these probabilities with certainty and use estimates with significant random noise. The uncertainty about risk may be incorporated in the regulatory assessment of the risk. Lichtenberg and Zilberman (1986) suggest that instead of using the expected value of risk estimates, risks will be presented by values that may be exceeded with a probability of .01 or .05 (the value that corresponds to the .95 or .99 of the cumulative distribution of estimated risks).

In addition to determining when and where to allow the introduction of GM varieties, the regulatory process should consider establishing conditions and practices that may be used to identify and reduce environmental risks. This may include refugia to control against resistance buildup, monitoring of impact on other species, etc. Furthermore, introduction of GM varieties may entail in some cases periodic evaluation of outcomes to improve information and then update regulatory decisions.

COMPARISON OF RISKS

The risks of GM varieties should be compared to the environmental risks without them. The use of GM varieties is the source of several environmental risk categories. They include buildup of resistance to Bt and other valuable organisms, damage to beneficial insects and other wildlife, and other "genetic drifts"—the spread of modified genes of wild varieties. However, these risks have to be compared to some of the environmental risks and side effects they eliminate. The introduction of Bt crop varieties reduces the use of pesticides and may increase yields. The reduction in use of some potent chemical pesticides reported in China, the United States, South Africa, and India (Qaim and Zilberman 2003) will likely reduce the negative side effects of these residuals. They include damage to beneficial insects and other wildlife, including fish and birds. While valuable species are harmed by modified varieties when they eat, say, Bt cotton in the field, pesticide residues can drift through the air and water, causing damage to regions farther away from the original location of applications.

Problems of resistance buildup may occur with excessive chemical pesticides, as it occurs with mismanaged use of GM varieties. The problems of genetic drift occur with many introduced varieties. Genetic materials of new varieties created by selective breeding may drift to traditional varieties in a manner similar to the drift of inserted genes to these varieties. Proposed varieties (including subsidization of in situ conservation of traditional varieties) designed to produce crop biodiversity may contain the side effects of both GM varieties and "modern varieties" that might have been used if GM varieties were not available.

The increase in yield associated with the introduction of GM varieties in some developing countries will likely generate substantial environmental benefits by reducing acreage needed for agriculture, slowing the process of land clearing and deforestation, and preserving biodiversity. Because of the high increase in U.S. agricultural productivity, the agricultural acreage in the United States in 2000 was smaller than it was in the 1920s. Actually, some reforestation of agricultural land occurred in the United States and other developed countries. Increased population growth and increased income (which will increase the demand for meats and thus the acreage needed for production of cereals) are likely to lead to increased farmland in developing countries if the GM varieties are not introduced. Thus, high yields with GM variety may reduce or reverse these outcomes.

A balanced perspective of the environmental risks with and without agricultural biotechnology will lead to identifying possible situations when introduction of agricultural biotechnology may actually reduce risks of environmental side effects versus situations when agricultural biotechnology may generate more negative outcomes. Policy intervention is needed mostly for the second-type of situations.

Economists from Sandmo on have demonstrated that risk aversion tends to reduce behavior and leads to under-investment, under-supply, and under-utilization of resources, and in many cases this is suboptimal. While many aspects of agricultural biotechnology possess risks that are hard to quantify and may be quite severe, our growing experience with these technologies provides us with lessons that show that within a limit the applications of these technologies are relatively benign. Thus far, most of the risks of these technologies have been carried out by the richest countries in the world, which at least in theory can afford to live with fewer risks than some of the citizens of the poorer nations. Shielding the citizens of the developing world from a technology possessing small risks—which at least in the past had containable risks but high potential for benefits—seems to us to be suboptimal.

USING RISKS WITHIN A DYNAMIC PERSPECTIVE

The risks of GM varieties should be used within a dynamic perspective. Technological change has an element of path dependency. Past experience builds expertise and provides a base for future development. Agricultural biotechnology is in its infancy, and the research and development activities that have brought us Bt and Roundup Ready varieties also generate process innovations, including the agribacterium, the gene gun, and many promoters that can build blocks of new innovations in the future.

The early applications of agricultural biotechnologies were pest-controlling technologies. These innovations were induced by the desire to reduce reliance on chemical pesticides in the United States (NRC 2000) and relative ease of operation. These technologies can be adapted to the needs of developing countries, and BT sorghum and cassava may have much value in some locations. The process innovations refined the development of these technologies and may be used in the future to pursue the objectives: development of drought-tolerant varieties, nutritionally enhanced varieties, etc.

Because of the high cost involved, the development of GM varieties has been a de facto "joint venture" of the public and private sectors. A moratorium on the use of the technology (unjustified on risk grounds as we argued above) will result in a substantial loss to investment and deter investors in agricultural biotechnology (for similar technologies) in the future, denying society the benefits of these investments.

As experience with agricultural biotechnologies accumulates, the technologies and their use will improve due to learning by doing (by producers) and learning by using (by farmers). Earlier introduction of these technologies to developing countries will generate the extra gain from both learning and development of specialized human capital.

The buildup of genetic knowledge and a database, resulting from continuous mapping of the genome of an increased number of varieties and accumulating knowledge on functional genomics, increases the potency of potential genetic modification. This new knowledge helps agricultural biotechnology become a major approach to identifying and introducing solutions to crop disease and production.

CONCLUSION

Agricultural biotechnology has been introduced and adopted in the United States, but its use globally has been subject to major controversy. Part of the debate about the future and value of agricultural biotechnology concerns its relevance and potential benefits to developing countries. It has been argued that the current applications of agricultural biotechnology, especially those that address pest-control issues, have limited value to developing countries, mostly because of the limited yield effect in the United States and the North. Furthermore, access constraints, because of intellectual property and registration requirements, and concern for environmental side effects and risks of agricultural biotechnology have been used to justify underemphasis of the role of agricultural technologies in developing nations.

This paper has shown that some of the concerns about the relevance of biotechnology for developing countries are unfounded. Using results from India, South Africa, and elsewhere, we have shown that yield effects of pest control in agricultural biotechnology may be substantial in developing countries with high pest damage and minimal use of alternative pest controls. We show that institutions, such as clearinghouses for IPR, can be introduced to provide developers of agricultural biotechnologies with access to intellectual property owned by corporations and universities in the North, thus substantially reducing their IPR costs. We suggest that registration requirements should be streamlined to reduce the regulatory costs for developing agricultural biotechnology. Our analysis suggests that agricultural biotechnology, if managed correctly, will not lead to substantial reduction in crop biodiversity but, rather, increase crop biodiversity in many cases. Albeit, many of the varieties that will be preserved and even reintroduced will be genetically modified. We also argue that while the introduction of agricultural biotechnology may lead to some risk and potential damage to beneficial insects, these risks have to be compared with the reduction in risk from pesticide use and expansion of agricultural production.

Agricultural biotechnology has significant potential for developing countries. Even current applications that reduce pest damage, as well as pesticide use, can increase agricultural productivity and lead to reduction of food prices in developing countries with humid and warm climates and significant unanswered pest problems. As the tools of biotechnology evolve, it will likely be used to help developing countries overcome some of the constraints in their agricultural production and to improve the quality and supply of their foods. The challenge facing policymakers is to develop the institutions that will enable developing countries to realize the potential of biotechnology.

REFERENCES

Altieri, M.A. 2001. *Genetic Engineering in Agriculture: The Myths, Environmental Risks and Alternatives.* Special Report No. 1, Food First, Oakland, California.

Ameden, Holly, and David Zilberman. 2003. *Adoption of Biotechnology in Developing Countries.* Working Paper, University of California, Berkeley, California.

Arrow, K., and A. Fisher. 1974. Environmental preservation, uncertainty and irreversibility. *Quarterly Journal of Economics* 88:312–319.

Dake, Karl, and Aaron Wildavsky. 1990. Theories of risk perception: Who fears what and why? *Daedalus* 119(4):41–61.

Graff, Gregory D., Susan E. Cullen, Kent J. Bradford, David Zilberman, and Alan B. Bennett. 2003. *The Public-Private Structure of Intellectual Property Ownership in Agricultural Biotechnology.* Working Paper, University of California, Berkeley, California.

Graff, Gregory, and David Zilberman. 2001. An intellectual property clearinghouse for agricultural biotechnology. *Nature Biotechnology* 19:1179–1180.

Hayami, Y., and V.M. Ruttan. 1985. *Agricultural Development: An International Perspective.* Baltimore: Johns Hopkins University Press.

Heller, Michael A., and Rebecca S. Eisenberg. 1998. Can patents deter innovation? The anticommons in biomedical research. *Science* 280(5364):698–701.

Lichtenberg, Erik, and David Zilberman. 1986. The econometrics of damage control: Why specification matters. *American Journal of Agricultural Economics* 68:262–273.

Marra, M.C., and G.A. Carlson. 1990. The decision to double crop: An application of expected utility theory using Stein's Theorem. *American Journal of Agricultural Economics* 72(2):337–345.

Moschini, G., H. Lapan, and A. Sobolevsky. 2000. Roundup Ready(r) soybeans and welfare effects in the soybean complex. *Agribusiness—An International Journal* 16:33–55.

NRC (National Research Council). 2000. *The Future Role of Pesticides in US Agriculture.* Washington, D.C.: National Academy Press.

Qaim. M. 1998. *Transgenic Virus Resistant Potatoes in Mexico: Potential Socioeconomic Implications of North-South Biotechnology Transfer.* Brief No. 7. ISAAA: Ithaca, New York. p. 48.

Qaim, Matin, Cherisa Yarkin, and David Zilberman. 2003. *Impact of Biotechnology on Agro-biodiversity.* Paper presented at the Allied Social Science Meeting, Washington, D.C., January 3.

Qaim, Matin, and David Zilberman. 2003. Yield effects of genetically modified crops in developing countries. *Science* 299:900–902.

Segerson, Kathleen. 1999. A unifying framework for policy analysis. In *Flexible Incentives for the Adoption of Environmental Technologies in Agriculture.* Frank Casey, Andrew Schmitz, Scott Swinton, and David Zilberman, eds. Norwell, Massachusetts: Kluwer Academic Publishers. pp. 79–95.

Sunding, David, and David Zilberman. 2001. The agricultural innovation process: Research and technology adoption in a changing agricultural industry. *Handbook of Agricultural and Resource Economics.* Bruce Gardner and Gordon C. Rausser, eds. Amsterdam: Elsevier Science. pp. 207–261.

Traxler, G., and J. Falck-Zepeda. 1999. The distribution of benefits from the introduction of transgenic cotton varieties. *AgBioForum* 2(2):94–98. Retrieved July 15, 1999, from the World Wide Web: http://www.agbioforum.missouri.edu.

Traxler, G., S. Godoy-Avila, J. Falck-Zepeda, and J. de J. Espinoza-Arellano. 2001. *Transgenic Cotton in Mexico: Economic and Environmental Impacts.*

Paper presented at the 5th ICABR International Conference on Biotechnology, Science and Modern Agriculture: A New Industry at the Dawn of the Century, Ravello, Italy, June 15–18.

NOTES

1. Department of Agricultural and Resource Economics, University of California, Berkeley, California.
2. Center for Development Research, University of Bonn, Walter-Flex-Strasse 3, 53113 Bonn, Germany.

VI

STATISTICS AND TRENDS IN WORLD AGRICULTURE

This section provides statistical information on trends in agriculture and hence factors that affect world agriculture.

25

FACTORS AFFECTING WORLD AGRICULTURE

B. Babcock, J. Fabiosa, H. Matthey, M. Isik, S. Tokgoz,
A. El Obeid, S. Meyer, F. Fuller, C. Hart, A. Saak and
K. Kovarik (Food and Agricultural Policy
Research Institute—Iowa State University)

The figures and text presented in this chapter are from the *FAPRI 2003 World Agricultural Outlook Briefing Book* produced by the Food and Agricultural Policy Research Institute (FAPRI) at Iowa State University. The *Briefing Book* is based on the *FAPRI 2003 World Agricultural Outlook,* which summarizes FAPRI's projections for world agricultural production, consumption, prices, and trade of major agricultural commodities. These projections are based on supply and demand models developed for major agricultural commodities produced and consumed in major countries and regions of the world. A number of assumptions underlie these projections:

- Average world weather patterns will prevail
- Existing domestic policies will remain unchanged
- International policy commitments made under trade agreements will be met

Conjectures on potential policy changes are not included in the process. Major macroeconomic drivers, such as economic recovery in major industrial countries, deepening financial crises in other countries, and growing strength of the dollar, are factored into the projections. These factors may have a significant impact on consumption, prices, and trade patterns. The FAPRI projections are updated annually to reflect changes in underlying economic conditions, assumptions, and supply and demand situations. These annual projections serve as an important baseline for the world agricultural situation, trade prospects, and food security.

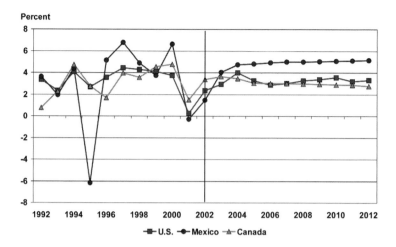

FIGURE 25.1 NAFTA Real GDP Growth Rates. A slowdown in the U.S. economy in 2001, at 0.25% GDP, affected the other NAFTA countries. Canada's slowdown was not as deep (1.5%) and its recovery came relatively earlier, reaching a peak in 2003. The projected long-run growth rates after full recovery is between 2.9% and 3.2%. Mexico will rebound as well and is expected to grow at rates between 4% and 5% after 2002.

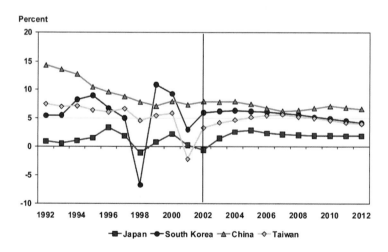

FIGURE 25.2 Asia Real GDP Growth Rates. Japan, which was in a recession in 2002, sees a weak recovery in 2003 and average growth of 2.2% thereafter. Despite the softness of Japan's economy, Asian economies are expected to grow 4% to 5% annually in the next decade. China is the only bright spot in Asia for 2002, with real growth above 7%.

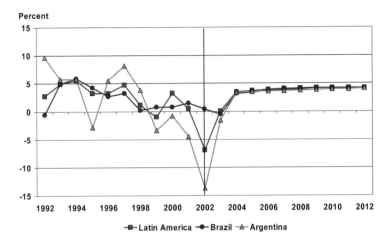

FIGURE 25.3 Latin America Real GDP Growth Rates. Debt burden, political instability, and unsustainable policies dampened Latin America's performance. With contracting economies since 1999, Argentina and Uruguay post positive growth beginning in 2004. Brazil seems to have avoided Argentina's problems but is expected to be in mild recession in 2003. The region is expected to grow at 4% the rest of the period.

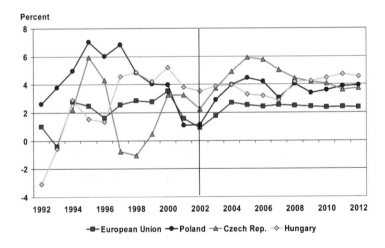

FIGURE 25.4 Europe Real GDP Growth Rates. The EU-15 region experienced moderate economic growth in 2001 but a slight slowdown in 2002, with an aggregate growth rate of only 0.97%. Beyond 2002, growth will accelerate at an annual rate of 2.4%. The larger acceding countries including Poland, Hungary, and the Czech Republic were affected by the EU slowdown. The other countries managed to post modest growth rates; the whole region grew at 4.1%. Much effort has gone into making the countries EU-ready.

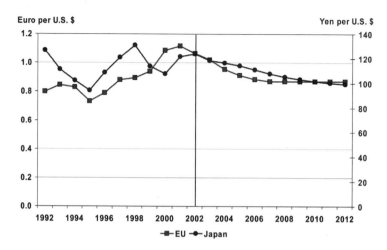

FIGURE 25.5 Exchange Rate Projections. The EU-15 currency depreciated by 7.4% relative to the U.S. dollar over the last six years. Over the next six years beginning in 2002, the euro is expected to appreciate by 3.5%. The yen depreciated in 2001 and 2002 by 12.7% and 2% relative to the U.S. dollar. Beginning in 2003, the yen appreciates moderately at a rate of 2.1% for the remainder of the projection period.

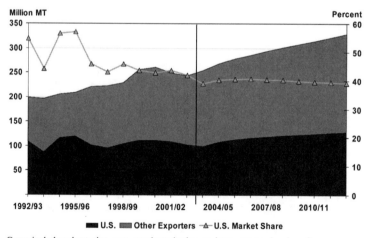

Crops include: wheat, rice, corn, sorghum, barley, soybean, rapeseed and sunflower

FIGURE 25.6 World Crop Trade and U.S. Market Share. Crop trade is projected to grow by 24.5%, adding 102 mmt over the baseline. Trade liberalization and increasing demand from developing countries (e.g., China) generate the demand side of this increase, whereas recovery in Canada and Australia combined with the more prominent role of Argentina generate the supply to meet this higher demand. In the beginning of the projection period, the highest increase in trade comes from oilseeds. In later years, grains trade increases faster, making up nearly 50% of the total increase in trade over the next ten years. The U.S. market share declines slightly, from 41.9% in 2002-03 to 39.1% in 2012-13.

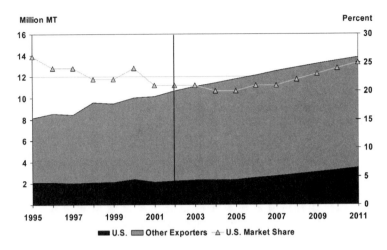

FIGURE 25.7 World Meat Trade and U.S. Market Share. With increasing beef imports over the next four years, the U.S. world meat market share decreases by 0.89 percentage points in the short run. But continuing growth in pork and broiler exports and a decline in beef imports in the second half of the decade allow the U.S. to regain 5 percentage points in its market share, taking it to a share of 25.18% of total meat trade in 2012.

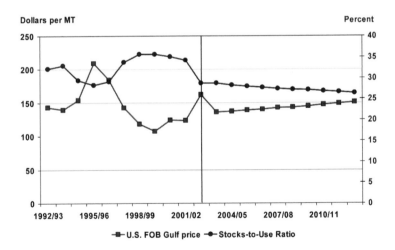

FIGURE 25.8 World Wheat Stocks-to-Use Ratio vs. Price. Because of less area, lower world stocks, and sustained demand, world wheat price increased considerably in 2002–03. With recovery in area and production, world wheat price decreases in 2003–04. The decrease in the stocks-to-use ratio maintains an upward pressure on wheat price after that. The Gulf FOB wheat price is projected to grow 1.1% annually after 2004–05. The stocks-to-use ratio steadily declines, reaching 26.4% by 2012–13.

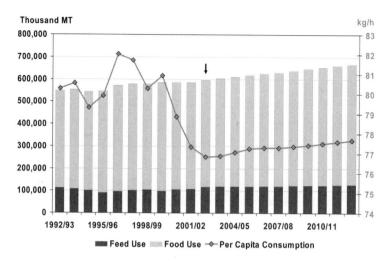

FIGURE 25.9 World Wheat Feed, Food Use, and Per Capita Consumption. Fueled by population growth and a slight increase in per-capita consumption, world food use is projected to grow by almost 11.1% over the next ten years. Most of this increase comes from Asian and Middle Eastern countries, in which a shift in diets and income growth boost per-capita consumption. Feed use is projected to grow 8.6%. The bulk of this increase comes from the EU, the FSU and the U.S., where most of the substitutions with other grains occur.

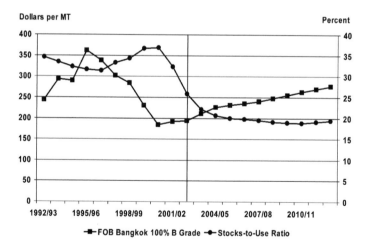

FIGURE 25.10 World Rice Stocks-to-Use Ratio vs. Price. International rice prices remained below $200/mt for the third straight year in 2002, but drought in India and reductions in China's area decreased world stocks to their lowest level since the mid-1980s. The tightening of world supplies is expected to raise rice prices 16.7% over the next two years. Rising global demand encourages prices to strengthen an average of 2.5% annually after 2004, with the Thai price reaching $277/mt by 2012.

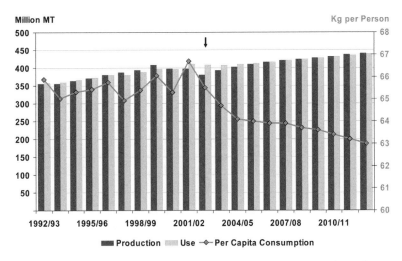

FIGURE 25.11 World Rice Production, Use, and Per Capita Consumption. The gap between global production and consumption is expected to close over the next two years as rice area recovers. Total world rice consumption increases 0.7% annually over the next decade, but average consumption per person gradually declines. Declining per capita consumption in a number of Asian countries—particularly China, India, Japan, and Indonesia—more than offsets consumption growth in other regions of the world. Urbanization, income growth, and diversification of diets are major factors contributing to the decline in Asia.

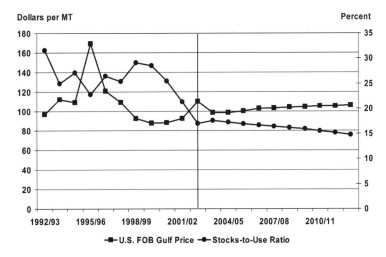

FIGURE 25.12 World Corn Stocks-to-Use Ratio vs. Price. In 2002–03, low world corn production led to an increase in world corn price along with a significant drop of world stocks. In 2003–04, the nominal Gulf FOB corn price decreases 11.6% because of recovery in production. Driven up by increasing demand from world markets and lower stocks, corn price is projected to grow 0.8% annually after 2003–04, reaching $106 per mt by 2012–13. The stock-to-use ratio decreases to 14.8% by 2012–13.

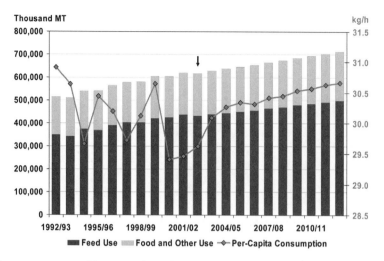

FIGURE 25.13 World Corn Feed, Food Use, and Per Capita Consumption. As a result of mildly increasing per-capita consumption and recovery in the livestock sector, the main increase in demand for corn comes from feed use. Feed use is projected to reach 499.7 mmt by 2012–13, increasing by 66.9 mmt over the baseline. Food use increases by 28.8 mmt over the next ten years.

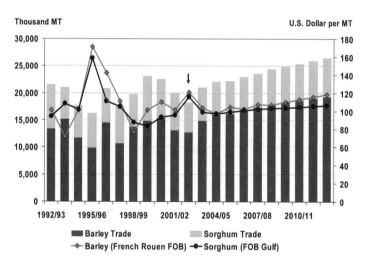

FIGURE 25.14 Barley and Sorghum Trade and Prices. World sorghum trade reached an all-time low in 2002–03 because of low production. This trend is projected to change in 2003–04, as area and production increase because of a high world price in 2002–03. Recovery in world sorghum trade is mild, reaching only 7.2 mmt in 2012–13. This higher demand is supplied mostly by the U.S. World sorghum price is high in 2002–03 but decreases 17% in 2003–04. World barley trade grows steadily at an annual rate of 4%, fueled by higher demand from China and Saudi Arabia. World barley price decreases 16% in 2003–04 and reaches $118.2/mt in 2012–13.

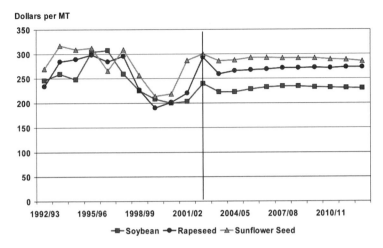

FIGURE 25.15 World Oilseed Prices. World oilseed prices climbed strongly in 2002–03 as demand grew faster than supply. Soybean prices are expected to weaken next year under the pressure of record supplies. High oil demand boosted sunflower prices in 2002–03 despite an expansion of world production. The continued decline in rapeseed production caused rapeseed prices to increase for the third straight year. In the long-run, all oilseed prices are expected to return to their historic relationships.

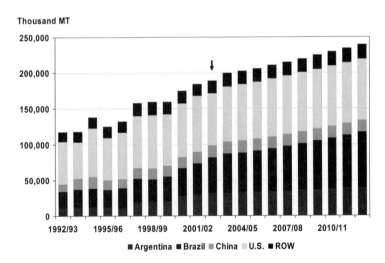

FIGURE 25.16 World Soybean Production. World soybean production reaches 239 mmt by 2012–13, an increase of 27% over the current year. The U.S. remains the dominant producer in the world, but its share drops from 39% to 36% during the baseline period. Brazil expands its production by 57%, raising its share to 32%. World soybean production stays very concentrated: the top three producer countries supply about 85% of all soybeans.

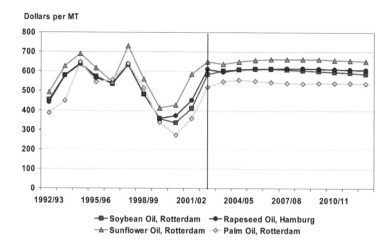

FIGURE 25.17 World Vegetable Oil Prices. Rising 2002–03 vegetable oil prices reflect the fact that consumption has grown considerably faster than production for the second consecutive season. Oil prices maintain their positive trend in the near future, driven by strong world demand. In the second half of the outlook, oil prices weaken slightly under the pressure of a stronger production response. Palm oil partially closes the gap to soybean oil but remains the low-price oil.

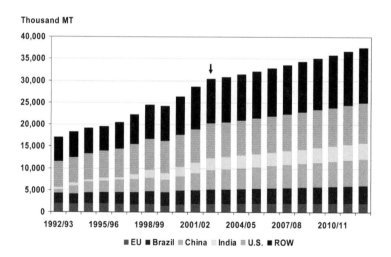

FIGURE 25.18 World Soybean Oil Consumption. World soybean oil consumption reaches 37.5 mmt by 2012–13. Per-capita consumption grows 1% annually, to 5.3 kg/year. The strongest total demand increases are in India and China, each growing around 3% annually. The U.S. remains the largest consumer, but its share drops slightly. World consumption falls short of production during the outlook period; stocks increase 28%.

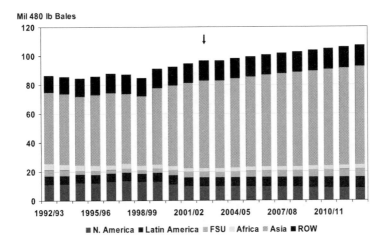

FIGURE 25.19 World Cotton Consumption. World cotton area fell significantly in 2002/03 to 76 million acres, with over half of the contraction coming from Asia. World cotton area is expected to rebound to 81 million acres in 2003–04. Production was 87.2 million 480-lb bales in 2002–03 which helped to support prices, resulting in a season average A-Index of $0.54/lb. Production rebounds in 2003–04 to 93.2 million 480-lb bales and shows continued growth throughout the period. World consumption keeps pace with expanding production, increasing an average of 1.2% annually.

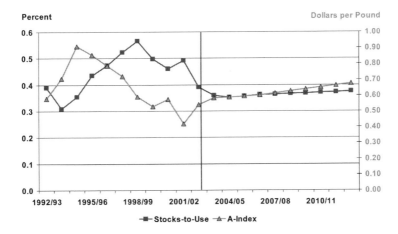

FIGURE 25.20 Cotton Stock-to-Use Ratio vs. Price. The Chinese continued their substantial stock reductions in 2002–03, with stocks falling to 8.94 million 480-lb bales. The stock reduction slows in 2003–04 and stocks stabilize at 6 to 7 million 480-lb bales, growing with consumption in China towards the end of the forecast period. A-Index prices are expected to reach a season average price of $0.58/lb in 2003–04 and increase throughout the forecast, with prices accelerating somewhat, after 2005–06.

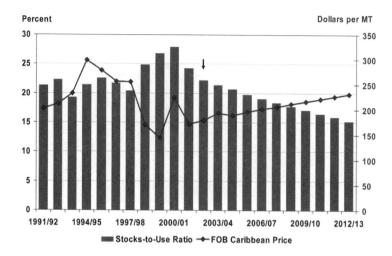

FIGURE 25.21 World Sugar Stocks-to-Use Ratio vs. Price. World harvested area increases between 2002–03 and 2012–13 by 4.7% for sugarcane and by 11.9% for sugar beets. World sugar price increases by 4.6% in 2002–03 leading to a 4% increase in sugar production. By 2012–13, price is expected to increase to nearly 10.5 cents per pound. The stocks-to-use ratio peaked in 2000–01 at 27.9% and is projected to decline to 15% by 2012–13. World trade increases by 12% by 2012–13.

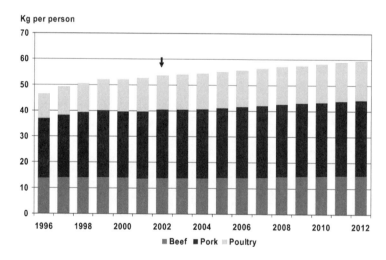

FIGURE 25.22 Per Capita Meat Consumption. Many countries reach full economic recovery beginning in 2003–04 (with the regional growth rate in the range of 2.62% to 4.38%). This causes per-capita meat consumption to increase by 0.38% to 2.18% annually, or 5.70 kg over the baseline, reaching a level of 59.46 kg per person per year by 2012.

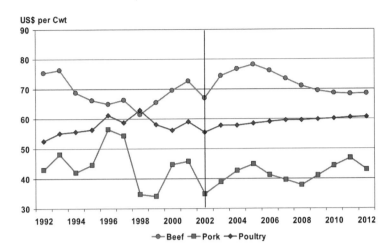

FIGURE 25.23 World Meat Prices. Recovering meat demand and rising feed crop prices strengthen world meat prices. The world beef price increases by 5.36% annually over the next three years, peaking at $78.23 per cwt in 2005. The pork price turns around in 2003 and reaches a peak of $44.86 per cwt in 2005. It cycles with another peak in 2011, 4.61% higher than in 2005. The poultry price increases by 0.89% annually over the next 10 years.

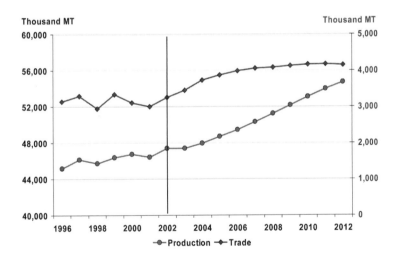

FIGURE 25.24 World Beef Production and Trade. A two-year decline in beef trade fueled by BSE and FMD reverses to an annual growth rate of 3.01% in the next decade. Beef production also recovers to a 1.51% growth rate, reaching 54.705 mmt in 2012. Recovery in major importing countries such as Mexico and Russia slightly reduces growth in trade in the outer period, ending with 4.14 mmt in 2012.

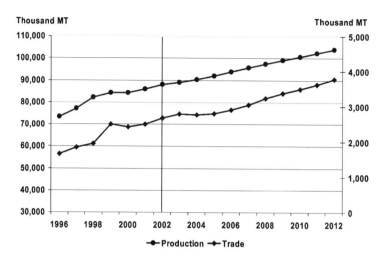

FIGURE 25.25 World Pork Production and Trade. Weak economies in Asia and SPS challenges in major exporting countries slowed down pork production and trade in the last two years. In the next decade, trade increases by 41.45%, reaching 3.78 mmt in 2012. Pork production increases at a rate of 1.73% (15.82 mmt), reaching 103.92 mmt in 2012. The pork price returns to an upward trend in 2003, reaching a peak of $44.86 per cwt in 2005, and another peak, 4.61% higher, in 2011.

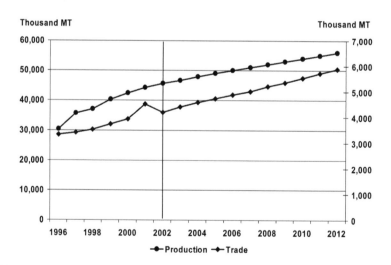

FIGURE 25.26 World Broiler Production and Trade. The world broiler market benefited from recent SPS challenges in other meats in the last several years. However, Russia's temporary ban on U.S. exports slowed trade in 2002. Recovery begins in 2003 and trade grows by 40.28% in the next decade, reaching 5.89 mmt in 2012. Total broiler production increases by 10.15 mmt, 2.13% annually, reaching 55.87 mmt in 2012. Strong demand helps to strengthen price in 2003, which grows by 0.89%, ending at $60.74/cwt.

U.S. Dollar per MT

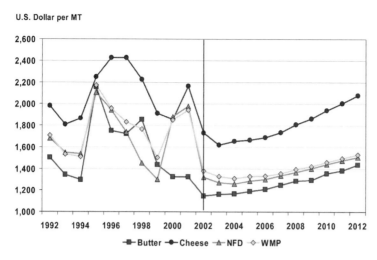

FIGURE 25.27 FOB Northern European Dairy Product Prices. World prices for NFD and WMP decreased by 36.4% and 34.8%, respectively in 2002. From 2004 onward, NFD and WMP prices rise an average of 1.9% to 1.6% annually. An increase in exports from the EU along with weak import demand contributed to the 20.4% decrease in cheese prices in 2002. Similarly, butter prices decreased about 14.9% in 2002. Butter and cheese prices rise steadily after 2004, increasing 2.3% and 1.9% annually, respectively.

Thousand MT

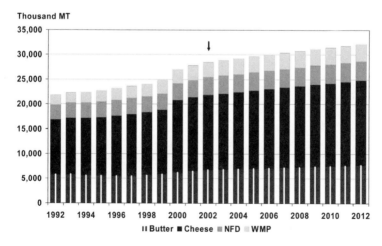

FIGURE 25.28 Dairy Product Output for Modeled Countries. Total butter production increases 14.4% over the baseline. The U.S. butter production decreases 3%, while butter production remains relatively constant in the EU and Japan. Total cheese production grows 13.3%, with the U.S., Australia, and New Zealand production increasing about 1.8%, 2.3%, and 2.9% annually, respectively. NFD production declines in the U.S. and the EU, while it increases in Mexico, Poland, Ukraine, India, and New Zealand. Total NFD output rises about 7.5% and production of WMP rises 15.2%.

Index

457